WATER MANAGEMENT IN INDIA

About the Author

P.C. Bansil heads the Techno Economic Research Institute, New Delhi, founded by him in 1984. He belongs to the first batch of Indian Economic Service and has worked with the FAO for about 15 years in Sri Lanka, Thailand, Rome, Zambia and Libya. He has authored 28 books and reports, besides a large number of reports and papers prepared for various governments. He has published over 400 papers in specialized technical journals. His work experience relates to the formulation, implementation and evaluation, and regional plans for agriculture, statistics, marketing, livestock, irrigation, fisheries, forestry, wildlife, national parks, tourism, agro-industries and social sectors like demography, health, education and housing, as well as project identification, formulation and evaluation in these fields.

Dr. Bansil was recently appointed by the Planning Commission as Chairman of the Tenth Plan Working Group on Animal Husbandry Economics and Statistics. He has just completed the study "Livestock in India 2000-2030" sponsored by the Food and Agriculture Organization of the United Nations. During his FAO assignments in Sri Lanka, Zambia and Libya as Senior Planning Advisor, he undertook micro level studies on issues relating to water..

WATER MANAGEMENT IN INDIA

P.C. Bansil
Techno Economic Research Institute
New Delhi

CONCEPT PUBLISHING COMPANY, NEW DELHI-110059

ISBN: 81-8069-097-0

First Published 2004

Published and Printed by
Ashok Kumar Mittal
Concept Publishing Company
A/15-16, Commercial Block, Mohan Garden
New Delhi-110059 (INDIA)

Phones : 25351460, 25351794
Fax : 091-11-25357103
Email : publishing@conceptpub.com

ROTARY INTERNATIONAL

One Rotary Center
1560 Sherman Avenue
Evanston, IL 60201-3698 USA

JONATHAN B. MAJIYAGBE
President, 2003-04

Foreword

Just the word 'water'—although the Drink of Life—may not attract our attention, but we get driven towards a glass of water anytime thirst grips our throat. This colourless, odourless, tasteless, natural substance is an important constituent of our diet but we seldom realize that proper utilization of water is necessary to keep us healthy and fit. It is interesting to know that water is one of the major structural components of the human body. All the tissues of the body including teeth and bones contain water. As much as 75% of the infant's body and 65% of an adult's body are composed of water. Water plays an important role in possessing and maintaining a beautiful and a healthy body. It acts as a medium for various metabolic reactions taking place in the body. The waste products of the body are excreted in the form of urine and perspiration and water is the medium for both of them. It is an important constituent of all the intracellular and extracellular tissue fluids of the body. There is a constant exchange of water in and around these cells to maintain the osmotic pressure. Water helps in maintaining the body temperature by distributing the body heat. It helps in the lubrication of various organs. Water is present in the mucus secretions of gastrointestinal, genitourinary and respiratory track.

The body suffers a loss of about 2 liters of water everyday in the form of urine, faeces, sweat, nasal and eye secretions etc. Some water is also lost with the air that is exhaled during breathing. If this lost water is not replenished in the body, it is likely to cause dehydration. Chronic low fluid intake is a common but often overlooked cause of mild dehydration and fatigue. Thirst is the most common noticeable sign of dehydration. The other symptoms of dehydration include rapid breathing, drying up of lips and mouth, dizziness and inadequate dark-coloured urine output. A loss of about 10 per cent of water from the body may cause serious symptoms like headache, fatigue palpitation and mental confusion. And if this loss of water increases to 15-20 per cent, it may prove to be fatal.

Appreciating the importance of water and keeping in mind the concern over the shortage of drinking water, the year 2003 has been declared as "International Year of Freshwater". Although 70 per cent of the world's surface is covered by water, only a fraction of that – 2.5 per cent – is freshwater. Less than one per cent of the world's freshwater resources are available for human use. Over 80 countries representing 40 per cent of the world's population are subject to serious water shortage and conditions might worsen in the next 50 years as population grows and global warming increases.

With these few words about the importance of water to sustain this Universe, I am glad that in this year of 'Fresh Water' at a time when all the International Agencies – UNDP, World Bank, WHO, UNICEF and FAO are ceased with the problem, Rotary has taken the responsibility of educating the public about managing this 'elixer of life'; hence this publication. This in-depth study by a Rotarian is the result of dedicated efforts for a cause so dear to Rotary. I congratulate the Rotary community in general and Rtn Sushil Gupta RID and PDG Rajan Dhingra, as well as Dr. P.C. Bansil in particular for this commendable effort. I am confident that the book will spread the Rotary message of the need for Managing Water judiciously.

December 2003

Jonathan B. Majiyagbe
RI President

Preface

This five letter word 'Water' conjures up different feelings from 'blue gold' to 'blue elephant'. This is the only commodity on earth without which no life is possible – whether human, animal or plant. Being available in plenty and having been considered as a 'free gift' of nature, no thought was ever given towards its proper management. With the ever increasing population, the demand for water has gone up disproportionately – now reaching a flash point. The whole world today is seised with the issues relating to various aspects of water, so much so that it is being predicted that future wars may be fought on 'water'. India is not far behind in attaining the status, of 'water stress'. Hence schemes from recharging ground water of Thar desert to interlinking of rivers are in hot pursuit.

It was my chance meeting some three years back with Rtn Sushil Gupta (now Rotary International Director) which served as a motivating factor to start thinking about water and its related problems. We have since then been toying with the idea of putting 'managing water' as a thrust area for Rotary. There will be no better service to humanity for Rotary than to highlight the emerging problems relating to water. To make people conscious of the need for a 'clean' potable glass of drinking water and find out ways and means to achieve this goal for every body, is possibly the biggest service which goes well with Rotary objectives. Now time has come when in this year of 'Fresh Water' governments all over the world are forced to provide regulations for 'right to safe water' as above all other 'rights'. This is because in the absence of this, there is no life and if, there is no life that will be dooms day. It is a matter of great privilege for me when I see light at the end of the tunnel – this treatise being released by nobody other than the Rotary International President, I am confident that the few encouraging words put by him in his 'foreword' will carry the message of this little effort to all concerned in the world. The problem of water is with everybody, everywhere and perhaps for all time to come; hence this book.

Actually the first draft of the book was prepared by me about a decade back when M/s Kirloskar Brothers asked me to present a paper on the subject which I read at an International Conference, organized by them. This, however, is a complete transformation of the original paper which had actually lost its relevance today. It will be preposterous for me if I take the total credit for this 'work' to myself. We have been working over it for over a year. The book would not have seen the light of the day, but for the untiring and dedicated labour put in by my colleague (comrade in arm) Mr. K. R Dasgupta. I have hardly any words to acknowledge his contribution which goes beyond any expression. First draft of the chapter on Interlinking of River Water was contributed by Mr. S.P. Malhotra, my old friend and Ex. FAO Commodity Specialist. For a support to a subject of this topical nature, there is always the need to go through various newspapers and literature being released everyday. We had to depend heavily on all such material and it has not been possible to acknowledge individually for all these references.

I must hasten to add that for whatever virtues the analysis may contain, I am indebted to all those who have helped me in this exercise. I, however, accept all responsibility for any error, misinterpretation, omissions or shortcomings and deficiencies in the treatment of the subject. The responsibility for the views expressed in this study is that of mine alone.

New Delhi
January 2004

P.C. Bansil

Contents

PART IV: CONCLUSIONS

Part I

SETTING

1

Introduction

"Agni and Water are givers and sustainers of life, they are affectionate mothers, givers of all, givers of life, they are have healing powers"

—Rig Veda

If there is magic on this planet, water has it. In its very ordinariness, water is extraordinary. From the awesome beauty of giant glaciers to the fragility of snow flakes, from the fury of floods to the silence of lakes, from the poetry of rain to the magic of mist, water is a many splendoured thing. The water of most of the rivers is considered holy. Every Srivaishnava has to chant several mantras and slokas while taking bath. River waters are treated Holy by the people in India and are utilized in various festivals/poojas/religious ceremonies. It is believed that Ganga, the heavenly river was brought down to the earth through the efforts of King Bhagirath, who underwent great penance to wash away the sins of his forefathers. Holy water of celestial Ganga is to be kept in the house of every Srivaishnava so that it could be drunk at the time of leaving this mortal world. Therefore, elementary rules against pollution of water are already formulated, which is the essence or simple rule for hyginene.

Water is the most important input for survival/ growth of not only plant, human beings, animals and other living beings on the earth but also economic development and environmental sustainability. At the same time, it is a dynamic component of the soil and plant system. This is the most scarce and precious natural resource. In arid areas water is known as Liquid Gold. Indeed few other resources affect so many areas of the economy or of human and environmental health as water. Water is in short a blessing for humanity.

The word ' Bhagwan' (God) stands for 'BH': BHOOMI; or Earth (first element),"A":AGNI or Fire (second element),"G": GAGAN or Space/ Sky (third element), "W":VAAYU or Wind (fourth element),"N": NEER or Water (fifth element). Ours is an ancient civilization with continuity extending over millennia. The earliest recorded thought found in Rig Veda begins with a vivid description of five elements (panchatatva)—earth, water, energy, air and space that constitute our universe and provide the life support for man and animal and the base for all forms of vegetation and all forms of human activity. These elements are defined as the manifestations of the absolute and man is ordained to conserve these elements for his eternal prosperity by scrupulously performing his duties prescribed in this regard in the various religious texts. No matter who we are, where we are and what we are, we all are dependent on water. We need it every moment in so many ways.

The ashrams of the ancient sages represent systems of ecological management marked by the most rational husbandry of natural resources, partnership between man and animal, sound balance between material satisfaction and spiritual and cultural pursuits and community life that ensured health, happiness and harmony for all the constituents. The Yagnyavalka Smriti spells out man's obligations towards protection of the plant and animal kingdom for his own survival and prosperity. Shankara, Madhwa and Ramanuja—all the three schools of philosophy are in unison in stressing the importance of conservation of natural resources and their judicious use by man for satisfying his material needs. No religion justifies the destructive attitude modern man has developed towards his environment. All urge man to treat nature as a god-given gift. Hinduism expresses the firm belief that the natural environment in which man is placed is created by the Lord. *Akasa* (space) is the first creation of the Lord, and the Brahman in the form of akasa created marut or *vayu* (air) and vayu in that form created *tej* or energy. The other element *apa* (water and *kshiti* (earth) are further creations. It is affirmed that the Lord alone, residing in these various forms of *akasa, vayu, tej, apa* and *kshiti*, created the further entities. The supreme is immanent in every one of these and these are not mere inert and independent entities.

Since Vedic times, water has been enjoying the most respectable and unique status. Melas and fairs are held in and around rivers. Since time immemorial, the Kumbh Mela, the greatest of the fairs is observed while standing in water. It has attracted people from all walks of life.

It is celebrated every 12 years on river Ganga (Prayag/Sangam). The Kumbh Mela has wielded a mesmeric influence over the mind and hearts of the Indian masses. Other water related festivals include the Pushkar Mela which is held every year in Kartika month in around Pushkar lake near Ajmer in Rajasthan. Water also finds its place in various festivals celebrated in different parts of the country like Makar Sankaranti, Shiva Ratri, Holi, Gangaur, Baisakhi, Ganga Dahshhera, Ganesh Chaturthi, Onam, Karwa Chauth, Chhath Puja and Aadiperukku.

Water had both to be managed and exploited for the survival of the human race and, therefore, with the first human settlements around 6,000 years ago began a two-fold struggle with water. On the one hand, people had to protect themselves against floods, on the other they had to ensure safe water supply for domestic use and irrigation. Hydro-technical installations were thus amongst the earliest technological achievements of mankind. Art of storing water in cisterns was developed well before the first cities came into existence. The earliest cisterns were hewn into the rock. Masonry cistern developed later. From about 2000 BC mortar began to be used for sealing these cisterns. By the middle of the first century AD, covered cisterns with storage volumes of upto 75,000 cubic metres were being built and fundamental elements of dam construction were known.

The earliest dams were constructed in Jawa (Jordan) around 3000 BC and in Wadi Garawi (Egypt) around 2600 BC. The great civilizations, in the valleys of Nile, Tigris, Euphrates, Indus and Huang Ho rivers, flourished only on the basis of their superior water management systems. Both the Indus valley and the Mesopotamian cultures had excellent water supply and sewage systems. The standard of their Hydro-technical infrastructure was not equalled even 2,000 years later by the Romans. The early Hindu texts, written around 800-600 BC, reveal a certain knowledge of hydrological relationships. The Vedic hymns, particularly those in Rig Veda, contain many notes on irrigated agriculture, river courses, dykes, water reservoirs, wells and waterlifting structures.

The Hydrological cycle and the water balance principles were thus known in earlier times through the wisdom and experience that was passed from one generation to the other. With water as the deciding factor for growth and prosperity, ingenious practices of water conservation, water harvesting, transportation of water from remote areas ensured meeting of water needs. These practices were given shades of ritualistic overtones and woven into religious and other socio-cultural rituals to ensure widespread practice at grassroots level. At the same time technological feats such as the Roman Aqueducts, the stepped wells of Adalaj exhibit a high level of engineering expertise. There are numerous examples of water harvesting through underground and surface tanks at the micro level, Baolis and Wells at meso level and canals, dams, aqueducts at macro level.

Recent Realisation

Water has been generating significant concern perhaps because irregularities in its availability tilt the balance of life. We can ill-afford to ignore the water crisis that has gripped many parts of the world or the dangerous floods that claim hundreds of lives every year in certain other parts. Hence the bid to spread awareness among every section of society on protecting this valuable resource is gradually gaining momentum. Water has the focus of environmental issues and this, combined with the fact that the year 2003 has been declared as the 'International Year of Freshwater', has led water to become the subject of much discourse.

Water is considered to be a renewable natural resource since it is continually being renewed through nature's hydrological cycle. Also it may be regarded as a unique resource, as the total amount of water available on the 'Global basis' remains constant. It does not imply that 'local' or regional sources of water can neither be exhausted by 'irrational use' nor rendered unusable due to large scale contamination. More and more water is being used for different purposes which tend to deteriorate the quality of water receiving bodies and simultaneously there is an increasing demand for 'good quality' water. And caught between growing demand for freshwater on one hand and limited and increasingly polluted water supplies on the other, many developing countries face difficult choices. Population continue to grow rapidly. Yet there is no more water on earth now than there was 2000 years ago, when the population was less than 3 per cent of its current size. Today 31 countries accounting for under 8 per cent of the world population face chronic freshwater shortages. By the year 2025, it is apprehended, 48 countries are expected to face shortages, affecting more than 2.8 billion people or 35 per cent of the world's projected population. However, enough freshwater is still available worldwide to meet needs for the foreseeable future if the water were evenly distributed. But water is poorly distributed across countries and across seasons. Competition for water is becoming more acute increasing the potential for conflicts between sectors and water wars between countries.

Water as Free Gift

If we think air and water are free gifts of nature, we must learn to honour the gifts and the gift-giver. Although water forms an essential ingredient of life, yet it is carelessly (mis) used and consumed. An increase in human population, developmental activities and changes in lifestyle have put tremendous pressure on water resources. We cannot afford to be content with the idea that these commodities will remain freely available to mankind forever. We are on the threshold of a water crisis. Ask those women—folk of the desert-affected remote villages of Rajasthan and Haryana, who fetch water in pitchers on their heads daily as a routine from as far as 8 km distance, the value of water. The situation in the urban areas is also not so happy, Come summer and there is low water pressure in the water supply pipes or no supply at all. There is a hue and cry everywhere.

Because of an easy access to the groundwater, people have been exploiting the hidden treasure of groundwater indiscriminately. This has resulted into the depletion of groundwater reserves which is reflected by the fall in the water levels by a few metres in the recent past in the areas of good quality groundwater. This is adding to the cost of extraction of groundwater. We can maintain the water levels of the groundwater by increasing artificially the groundwater bodies through rainwater harvesting. There is a large variation in the precipitation in the country spatially and temporally. Eighty per cent of the precipitation occurs during the monsoon period restricted to only about three or four months. The period of the monsoon is also not similar everywhere in the country. But generally it is from June to October. Spatially, the amount of rainfall varies from about 10 cm in Rajasthan to about 1,000 cm, in the North Eastern States of Mghalaya and Tripura. The average annual rainfall of the country is 110 cm. As much as 70 percent of the precipitation goes waste as runoff and ultimately reaches oceans depending upon the amount and intensity of rainfall, soil characteristics, topographical gradient and subsurface lithology etc., It is the need of the hour that we tap this run off having a large scope.

Water and the Economy

Water plays a vital role in the economy, through its use in agriculture and industry, as well as supplying energy through hydropower. A third of the world's harvested crops are produced by irrigated agriculture. Much of the natural environment, from coastal ecosystems to wetlands, depends on water. Degrading the quality or supply of that water can have disastrous effect on the environment and on biodiversity. Water resources planning is a scientific tool aimed not only at optimal utilization and management but also to make developmentalists aware of environmental and social equity considerations as well as to make the environmentalists aware of economic development considerations.

Irrigation and drainage have a proud tradition and an important role to play in the developing world. Irrigation is the acid test of development. To build a successful irrigation project is like "meeting a payroll" in business, or winning the decathlon in sports.

Irrigation, which consumes far more water than any other use, has generated enormous benefits. By helping raise farmers' yields and stabilize food production and prices, irrigation has been key to achieving food security in many parts of the world. About 250 million hectares are irrigated worldwide today, nearly five times more than at the beginning of the 20th century. Yet inappropriate water and agricultural policies and poor irrigation management have also lowered groundwater tables, damaged soils, and reduced water quality. Moreover, growing populations with rising incomes will further increase the demand for irrigation water to meet food needs.

Thus, without water, we cannot grow food. It takes a remarkable amount of water to produce crops—one to three cubic metres to yield just one kilogram of rice, for example. And as population grows, water needs for agriculture will only increase. At the same time, there is growing competition for water from municipalities and industry. Currently, globally, industry claims about 20 per cent of total water withdrawals and municipal users the remaining 10 per cent. But of the amount they claim, only 10 to 20 per cent is actually consumed. We will need to reuse a greater portion of the 80 to 90 per cent that is wasted. And before it is reused or returned to rivers and aquifers, water must be treated to remove pollutants. But as demand for water by all users grows, groundwater is being depleted, other water systems are becoming polluted and degraded, and developing new sources of water is getting more costly. Many people are concerned that water will be the main obstacle to growing enough food in the coming decades.

Protecting Precious Resources

Investment in irrigation projects has decreased even though increased production from irrigated lands is likely to be decisive in meeting future food needs. Efficiency of water use in agriculture, industry, and urban areas is generally low. Degradation of land and water resources through water logging, salination, and ground water mining are mounting while excessive use of water in some locations causes lack of access to water elsewhere[1]. In many locations, water is still treated as a free good with little or no clearly defined property and user rights. Only a small portion of applied water and fertilizers are utilized by various crops in different cropping systems and the major portion of both the inputs is lost through leaching which results in deterioration in quality of underground water. Around 50 per cent of the increased demand of water could be met by increasing the effectiveness of irrigation. About 35-60 per cent of applied nitrogen is lost through leaching as NO_3-N. Another important fact to be noted is that of the total rain water (nearly 4000 billion cubic metres) received in the country, only approximately one-third is recharged into the aquifers while two-third is lost due to run-off and evapo-transpiration. There has been under-utilization of created irrigation potential. High conveyance loss of water, lack of equity, adequacy and reliability of water supply, wastage of water by over-irrigation by the farmers, excessive exploitation of ground water in fresh and marginal zones, increase in operational and maintenance (O & M) expenditure coupled with poor maintenance because of lack of funds with the agencies, limit increase in irrigation efficiency. Excessive weed growth and silting in the carrier channels, traditional cropping patterns are some of the banes of our water irrigation sector. The scope for further development of new irrigation is limited and is a costly solution. There are problems of water-logging and salinity in various pockets. The ground water table is declining sharply in almost one-third of the country. There is a massive increase in power consumption for extraction of underground water and in the cost of drilling and running of tube wells.

MANAGEMENT OF WATER RESOURCES

UN Commission for Sustainable Development, Second Session, New York, 1994 noted as follows:

> "As the crisis approaches and as water resources become scarcer, the risk of conflict over them will become greater. After 2025 AD climate change could also make conditions worse if precipitation amounts decrease in the major food producing regions and evaporation rates increase. The bulk of the increase in food production has to come from irrigated lands and this, in turn, will require more money to be spent on long distance water transfers, dams and the like, should the resources be available. The increasing size and number of cities will create a much bigger pollution load unless sanitation systems are provided. Urgent and decisive action must begin now if impending water crisis of a national proportions later in the 21st Century—are to be avoided during the next 30 years".

In view of overall shortage of fresh water at the macro level and actual scarcity and water stress in many arid and semi-arid areas of India, there is an urgent need for conservation and better management. Effective strategies must consider not only managing the water supply better but also managing demand better. Needless to say, when communities manage fresh water resource efficiently, they also manage other natural resources better, improve sanitation, and reduce diseases. At the national level, especially in water short regions with dense population, adopting water-shed or river-basin management perspective is a needed alternative to uncoordinated water-management policies by separate jurisdictions. At the international level countries that share river basins can fashion workable policies to manage water resources more equitably.

In India, government pricing policies, particularly with reference to electricity for pumping groundwater, often result in inefficient and unsustainable exploitation of precious water resources. On the other hand, private ownership of groundwater resources leads to the emergence of water lords and inequitable water markets. There is a loud demand that a national policy relating to access to water for all should be designed in such a manner that, first, irrigation does not result in long-term harm to soil health and, second, gender and social equity in sharing the available water is ensured. Participating management of irrigation water resources, including systems of rational distribution of water, will help to foster the equitable and efficient use of water.

Pricing policies should signal the inter-generational equity aspects of water use. No doubt, there is an urgent need for an adoption of an integrated approach to water management. Demand management may also find a major place in the national water policy. The long-term vision may focus on integrated management of water resources.[2]

Many parts of India are facing environmental catastrophe because of over use of water. For example, emerging water crisis and its poor management in Punjab have been a cause of great concern. A recent study of the Punjab Agricultural University on water situation in the State highlights that as against the requirement of some 4.90 million hectare metres of good quality water, only 3.12 million hectare metre water is available. In the past four decades there has been a 76 per cent increase in the state's water requirement for agriculture due to the cultivation of water-intensive crops like paddy. The underground water table has been going down. Lack of recharging potential due to negligible flow and no storage in the rivers for nine months in a year is alleged to be the principal cause of falling underground watertable. What would have happened slowly anyway has been markedly accelerated by the cultivation of rice. Hence, as agriculture intensifies, attention must be paid to the environment. Improper fertilizer use can contaminate surface water and groundwater. And when irrigated soil is not properly drained, land becomes waterlogged. Salinization can also occur as naturally occurring salts concentrate in soil, leaving it infertile. About 10 per cent of irrigated land worldwide is damaged by salt build-up, a waste of farmland that could be used to increase food production. Further, removing water from rivers and lakes can jeopardize wetlands, an important source of inland fisheries and a natural filter that cleanses wastewater.

Tools to Improve Irrigated and Rainfed Production

Though India and other developing countries depend on both irrigated and rainfed crops to feed their people, much of the increase in food production will need to come from irrigated land. Finding enough water to support that increase requires producing more crop per drop. The most common forms of irrigation—surface irrigation in which waters flood fields and sprinkler irrigation which mimics rainfall-can waste water. More efficient are localized methods such as drip irrigation, which put water only where it is needed. Rainfed agriculture, which produces more food overall than irrigated agriculture, benefits from practices to maximize the collection of rainwater. Water harvesting, collecting water in structures ranging from small furrows to dams, allows the farmer to conserve rainwater and direct it to crops. Water harvesting can boost yield two to three times over conventional rainfed agriculture. Introducing improved varieties, improved cropping patterns and using minimum tillage methods which conserve water further increase yields.[3] Hence, proper water management is not only essential for adequate availability of quality water for agricultural, industrial and domestic uses, but also for protecting the ecosystem.

The Third World Water Forum (WWF-3) held at Kyoto (Japan) attended by about 6,000 delegates drawn from 166 countries along with several head of states from African and Asian countries, and more than 100 Ministers of Water Resources and Agriculture stressed the need for forging and promoting new partnerships in water services.

In India, the break—up of the joint family system has an adverse effect on the environment, including on freshwater conservation. As families break up into smaller, independent nuclear units and start living in separate houses, their freshwater consumption shoots up. This phenomenon is being witnessed even in countries like Portugal, Spain, Italy, etc., where the population is actually declining.

Despite the gradual worldwide acceptance of the fact that we face a catastrophic freshwater crisis in the near future, many water experts are of the view that we are possibly already too late to halt the rapid despoliation of our water resources. In the long run, probably the only thing that can turn things around in that Holy Grail of cheap, abundant, non-polluting energy, thermonuclear power. Only thermonuclear reactors, using the enormous hydrogen reserves in the oceans to produce virtually limitless power and giving of inert helium as the only " smoke" can provide energy for massive seawater desalination plants. However, researchers say they are still at least four decades away from viable fusion reactors of this size. So far, the only desalination plants of any consequence exist in the oil-rich Persian Gulf emirates and kingdoms, consuming vast amounts of petroleum to produce pure water from seawater.

Environmental Demand for Water

Of late, environmental demands for water have attracted increasing demand. Environmental water requirement is defined by IFPRI as the amount and quality of water required to protect an ecosystem to enable ecologically sustainable development and water resource utilization. Extreme changes in ecosystems may occur if water available for environmental uses falls below a certain threshold. Many countries are beginning to incorporate these concepts into their water resources management strategies to reserve a certain quantity of water for environmental or ecosystem uses.

Water reserved for the environment can help regulate pollution and sustain the riparian ecosystem. Sufficient in-stream water availability can help to temper water pollution through dilution of contaminants in the watercourse. Water availability is critical to the maintenance of local flora and fauna in a watershed (particularly fish and other aquatic species). Water is also needed to maintain existing wetlands, which can have additional positive impact on the watershed because they can recharge aquifers, digest organic waste, and store runoff.

FRESH WATER RESOURCES

In view of the immense importance of Fresh Water for human existence a detailed discussion on the various aspects of Fresh Water Resources is called for. Although about three-fourths of earth is water, the estimated volume of freshwater in our rivers, groundwater, snow and ice, is about 2.5 per cent only, the rest being sea/salt water. Most of the freshwater are either in the form of ice and permanent snow cover in Antarctic/Artic regions (about 69 per cent) or is stored underground in the form of deep underground basins/aquifers, soil moistures (30 per cent). Total usable freshwater supply to ecosystem and humans from river system, lakes wetlands, soil moisture and shallow groundwater is less than 1 per cent of all freshwater and only 0.01 per cent of all the WATER ON EARTH. As per World Health Organization estimates only 0.007 per cent of all water on earth is readily available for human consumption globally. This indicates that Fresh Water on earth is finite also unevenly distributed.

How Freshwater Resources are Affected?

Whatever its form, every drop of the world's water is locked into the hydrological cycle. However, the speed of movement of water through different phases of the hydrological cycle varies considerably. The average time a drop of water stays in the atmosphere is about eight days and in a river about 16 days. But such duration can run into centuries for a glacier and tens of thousands of years for water moving sluggishly through a deep aquifer. Though water drops are continuously recycled, freshwater available in form of lakes and river storage (which is about 0.3 per cent f all fresh water resources) are renewable.

Human actions modify the hydrological cycle and can seriously pollute available freshwater. Climate change also affects the hydrological cycle significantly thereby affecting fresh water production and its distribution.

Major Freshwater Resources of the Country

The River Systems

The major river systems of India viz. the Himalayan river system which is snow and rainfed and the Peninsular river system which is totally rainfed are the main source of water. The major part of Himalayan river system comprise the Indus and the Ganga-Brahmaputra-Meghna (Barak) system. The Ganga, rising from the snow capped Himalayan mountains, flows through the great Indo-Gangetic plains. The Brahmaputra rises in Tibet where it is known as the Tsangpo and runs a long distance until it enters into India in Arunachal Pradesh under the name of Siang or Dihang. The Ganga and the Brahmaputra join in Bangladesh and continue to flow under the name Padma forming the Sunderban delta.

The Indus, which is one of the great rivers of the world, rises near Mansarover in Tibet, flows through India, thereafter to Pakistan and finally falls into the Arabian Sea near Karachi. Its major tributaries the Jhelum, Chenab, Ravi, Beas and Sutlej originate in India and after flowing through the Punjab plains-join the Indus.

The important river systems in the Deccan under peninsular river system are the Narmada and the Tapi, which flow westwards into Arabian Sea. The east flowing rivers of the Deccan, the Brahmani, the Mahanadi, the Godavari, the Krishna, the Pennar and the Cauvery fall into Bay of Bengal. There are numerous coastal rivers, which are comparatively small. While only handful of such rivers drain into the sea from the east coast, there are as many as 600 such rivers on the west coast.

Main source of freshwater is the precipitation including snowfall, averaging to about 4000 billion cubic metre (b.cu.m) received on the Indian land mass, and the flows, about 390 b.cu.m received from other countries. The estimated run off 1869 b.cu.m.

Due to large spatial and temporal variability in the rainfall pattern the water resources distribution in the country is highly skewed. The Ganga-Brahmaputra—Meghna basin covering 33.5 per cent of the country's area contributes as much as 60 per cent of the water resources whereas the west flowing rivers with only 3.5 per cent of area contributes as much as 11 per cent of the water resources.

Ground Water

Based on large volume of hydro-geological and related data generated by Central Ground Water Board and State Ground Water Departments the replenishable ground water resources in the country have been estimated as 432 b. cu. m. In the alluvium plains of the Indo-Gangetic valley, ground water depths measure upto 450 metre. The coastal aquifers also have similar depth range of ground water availability. Inland river basins in the country have recorded shallow over depth within the range of 100-150 metres.

Static ground water resource also sometimes known as "fossil" water, is available in the aquifer zones below the zone of water level fluctuation, countrywide its availability has been assessed as 10,812 b.cu.m. Basis being the depth of availability of ground water and the productivity of deeper aquifers. For purposes of sustainability, development of groundwater resources needs to be limited to replenishable depths of naturally occurring ground water, available in sub-surface domain. Ground water is widely dispersed. It is an important source of water for drinking and irrigation. Ground water contributes 51 per cent of the irrigation potential of the country through more than 4 million dug wells, 5 million shallow tube wells and some ninety thousand public tube wells.

Freshwater Resource Assessment

Water resource assessment is the determination of its source, extent, dependability and quality. On the above parameters are based our evaluation of its utilization. Planners and decision makers require such information for meeting the expected demands. Existing and future uses of water are to be determined by giving due consideration to its quality and the ecosystem needs of the aquatic environment all being legitimate user of the resources.

The river basin is recognized as the appropriate unit for planning and development of our water resource. Measurement of its quantity and quality, and of other characteristics of the environment affecting water, are an essential requisite basis for adopting effective water management strategies.

Measuring on a regular basis the hydrological elements, which control water resources, is necessary to determine how much water is available for use. These elements include precipitation, evaporation and river flow, as well as the water stored in soil, aquifers, reservoirs and glaciers. The water's quantity, quality and biological characteristics are to be measured regularly.

Despite the large number of hydrological stations, the network needs additional stations. Indeed, national networks in several developing countries are declining to the extent that many nations are less able to determine their water resources now than was the case in 1967. Hydrological information is vital for investment purposes, as well as for advancing scientific understanding. Keeping this in view the HYDROLOGY PROJECT was initiated in eight peninsular states of the country with a view to upgrading the water resources assessment capabilities, capacity building of the concerned state agencies and to create a national data bank for free exchange of all data with access facilities for genuine users.

Importance and Need of Freshwater

No matter who we are, where we are, and what we do, we are all dependent on water. We need it everyday, in so many ways. We need it to stay healthy, for growing food, vegetation, transportation, irrigation, industry and its sheer life giving properties. However, despite the importance of freshwater resources in our lives and well-being, we are increasingly beginning to take this resource as being infinite, and for granted. In today's world, much water is wasted, used inefficiently and polluted through abusive use. The per capita availability of fresh water is fast declining all over the world. If the present Consumption pattern continues, two out of every three persons on earth will live in water stressed conditions—moderate or severe water shortages—by he year 2025. In India, the per capita average annual freshwater availability has reduced from 5177 cubic metres from 1951 to about 1869 cuic meters in 2001 and is estimated to further come down to 1341 cubic metres in 2025 and 1140 cubic meters in 2050.

FRESH WATER YEAR

India has declared 2003 as Year of the Freshwater in keeping with the United Nations Resolution declaring 2003 as international Freshwater Year.

The Mission: Fresh Water for All

The objective of adopting 2003 as the year of Fresh water are:

- Increasing awareness among stakeholders regarding scarcity value of freshwater
- Conservation and efficient use of Freshwater
- Freshwater preservation—quality and its ecosystem
- Augmentation of freshwater resources
- Community partnership for informed decision making

2003 is a year of opportunity, wherein we can rededicate ourselves and focus attention on protecting and respecting our water resources as an individual, a community, as a nation and a global family. By protecting our freshwater, we help to ensure our future and our planets long term prospects. Individuals, Institutions, Governments each has a role to play, " Together We Can Make A Difference".

Quality of Water—Consequences of Over Use

Human health is dependent on wholesome and reliable supply of water and safe sanitation. It is estimated that at any given time about half the people living in developing countries are suffering from water-related diseases caused directly by infection, or indirectly by disease-carrying organisms that breed in water, such as mosquitoes. Diarrhea, infections by parasites, river blindness and malaria are among the most widespread of these diseases. Health concern also arises due to exposure to various chemicals in drinking water; for example, high levels of nitrates, effect of which are not easy to quantify. Pollutants can build up in shellfish to the point that they harm the people who eat them; eating seafood contaminated by mercury from industrial discharges results in Minimata disease. Also due to over extraction of ground water, several parts of the country faces problems relating to arsenic, fluoride, salinity.

The effects of pollution on wildlife can also be far reaching. These include death, eggshell thinning, population decline, reduced success of hatching, birth defects and a range of other health hazards for the bird, fish and other forms of life which live in the rivers, lakes, wetlands and deltas. Changes in the conditions of the aquatic environment arising from human activities can endanger the different species living there, in some cases leading to the decline and extinction.

Over use and pollution of the world's fresh water resources are a recent development. The long-term consequences are not well known. Already, however, they have taken a heavy toll on the environment, and they pose increasing risks for many species. Indeed, the supply of water for human use entails not only the problem of quantity but also one of quality, since practically all human activities impair the natural quality of water. Beyond a certain degree of pollution water becomes unusable. The result is a reduction in the supply of usable water and problems with the disposal of the polluted water. The most important causes of pollution are:

- The discharge of untreated or insufficiently treated industrial and domestic water (bacterial, organic an heavy metal etc. pollution) as it has happened in Yamuna;
- Storage of domestic and industrial waste on insufficiently safe dumps (pollution by seepage water);
- Irrigation (discharge of salt etc. into groundwater reservoirs and surface waters through unsuitable irrigation and drainage methods);
- Excessive use of fertilizers and inefficient use, unsuitable storage and disposal of pesticides (discharge of nitrate, pesticides, herbicides, etc.);
- Deforestation and ploughing-up of grasslands (pollution of outlets with nutrients and suspended matter);
- Overuse of groundwater (salt water contamination of groundwater bodies in coastal areas, subsidence and consequent changes in the behaviour of surface runoff, rising salt content and ultimately the drying up of lakes);

- To these direct kinds of pollution of the groundwater with harmful substances must be added indirect harm done by atmospheric pollution, since the substances released into atmosphere (such as nitrogen monoxide) enter the water via precipitation.

While agriculture remains the biggest source of water pollution, wastes from industries and municipalities have increased enormously in recent decades. Industrial pollutants, such as wastes from chemical plants, are often dumped directly into waterways. Oil and salts are washed off city streets. Heavy metals and organochlorines are leached from municipal and industrial dump sites[4]. In fact, all of India's 14 major rivers are badly polluted. Together they transport nearly 50 million cubic meters of untreated sewage into India's coastal waterways every year. The city of New Delhi dumps almost 200 million litres of raw sewage and 20 millions of industrial wastes into Yamuna River every day as the river passes through the city on its way to the Ganga[5]. However, there has been a visible reduction in the levels of solid waste in the river during the last couple of years and the government efforts by putting wire meshes, using cleaning dredges and contribution made by treatment plants is beginning to show positive results.

When human activities damage fresh water ecosystems, they ultimately damage the salt water environment as well. Oceans are connected to the land through the complex network of rivers, streams, and lakes that comprise watersheds. Managing freshwater resources will help protect the coastal waters. Protecting the coastal environment, rather than treating oceans and seas as garbage dumps, would help avoid a possible ecological disaster in the future.

Water Policy

Policy reforms are needed to provide secured water rights vested in individuals or groups of water users that increase incentives for investment, improve water use efficiency, reduce the degradation of the environment, and encourage flexibility in resource allocation. In some countries and regions, these rights should be tradable, providing further reason to conserve water. Irrigation infrastructure and management should be turned over to water user associations where well-defined water rights provide incentives for user groups to economize on water use. Governments should reform distorted price incentives and reduce or remove subsidies on water to prevent overuse or misuse. In estimating the true value of water, one must include such issues as the time spent transporting it and the health benefits that accrue from access to clean water[6].

The precise nature of water policy reform will vary from country to country, depending on underlying economic conditions and institutional capabilities, relative water scarcity, and level of agricultural intensification. Water policy reforms must transcend national boundaries. In many regions, long-term solutions will require cooperation between countries that share scarce water resources.

In India, national water policy was adopted for the first time in 1987. The revised National Water Policy has been adopted by the National water Resources Council in its 5th meeting held on 1st April 2002. The National Water Policy—2002 recognizes water as a precious national asset. It embodies the Nation's resolve that planning, development and management of water resources would be governed by National Perspectives. The National Water Policy, 2002 emphasises integrated water resources development and management for optimal and sustainable utilization of the available surface and groundwater, creation of well-developed information system, use of traditional methods of water conservation, non-conventional methods for water utilization and demand management. The revised Policy integrates quantity and quality aspects as well as environmental considerations for water through adequate institutional arrangements. Besides, 'ecological needs' have been assigned due priority in water allocation. Greater emphasis has been laid on water quality aspects. The policy also stresses involvement of people in project planning and participatory approach in water resources management.[7]

REFERENCES

1. Rosegrant, M.W., X. Cai, and S.A. Cline: Global Water Outlook to 2025: Averting an Impending Crisis, Research Report, Washington, D.C, IFPRI, 2002.
2. Swaminathan, M.S.: Innovations in Managing Monsoons and Water Resources; lecture delivered at Indian Science Congress, 88th session held on 3-7th January, 2001 at Indian Agricultural Research Institute, New Delhi—110 012.
3. Water-source of food security—theme of the World Food Day held on 16th October, 2002 under the auspices of FAO.

4. Dirt Poor: A survey of development and the Environment, *Economist*, March 21, 1998.
5. Harrison, P.: The third revolution, Environment, Population and a sustainable world. London, I.B. Touris, 1992, Page 305.
6. Meinzen—Dic, R.S. and M.W. Rosegrant, eds. Overcoming Water Scarcity and Quality Constraints, 2020 Vision Focus 9, Washington, D.C: IFPRI, 2001.
7. Ministry Water Resources: Annual Report 2002-2003.

2

Global Scenario

"Fierce national competition over water resources has prompted fears that water issues contain the seeds of violent conflict."

—Kofi Annan,
U.N. Secretary General

"The Wars of the next century will be on Water."

—Ismail Serageldin,
Vice President, World Bank

Introduction

Water is an abundant natural resource as three fourth of the surface of earth is covered with it. Total water resource on earth is estimated to be about 1360 million cubic kilometers which is 0.25 per cent of the planet's mass and if spread evenly over its surface, it would cover the planet to a height of 2.7 kilometers. But more than 97 per cent is in the form of ocean and seas, 2 percent is locked in ice-caps and glaciers and a large proportion of remaining 1 per cent lies far too deep in the ground to exploit. Thus, only 0.2 million cubic km is fresh water of rivers, lakes, swamps and reservoirs; and 23.4 million cubic km is ground water which is mostly saline. It is only this limited quantity which is available to meet the water demands of human and livestock world over. Figures 1 and 2 indicate the total water availability and distribution of fresh water on earth.

More than half the world's people rely on water originating in mountains for drinking, growing food, producing electricity and sustaining industries. Yet while human demand for water increases,

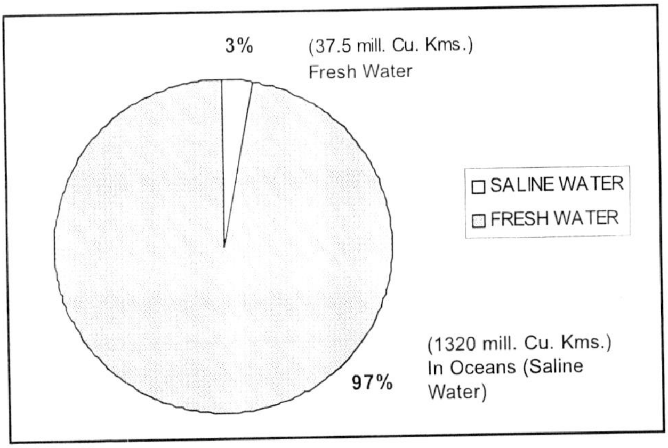

Of the total water on earth only 3 per cent constitutes fresh water. Rest is saline water in the Oceans.

Fig. 1: Total Water on Earth

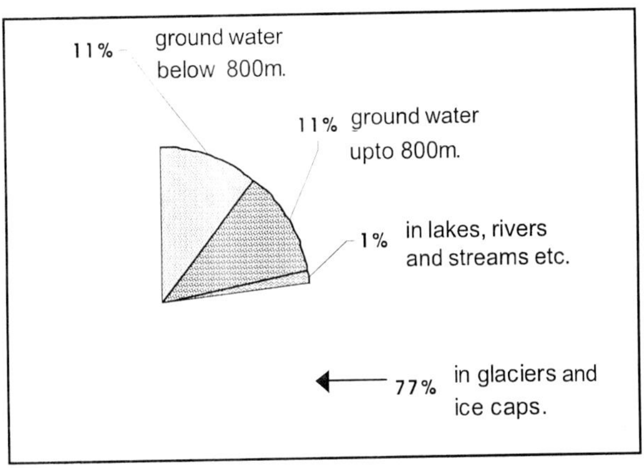

Fig. 2: Distribution of Fresh Water on Earth

- 11 per cent of the total fresh water on earth is ground water available upto a depth of 800m which can be extracted for use.
- Mindless extraction and over exploitation of very small quantity of this precious nature resource has caused a rapid depletion and deterioration in this quantity and quality both.

human activity threatens its future availability. Deforestation of mountain woodlands, mining, urban sprawl and global warming are taking a toll on mountain watersheds. By celebrating International Year of Mountains (2002), FAO is working to increase awareness about the importance of mountains and mountain communities.

Water undergoes a permanent cycle. Every year about 566000 km^3 evaporates from the earth's surface, 80 per cent of it over the sea. The same amount falls back into the earth as precipitation, but only 110000 km^3 over the land masses, of which 71,000 km^3 evaporates directly again. Thus, the average amount of water running off the land masses each year as a renewable resource is about 39,000 km^3, Of this renewable resource, 64 per cent or 25,000 Km^3 flow off very quickly. This part is known as the unstable runoff. The remaining third, in other words about 14000 km^3, seeps into the ground and becomes groundwater, which later forms the "stable" runoff. The latter is also known as fair-weather or groundwater runoff[1].

From the above scenario of global water supply, it may appear that there is no lack of fresh water worldwide. But it must be taken into account in individual regions of the world, because of the spatial and temporal variations in precipitation, the potential usable water supply is very small.

The most frequently used criterion for assessing the availability of the renewable water supply is per capita supply, a supply less than 500 m^3 per capita per annum being regarded as the critical lower limit, 1000 m^3 per capita is very low, 2000 m^3 as critical. According to a study carried out by the World Water Research Institute, it is especially the countries which are in or near semi-arid and arid regions that are in a critical situation. Based on a supply of less than 500 m^3 per capita per annum, the most endangered regions are North Africa with 3 and the Middle East with 8 countries at threat[2]. The World Bank estimates that by the year 2025 one person in three, in other words 3.25 billion people, in 52 countries will have conditions of water shortage.

As per the World Wide Water Development Report, by the middle of this century, at worst seven billion people in 58 countries will be faced with water scarcity, at best two billion in 48 countries will be faced with water scarcity, depending on factors like population growth and policy making. Climate change will account for an estimated 20 per cent of this increase in global water scarcity.

The report calculates that there is an estimated 12,000 km^3 of polluted water world wide, which is more than the total amount contained in the world's ten largest river basins at any given moment. Therefore, if pollution keeps pace with population growth, world will lose 18,000 km^3 of freshwater by 2050. The report ranks 122 countries according to the quality of their water as well as ability and commitment to improve the situation. Belgium is considered the worst in respect of quality of its groundwater, combined with heavy industrial pollution and poor treatment of wastewater. It is followed by Morocco, India, Jordan, Sudan, Niger, Burkina Faso, Burundi, Central African Republic and Rwanda. The list of countries with the best quality is headed by Finland followed by Canada, New Zealand, United Kingdom, Japan, Norway, Russian Federation, Republic of Korea, Sweden and France. Many policy makers and advocates have called for the recognition of water as a human right, as an essential step in ensuring that action is taken on behalf of those, who are suffering from lack of access to clean water supplies.

Integrated rural water development is needed to improve water use in both rain-fed farming and irrigated lands. The global irrigated area is expected to increase annually at about 0.6 per cent in the coming 30 years. But 70 per cent of the increase in cereal production is expected to come from these irrigated areas. There is significant scope for increasing rain-fed production through suitable water harvesting systems, but appropriate investments and policy changes are needed. In chapter 18 of Agenda 21 of the International conference on water and the Environment held in Dublin in 1992 also says:

> "Sustainability of food production increasingly depends on sound and efficient water use and conservation practices consisting primarily of irrigation development and management with respect to rainfed areas, livestock water supply, inland fisheries and agroforestry. Achieving food security is a high priority in many countries, and agriculture must not only provide food for rising populations, but also save water for other uses. The challenge is to develop and apply watersaving technology and management methods and through capacity building, enable communities to adopt new approaches, for both rainfed and irrigated agriculture."

Living on a Spoonful of Water

It is from this spoonful of water that the entire world has been trying to quench its thirst. And a spoonful just does not seem enough. It can easily spill into that jug of non—usable water and leave us with precious little to quench our thirst. Does the idea of a thirsty, drought—stricken mankind seem too far fetched? It really would not appear so if we were to take a look at some of the facts and figures that environmental agencies have presented before us in recent times. Here are some facts that should shake us out of our complacency. As stated earlier, the fresh water available amounts to less than one-half of the 1 per cent of all the water on the earth. Fresh water is renewable only by rainfall, at the rate of 40,000 to 50, 000 cubic kilometers per year. Due to intensive urbanization, deforestation, water diversion and industrial farming, however, even this small finite source of fresh water is disappearing fast. If the present trends persist, the water in all river basins on every continent could steadily get depleted.

Global consumption of water is doubling every 20 years. According to a United Nations study, more than one billion people on the earth already lack access to fresh drinking water. Forty per cent people live in countries where water is scarce: by 2025, this is expected to rise to 66 per cent. The greatest environmental disaster afflicting the planet is the scourge of dirty drinking water. Dirty water kills 2.2 million a year in developing countries. Most of the victims are children. In fact, the water crises the world over has reached such immense proportions that the international community took it us as one of the main issues at the Earth Summit held at Johannesburg. The summit focused on "Water—Its Rapid Depletion and Deterioration". The Earth Summit had itself been held in a land threatened by famine, estimated to affect nearly 10 million people across southern Africa. In Malawi, Mozambique, Zambia, Swaziland and Lesotho, the main cause of famine is drought. On the other hand, floods have in recent times been the cause of widespread misery in much of Central Europe.

The Water Kingdom

Water is life. Natural and human systems depend on it. It is needed every day in many ways—for growing food, to stay healthy. For plants and animals, for irrigation and industry. All this is in theory, but the real picture is not exactly like this. The statistics worldwide are quite disturbing. One person in six lives without regular access to safe drinking water. Over twice that number (2.4 billion people) lack access to adequate sanitation. Water related diseases kill a child every eight seconds, and are responsible for 80 per cent of all illnesses and deaths in the developing world. This is just the tip of the iceberg. Water is not easily accessible to all. For a good proportion of people in the developing world, to have a drink or a pot of soup means walking miles for a bucket of water. The scarcity is increasingly becoming a raising issue and has invited attention. Even the theme of the World Environment Day in 2003. "Water-Two Billion People are Dying for It!"—highlights the centrality of water to human survival and sustainable development.

One of the most important issues facing the mankind today, is availability of clean freshwater. And it is likely to take still serious dimensions as the growing demands are outstripping supplies with pollution and contamination of rivers, lakes and streams. In recognition of importance of water resources to the planet's future, United Nations General Assembly has proclaimed 2003 as the International Year of Fresh water. Taking United Nation's figures in view, fresh water is the single most precious element for life on Earth. It is essential for satisfying basic human needs, health, food production, energy and maintenance of regional and global eco-systems.

In the words of Kofi Annan, UN Secretary—General, Lack of access to water—for drinking, hygiene and food security—inflicts enormous hardship on more than a billion members of the human family. Water is likely to become a growing source of tension and fierce competition between nations, if present trends continue, but it can also be a catalyst for co-operation. The International Year of Fresh water can play a vital role in generating the action needed—not only by civil society communities, the business sector and individuals all over the world. Notwithstanding the bad water situation, provision of water services has risen across the developing world during past 20 years. However, these gains have largely been cancelled out by population growth. Result is that many parts of the world now face the spectre of water scarcity because of climatic change, pollution and over consumption. The challenge faced by the world bodies is to provide water services to all, especially the poor, to maximize water productivity, especially in agriculture; and to ensure that rivers and groundwater aquifers that are shared between two or more countries are equitably and harmoniously managed.

What Needs to be Done

- Changing behaviour in water use, sanitation and hygiene.
- Mobilizing the energy and participation of communities, particularly women groups.
- Setting national targets and plans to generate investment.
- Putting in place policies and regulatory frameworks for water management that take into account both public health and eco-system needs.
- Forming partnerships between private companies, bilateral donors, development agencies, banks, civil society and local communities.

Currently, it is estimated that approximately $30 billion per year is spent on meeting drinking water supply and sanitation requirements worldwide. And that an additional $14 to $30 billion per year would be needed to meet the targets. Koichiro Matsuura, Director—General, UNESCO, once commented, " Of all the social and natural crisis we humans face, the water crisis is the one that lies at the heart of our survival and that of our planet Earth. No region will be spared from the impact of this crisis that touches every facet of life, from the health of children to the ability of nations to secure food for their citizens. Over next 20 years, average supply of water world wide per person is expected to drop by a third".

Water Worries—World Over

The Government in China, the world's most populous country, planned to invest more than $30 billion over the next few years to fight water pollution and help relieve shortages. It is a country that still lacks water resources, and the problem of water pollution remains severe, Presently top priority of the Government of the country is to ensure clean drinking water for her people. China is working on a multi-billion dollar diversion project aimed at channeling water from the flood-prone south to the thirsty north. But environmentalists have raised concern over the Three Gorges Dam—the world's largest hydroelectric project which China began filling up after a decade of work and which will flood hundreds of abandoned towns, villages and factories along the Yangtze river.

Dead Floods

In Bangladesh, where water can often be both a blessing and a curse, the government launched a tree-planting drive that it aimed to turn the country into a "garden of green[1] by 2,015. Although floods have killed hundreds of thousands of Bangladeshis over the years, the country still faces acute water shortages. The biggest problem in Bangladesh is arsenic contamination, which has forced authorities to seal thousands of wells across the country. Health officials say, one-third of the country now had arsenic contamination in wells, while rivers—many drying or polluted by chemical waste were no longer a source of safe drinking water.

Globally two billion people are dying for water. Water—related diseases kill a child every eight seconds. One person in six lives without regular access to safe drinking water. Over twice that number—2.4 billion—lack access to adequate sanitation. In India which is often gripped by a heat wave in summer, the water scarcity heightens the misery of its people. Excepting most of Africa and Latin America, the rest of the world is confronted by a scarcity of natural water resources. This is awful considering that the International Water Management Institute has predicted that the use of water worldwide will increase by 50 per cent over the next 30 years and 4 billion people (half of the world's population) will live under conditions of severe water stress by 2,025. It is also likely that conflict over water will worsen, especially in areas already suffering from water stress like in South and West Asia, and North Africa. The battle over direct economic use and environmental needs is also likely to worsen, with the exploitation of water resources reaching dangerous levels in many countries. China, for example the most water-stressed country, exploiting 44 per cent of its usable water.

Consumption

It was estimated that world-wide water use increased more than tenfold during the last century. Water exploitation rose particularly sharply in the second half of the 20th century. Disproportionate increase in water exploitation took place during this period in Asia and Europe[3]. Total agricultural water requirement was also estimated to have increased 6.5 times during this period. There were at the same time, changes in the pattern of water requirements. In 1900, agriculture accounted for nearly 90 per cent of total water requirements, but as at present, the corresponding figure is likely to be around 70 per cent. Industrial water use, which was about 6 per cent of the total water consumption at the beginning of the last century is estimated to have increased four-fold, to nearly 20 per cent now. There is thus growing competition for water in municipalities and industry.

Agriculture is thus currently the largest water-using sector worldwide. Within this sector by far the largest share is used by irrigated agriculture, which in some regions at least has contributed to the rapid rise in the amount of water used in the last 40 years. The gross irrigated area has increased world-wide from 94 million hectares in 1950 to about 270 million hectares today. More than 70 per cent of the irrigated area is located in the developing countries. It was only the increase in irrigated area, though, that made it possible in densely populated countries for food production to keep up with population growth in the past few decades. Over 30 per cent of the world's food production takes place on irrigated areas, which account for 17 per cent of the arable areas. In Asia irrigated agriculture accounts for 60 per cent of food production and in some countries, such as Pakistan, the figure is 80 per cent. In India, agriculture sector as a whole accounts for between 90 and 95 percent of surface and groundwater.

The data released by the World Bank, together with figures from the United Nations which indicate that water consumption doubles every 21 years, gives an idea of the international tensions for control of water resources that could unleash. It is not a coincidence that conflict over the control of these resources is, even today, one of the biggest obstacles to peace treaties between Israel and its Arab neighbours. At least 40 per cent of the world's population live near 250 sources of water which are the object of frequent and heated conflicts. According to the World Bank, the availability of water per person in the West Asia and North Africa may be reduced by 80 per cent by 2025. And there may be a 30 per cent drop in Egypt, a 40 per cent reduction in Nigeria and a 50 per cent decrease in Kenya. According to the FAO, scarcity of water in the third world is not due to a decrease in rainfall, but to a growing demand from agriculture and tourism.

A paper entitled 'Global Water Shortage' the Congressional Quarterly Researcher concentrates largely on West Asia where Israel, Syria and the new Palestenian entity have a problem over the scarce waters of the river Jordan. Turkey, Syria and Iraq are also involved in a dispute over water resources in the Tigris and Euphretes river basins.

Outside West Asia, water rights are a contentious issue in the Ganges river basin shared by India and Bangladesh, the Danube river, where Slovakia and Hungary are arguing over a planned Colorado river, whose waters, are so depleted by diversions in the American southwest that hardly a trickle of polluted saline water reaches Mexico.

An FAO study of 93 developing countries indicates that a number of water-scarce nations are already withdrawing water supplies faster than they can be renewed. Ten countries are in a critical state meaning that satisfying the needs of agriculture requires them to withdraw more than 40 per cent of their total renewable water resources. Another eight are water stressed—satisfying the needs of agriculture requires them to withdraw more than 20 per cent of that total.

A noted Swedish Hydrologist, Main Falkenmark, has calculated that 100 liters per day (36.5 cubic meters per year) is a rough minimum per capita requirement needed to maintain good health, while 5 to 20 times that amount tends to be needed to satisfy the requirements of agriculture, industry, and energy production. Falkenmark has suggested that when annual renewable fresh water availability falls below 1,700 cubic meters per person, a country will experience periodic water stress; below 1,000 cubic meters, a country will experience chronic scarcity. These benchmarks are rough but other factors such as climate and level of economic development influence water stress and chronic scarcity. Using medium-range UN population projections, some 34 countries will have less than 1,000 cubic meters of annual renewable water per capita by 2025. India by that time will fall in the category of 'Periodic Water Stress'[4].

Water Imbalances

World Water Day held on March 22, 2002 passed with barely a flicker of interest. A UN conference in Beijing had warned that we were silently but surely heading towards "water shock" which would dwarf even the oil crisis. Water is indeed as precious as oil: it is mined like any other natural resource and has always been the world's largest extractive industry. The new crisis arises because in the last two decades, for the first time in human history, more is being taken out across the globe than nature is putting in. Some 80 countries and 40 per cent of the world's population are already experiencing "water stress". Acquisition and control of water resources has driven communities to migrate since pre-historical times. It has generated increasing conflict as the world's river basins became fully colonized; almost half of the world's population now lives in states which have to share their water resources with their neighbouring states. From the Nile to the Rio Grande via the much disputed Jordan, most of the world's best-known rivers are commodities in contention.

Yet to regard the water crisis as a problem of natural shortage or inter-state conflict is to miss the point. Whether water is sufficient or not must depend in the end upon the extent of demand. Cases of unanticipated drought are relatively rare-though global warming is likely to generate more such cases as distribution patterns are altered by climatic change. The real problem is consumption rather than supply. Crop irrigation consumes 90 percent of all water uses in poor countries. Nearly half of that is wasted and never gets to the plants. Intensive agriculture is particularly greedy for water which does reach the fields. In the arid and semi-arid zones, where much irrigated farming goes on, loss by evaporation is greatest. Typically, only between 35 per cent and 50 per cent of withdrawn for irrigated agriculture even reaches the crops. Most soaks into unlined canals, leaks out of pipes or evaporates before reaching the fields. The revolution in urban third-world consumption habits, hailed as part of the great economic miracle, soaks up more water. The effort to emulate the unreal standards of international tourism, with hot water round the clock and fertilizer-rich golf courses, is phenomenally wasteful.

The recent Beijing conference on managing resources for large cities and towns was part of the run-up to the UN Habitat II Conference. This initiative, which stems from the Rio conference, places the responsibility where it properly lies upon our urban societies. The old injunction not to spend money like water needs updating. Whether directly through our taps or indirectly to satisfy an unsustainable lifestyle, we must stop spending water like water.

Policy-makers have yet to find a way to forestall the growing depletion of fresh water supplies in part because of the strictly local nature of water shortages. While West Asia faces shortages, Iceland and Canada have more water than they can use. Food grown to supply the city competes for water with the city itself, and failure to control waste disposal leads to the pollution of water sources. Extraordinary coliform levels have been reported-more than three million per 100 millilitres from canals in Jakarta and ponds in Nigeria, and 7.3 million down stream from Bogota[5]. In fact, about 2.3 billion people in the world suffer from diseases that are linked to water.

The world's population is growing rapidly, especially in urban areas. This would further stimulate demand for food and hence for irrigation as also potable water. Global demands for farm products is likely to increase three-fold in the next 20 years. More food, fuelwood, fodder and fibre would have to be produced from less land, water and energy per unit of output. At the same time, the demand for water by households and industries further stimulated by economic growth, is rising. These factors imply escalating competition for irrigated agriculture, which already takes a hefty share of available water-something like 70 per cent at the international level.

Water shortage is beginning to emerge as a problem even in countries traditionally seen as well supplied, lending support to the view of those hydrologists who predict a world water shortage. They argue that the use of the term 'drought ' has tended to mask a real situation of endemic scarcity. It is feared that by 2025, more than 2.8 billion people will live in 48 countries facing water stress or water scarcity. Of these 48 countries, 40 are either in the Near East and North Africa or in sub-Saharan Africa. Over the next two decades population increase alone-not to mention growing demand per capita—is projected to push all of the Near East into water scarcity. By 2050 the number of countries facing water stress or scarcity will rise to 58, and their combined population to 4 billion people—40 per cent of the projected population of 9.4 billion[6].

The significant novelty lies in the fact that northern temperate countries like Britain are also facing now an apparent water resource deficit of the kind that has impeded development in more arid regions of the southern hemisphere. The typical Northern response to shortage has been to develop new sources with more dams and reservoirs. According to a report by the World Bank[7], at least 40 per cent of the world's population live in houses that has no potable water. And 80 countries have difficulties providing water, leading to both economic and public health problems. The international financial body's report warns that without immediate rationing measures, this century's wars will focus on that clear liquid which so many still take for granted.

Clearly, the availability of good-quality water is central to economic progress and to maintaining a natural environment that can help to sustain such progress; yet in most countries, water resources are not being managed in an efficient and sustainable manner-a situation that cannot continue. The amount of fresh water available to any country is limited by nature. Since nearly all the easily available sources of water have now been developed, or are in the process of development, the unit costs of future projects would only be higher. For example, a recent review of domestic water supply projects indicates that the cost per cubic metre of water for the next generation of projects is often two to three times higher than that for the present generation.

As human activities increase, more and more waste products are contaminating available sources of water. Among the major contaminants are untreated or partially treated sewage, agricultural chemicals, and industrial effluents. These contaminants are seriously affecting the quality of water especially for domestic use. Already, many sources of water near the urban centres of developing countries have been severely contaminated, thus impairing their potential use. Unsafe water can be fatal. Water borne diseases are implicated in the deaths of 3 million people annually-mostly children-and cause over a billion episodes of illness a year. Yet insufficient progress has been made in supplying adequate water and sanitation, especially to the poor. Over 1 billion of the world's 5.4 billion people lack access to safe drinking water, and 1.7 billion lack access to adequate sanitation. There are also wide variations in the consumption levels. For instance, the daily water consumption per person in Papua New Guinea is of half a litre whereas in Uganda, every person consumes 3.5 litres of water each day[8].

CONFLICTS OVER WATER

The fears expressed by various world high officials regarding impending conflicts cannot be brushed aside lightly since fresh water use has already reached or even exceeded natural limits in many countries. Further, with the projection that by the year 2025, two thirds of the world population is likely to live in countries with moderate to severe shortage of water, these statements have to be viewed with greater concern, all the more so since conflicts over water have become more common among competing water users. A burgeoning population, mis-use of water resources and poor management practices have been causing depleted supplies, falling water tables, shrinking in-land lakes and stream flows diminished to ecologically unsafe levels, leading most parts of the world to the threshold of water famine.

There are about 260 international rivers covering almost one half of the land surface of the globe and an innumerable number of shared aquifers providing fresh water to the people within these basins.

Unfortunately, many of the countries are in conflict with their neighbouring countries on sharing of water from the common source, overlooking the fact that water does not recognize political boundaries and it eludes legal generalizations. The countries in North Africa and the Near East, not only face mounting internal competition for limited supplies of fresh water, but also find themselves with their neighbours over water rights. For example:

- Water has been the center of a continuing controversy between Israel and Jordan. In May 1997 a ceremony to create a joint "peace park" on the site where seven Israeli school girls were killed by a Jordanian border guard was canceled after Jordan charged Israel with delaying implementation of a water agreement in the Jordan-Israeli Peace Treaty of 1994. Under the treaty Jordan was to receive an additional 50 million cubic meters of water a year from Israel, mostly from the Yarmuk River, one of the River Jordan's main tributaries. The treaty did not state who would pay for the water and its transportation, however. Since the May 1997 "peace park" crisis, Israel has offered to foot half the bill, and the Israeli cabinet has approved a plan to supply Jordan with 50 million cubic meters of water, constituting the second transfer under the treaty.
- Israel already has used its military power to maintain access to the Jordan River. In the early 1960s Israeli soldiers halted a Syrian-Jordanian scheme to divert the river for irrigation. Later, Israel occupied vital sections of the Jordan River's headwaters, thus making sure that most of its flow would be available for Israeli towns and farms.
- Egypt has threatened Ethiopia with war if it carries out plans to divert more water from the Blue Nile for agricultural use. The Egyptian government sees this issue as one of life or death. Without the Nile's nourishing waters, Egypt could not exist as a nation, since it depends on the Nile for 98 per cent of its freshwater needs.
- The Southeast Anatolia Project in Turkey is one of the largest irrigation and power generation schemes in the Near East. This vast complex of dams, canals, and irrigation systems began operating in July 1992. By early in the 21st century it

is expected to divert at least half of the flow of the Euphrates River—some 4 trillion gallons of water a year—into Turkish dams and irrigation canals. This diversion will leave the downstream countries, Syria and Iraq, with less than half of the stable flow they now have access to. Syria also is planning to take some 3.5 trillion gallons out of the Euphrates before it enters Iraq, thus depriving Iraqi farmers of badly needed irrigation water—water that people in the area have had access to for the past 6,000 years. The entire region is set for a potentially ruinous conflict over limited water resources.

- The ambitious but politically ill-advised Greater Thal Canal (GTC) project in Pakistan is threatening to snow ball into a major problem for the Pervez Musharraf regime and the Pakistan press has warned that a "war for water" among the different provinces was looming large. Strong resentment prevails in Sindh province over the construction of which was laid down by President Musharraf on August 16, 2001. The Dawn editorial commented on July 2003 " A war for water among the country's federating units is looming large and the preparations for a showdown, perhaps final, seem to be in their final phase. The ties, especially between Sindh and Punjab, were marred, what the people of Sindh now strongly believe by the pro-Punjab policies pursued by the Centre in complete disregard to the due rights of the smaller provinces.

The Central Government has been pushing the Rs. 30 billion canal project despite the fact that the Sindh Assembly has passed a resolution against it. In protest against the construction of this canal, a Thal Canal Action Committee (TCAC) has been formed which has started demonstrations and fast all over the province. Reporting this Urdu Friday Special said that all national and regional parties are opposing this project alleging that the government prepared a plan aimed at rewarding former military generals. It said the Sindh land, which had already turned into a desert due to scarcity of water, would not get its due share of water because of this project.

Founder leader of the Muttahida Qaumi Movement (MQM), Altaf Hussain has asked Gen Musharraf and Prime Minister Mir Zafarullah Khan Jamali to stop work on the Thal Canal project forthwith. He also proposed that a round table conference be convened comprising Sindhi intellectuals, experts and all parties because the people of Sindh had strong reservations about Thal Canal adding that Sindh Assembly had also adopted a unanimous resolution against it. Feudal lords had become a part and parcel of the establishment and pointed out that the people of Sindh had been continually deprived of their fundamental rights, even though Sindh contributed 70 per cent of the national exchequer. He regretted that the people of Sindh were being deprived of their rights for the last 55 years although they had been paying 70 per cent revenue. He alleged that in Sindh lands were being distributed among retired Generals while Sindhi youth were not getting employment.

Anti-Thal Canal rallies have also been held in different parts of Sindh including Hyderabad, Badir, Zeel Park, Puno Aquil, Dadu Larkana and Nawab Shah. A Sindhi leader Dr. Qadir Magsi, who had been addressing the rallies had gone on record saying that due to shortage of water, the opportunities of employment had vanished completely and Sindhi youths were compelled to commit suicide. He alleged that all resources of Sindh were being diverted to Islamabad and Lahore. The Greater Thal Canal is a 23—mile floodwater canal (agreed in the 1991 inter provincial water accord) branching of from the Chashma-Jhelum Link Canal and taking water from Indus only during the flood season. The Rs. 30 billion canal is to operate only three months in the year to irrigate the arid areas of Bhakar—Layyah—Khushab benefiting a population of half a million. Pakistan-watchers say that the Sindhis simply don't trust the Punjabis. The Sindhis think that the Chashma Link Canal has been misused by the Punjabis. They are certain that the canal will skim off water that really belongs to the fast desiccating Sindh province.

Avoiding International Conflicts

An important part of any international water management strategy is to help countries that share river basins fashion workable policies to manage water resources more equitably. If the countries conclude proper treaties and establish water regimes, they become resistant enough to survive conflicts between otherwise hostile reparians engaged in skirmishes—or other issues. For example, Indus Water Treaty survived two wars between India and Pakistan. There are many more such treaties entered into by some of the riparian states during the last century to utilize the water resources of their common rivers, thereby averting potential threats of war. The agreement signed in 1946 between

Uruguay and Argentina regarding the Salto Grande project providing for equal rights of hydel power generated; the Itaipu Treaty of Paraguay and Brazil for the hydel development of the Parana River forming the boundary between the two; the 1959 Egyptian—Sudanese Agreement on the Nile waters emphasizing on the existing uses, etc. are a few other successful agreements.

The Essence of Survival

Article 25 of the Universal Declaration of Human Rights says: "Water is the most important single element needed in order for people to achieve the universal human right to a standard of living adequate for the knowledge that clean water is a universal right, it is the responsibility of every one to ensure the purity of water".

The theme for World Environment Day 2003 is "Water—Two Billion People Are Dying For It". It was chosen by the United Nations Environmental Programme to help safeguard our most precious resource. The theme has also been chosen to support the U.N. International Year of Fresh Water, 2003. It is not too late for us to focus our attention on protecting and respecting our water resources. Because by protecting our fresh water, we help ensure our future and our planet's long term prospects.

While these treaties/agreements were entered into under certain situations, there are also such treaties/agreements signed under compulsion. For example, in 1927 the former Soviet Union forced Turkey to enter into a treaty on equal sharing of the resource base on the river streams and brooks coinciding with the line of the frontier between the two countries. The U.S. and Mexico had signed a treaty on the Rio Grande in 1909 with the U.S. holding out the Harmon Doctrine that its rights as an upper riparian state were unlimited irrespective of whatever effect the unbridled exercise of these rights might have on the flow of the rivers in Mexico, the lower riparian state. However, while entering into a treaty on the water of the Columbia river with Canada (the upper riparian) in 1961-62, the U.S. (the lower riparian in this case) repudiated the same doctrine, since the application of this theory in this situation would be disadvantageous to it. As the demand for fresh water in these countries increased tremendously since the last few decades of operation of these treaties, disgruntlements is brewing among the affected parties with the feeling that their rightful share on these waters has been curtailed by past treaties/agreements. Since the beneficiaries may not like to compromise on their existing rights accumulated in the course of last many years, when these documents have to be revalidated for a further period and freshly concluded in this century it may prove to be a difficult task to continue with the provisions in the existing treaties/agreements. The resulting uncertainties may lead to an environment conducive to trigger conflicts since the international law in the present form is ineffective to arrest such a situation and the international institutions are not adequately equipped to resolve water disputes.

The Convention on the Non-Navigational uses of International Water Courses adopted in 1997 by the U.N. General Assembly sets out many important principles for cooperation and joint management of such water courses though no practical enforcement mechanism has been provided to back up the decisions of the final arbiter—the International Court of Justice. Further, it provokes further disputes by institutionalizing the two conflicting principles in the upstream-downstream uses of an International Waterway by emphasizing both the "equitable use" and the "obligation not to cause significant harm". This would enable the upstream riparian to stress on the principle of equitable use" while the downstream riparians could demand the protection of their pre-existing uses under the principle of "no significant harm". However, the Convention has been ratified by only 6 signatories so far against the requirement of 35 to make it a law.

Even within the nations, the upstream-downstream conflicts have caused embitterment among people. In India, in the case of the Cauvery river, legal battles are being fought by the upper and lower riparian States of Karnataka and Tamil Nadu, the former insisting on the principle of "equitable use" while the latter emphasizing on the principle of "no significant harm" to protect their existing rights.

With the scenario of water being a non-substitutable resource, for which legal principles are vague and contradictory, which is becoming scarce with population growth and developmental activities, there is a compelling argument that the plausibility of water war in this century is not a myth but can be reality.

WATER FOR IRRIGATION AND PROBLEMS THEREOF

A review of the growth in world irrigated area shows that, from the beginning of irrigation until 1950, the area irrigated expanded until it reached 94 million hectares. Between 1950 and 1978, irrigated area expanded much more rapidly than population, averaging 2.8 percent per year as it increased from 94 million hectares to 206 million hectares. Since 1978, however, growth in irrigated area has slowed down. Going from 206 million hectares in 1978 to 241 million hectares in 1991, it expanded at 1.2 percent per year. As a result of this, irrigated area per person has fallen by 6 percent over this period. World irrigated area estimated at the end of 2002 is around 280 million hect. leading to further fall in per capita irrigated area.

A substantial share of the area now being irrigated is by the unsustainable use of underground water. In some cases, as in the southern Great Plains of the United States and in Saudi Arabia, irrigation is based on the use of fossil aquifers. In the southern Great Plains, specifically in the area of the Texas panhandle, some irrigated land has already gone back to dryland farming as the shallow southernmost fringe of the vast Ogallala aquifer is depleted[9]. In addition, irrigation in many areas is based on over-pumping of underground water, pumping at a rate that exceeds the rate of aquifer recharge. In the world's three leading food-producing countries, viz China, the United States, and India, over-pumping is now common practice. The USDA, for example, has estimated that a quarter of U.S. cropland is irrigated by lowering underground water tables[10].

In China, irrigation is threatened by a growing scarcity of fresh water throughout much of the country's northern regions[11]. Few realise how tight the water supply has become throughout northern China. In late 1993, the Chinese Minister of Water Resources, Niu Maosheng said,"in rural areas, over 82 million people find it difficult to procure water. In urban areas the shortages are even worse. More than 300 Chinese cities are short of water and 100 of them are very short". The aquifer under Beijing has dropped from 5 meters below the surface in 1950 to 50 meters below in 1993.[12]

Largely as a result of poor design, irrigated land in many parts of the world is plagued with water-logging and salinization. Roughly 10 percent of the world's irrigated area appears to be suffering from salinization serious enough to markedly reduce yields. Another 30 percent may be moderately affected[13]. With a sizable fraction of the world's irrigated area based on the use of unsuitable underground water and with the productivity of a gradually expanding share of the irrigated area adversely affected by water-logging and salting, a substantial investment in new irrigation will be needed merely to maintain the current contribution of irrigation to world food production. Even under the most optimistic scenario, the irrigated area per person will be further declining.

Thus, water used for agriculture must be managed wisely. As agriculture intensifies, attention must be paid to the environment. Improper fertilizer use contaminate surface water and groundwater and end in a waste of farmland that could be used to increase food production. Removing water from rivers and lakes can jeopardize wetlands, an important source of inland fisheries and natural filter that cleanses wastewater.

Reallocation of water made available from one user sector to another is today seen as one of the most effective ways of combating critical shortages. The idea is first and foremost reallocation between irrigation and other sectors of water use. Irrigated agriculture, as mentioned, is the largest water use sector worldwide, at 70 per cent of use. By transferring a relatively small part of the water used in agriculture, the water supply for the other sectors could theoretically be significantly improved. In Morocco, for example, by transferring 5 per cent of the currently used irrigation water the usable water supply for communal drinking and general purpose water could be almost doubled. Whether such reallocation measures would be socially, politically, economically and technically feasible would have to be examined in detail in each individual case. In humid regions, where water use by irrigated agriculture is significantly lower than that for municipal business and industrial purposes, corresponding relocation measures are usually ineffective regarding the water supply situation and are at best only of local significance.

Virtual Water

A new concept of virtual water-the amount of water needed to create goods is gaining momentum among the water-scarce countries to determine their agricultural and industrial production strategies. When one consumes one kilogram of grain, one is also consuming 1,000 litres of water needed to grow that grain; when one consumes a kilogram of beef one is consuming 13,000 litres of water needed to produce that amount of meat, and this is the hidden or virtual water. It is this unconscious behaviour that causes humans to consume so much water. The contrast in water use can be noticed between continents. In Asia, people consume an average of 1,400 litres of 'virtual water' a day, while in Europe and North America, people consume about 4000 litres. About 70 per cent of all water used by humans goes into food production.

The magnitude of this variation demonstrates that diet is very important for water consumption. If the entire world consumed as much virtual water as do people in North America, the world would need 75 per cent more water than it currently uses for food production. No practical way will probably ever exist to trade large enough volumes of water like other products, simply because its weight and volume make this prohibitively expensive. However, the country that opts to be a net importer of 'virtual water', as opposed to real water, can relieve pressure on its own water resources. Among the biggest net exporter countries of 'virtual water' are the U.S., Canada, Thailand, Argentina, India, Vietnam, France and Brazil. Some of the largest net importing countries are Sri Lanka, Japan, the Netherlands, South Korea, China, Spain, Egypt, Germany and Italy. Unconsciously, through food imports, many water scarce countries have already eased tensions over their water problems, so that 'virtual water' imports are already playing a role. Now, virtual water trade is becoming more of a conscious decision by countries.

Many countries could resort to 'virtual water' trade in order to achieve a sufficient food supply for their people, but many Governments do not want or simply cannot afford to become dependent on global trade. This is crucial for countries like India and China. They feel that because they have such large populations, the world market would not be able to supply their food demands in any crisis and so, as much as possible, they want to take care of their own food needs. The governments should start thinking in a different way; at the regional level, they should begin to think how they could share the benefits of water, instead of sharing the water.

Ensuring Equitable and Fair Access to Water

One way to improve access to water may be to treat it as an economic good as well as a social right. A pricing policy that makes wasting water expensive is one of the best incentives to save water. And smallholder farmers with few resources must be treated fairly. Special attention must be paid to women, who represent the majority of the developing world's farmers, yet are often excluded by tradition from owning land and managing water. In conjunction with the United Nations Funds for International Partnerships, FAO is embarking on an ambitious three-year project to strengthen women's ability to manage irrigation and water resources in Cambodia, Nepal and Zambia. With the assistance of local non-governmental organizations, women in these countries are learning how to be more fully involved in water management decisions and actions, improving food security for their families and communities.

In water-short countries national governments need to give water resources management their highest priority. Crafting and implementing a national water strategy is essential to sustainable development. Such a strategy should include four elements:

- Adopting a watershed or river-basin management perspective, especially in water-short regions (also appropriate as an international response, since watershed frequently cross national boundaries);
- Instituting a workable water infrastructure so that national, regional, and local water needs can be met within the context of a national water policy;
- Enacting and enforcing water legislation and regulations that conserve water and value the resource properly according to type of use; and
- Connecting water management to the needs of agriculture, industry, and municipalities, and meeting public health requirements for proper sanitation and disease prevention.

WATER POLICY REFORMS

A crucial precondition for all the measures mentioned above in the framework of supply and demand management is a reform in water policy. Saving water especially depends on such reforms, though it will not be easy to carry out the water policy reforms needed because in the general view and as a matter of cultural and religious tradition water is regarded as a free natural asset, and ultimately also because such reforms often affect interests which profit from water policies as they stand.

The water policy reforms needed differ from country to country, and they depend on the degree of economic development, institutional capacity, and on the degree of water shortage and the intensity of agricultural production. Taking these influencing factors into account a specific and practicable water policy should be worked out. According to Rosegrant[14] the key elements of a comprehensive water policy reform should be;

- legal and social safeguarding of water rights and if need be the introduction of trading in these rights;
- management of the irrigation systems by water users and/or user associations;
- creation of price incentives for saving water;
- introduction of suitable technologies;
- environmental safeguards;
- encouragement of international cross-border cooperation.

The Food and Agricultural Organization of the UN also sees the reform of water resource policies as one of the fundamental preconditions for sustainable use of water resources and has recently begun a number of initiatives to encourage the necessary reforms.

Further, policies to guide agricultural water use are needed on the international, national and local levels. International agreements are especially useful when major source of water cross more than one country. For example, with technical assistance from FAO, the ten nations bordering the Nile River are implementing the Nile Basin Water Resources project as part of the Nile Basin Initiative, helping them to access and manage water equitably. The United Nations has developed a framework convention, now ratified by ten states, to guide negotiations concerning the management of freshwater. It will come into force when 35 nations ratify it.

Local tradition of water management and access must also be considered to ensure they do not conflict with proposed government policies. To back up new water policies, investment is required. National governments will need to establish the conditions to attract such investment. Innovative forms of credit will also be needed to help the poor take advantage of improved water management techniques.

THIRD WORLD WATER FORUM

There is hope and despair in meeting the human need for water; this is the message from the third meeting of the World Water Forum (WWF) which was held in Kyoto during 2003. There have been local, national and international exercises, which have proved successful in harnessing water for direct and indirect consumption. But these are as yet too few to make a difference to the direction of a larger trend, which is one of a spreading water stress in the world. The third WWF meeting, organized by the World Water Council (a forum which counts among its members U.N agencies and civil society orgnisations) heard the messages from participants from more than 160 countries. But despite the welter of promises made in Kyoto by government and non-government agencies, the larger message is one of the world falling back in solving the water problem. If the situation at the time of the first and second WWF meetings in 1997 and 2000 is compared with that in 2003, the conclusion cannot be otherwise. Too little is being done to follow up words with successful action on the ground.

The meetings in Japan heard that the demand for water was increasing three times faster than the global growth in population. The WWF conference was also presented with a gloomy prediction that the population experiencing water stress would rise from 450 million today to as many as 2.7 billion people, or a third of the world's population by 2025. But both these are business—as-usual projections, which assume that the norms of water extraction and consumption will continue as before and cannot be changed. This need not be so. Let us consider, for example, the cliché of a doomsday prediction that the wars of the 21st century will be fought over water, not oil. Yet, a resolution of international water disputes has been the rule rather than the exception over the past half-century. Violence between countries has erupted in 37 instances but as many as 200 bilateral and multilateral treaties on a sharing of waters have been signed. A notable example, as already stated earlier, has been the Indus Water Treaty that has been adhered to by India and Pakistan through two open wars and decades of tension. But there is no denying the existence of a multifaceted problem in water. One facet is that the world still has 1.2 billion people who do not have access to drinking water, while there are 2.4 billion people who have to make do without proper sanitation. Another facet of the water crisis is that sources of freshwater are drying up around the world, while pollution is damaging many others. The problem is, however, that the dominant approach—worldwide—to tackling the crisis in water is essentially one of augmenting supplies without any attention being paid to forms of use of water.

It is not a new set of observations but the discussions at the WWF meet highlighted three sets of practices which have blighted national efforts to tackle the water crisis. Inefficient and wasteful use of water in agriculture, industry and homes (the latter more common in the advanced countries) is one of them. The second is the pollution of water sources. The third is unsustainable extraction of groundwater resources. Action on all these fronts supplemented by exploitation, storage and transport of water will provide the solution, instead of the pure supply-side approach to 'production' of water that continues to rule national programmes, especially in developing countries. A bouquet of solutions that was suggested at the WWF meet included improved governance in management of water resources, capacity building and additional financing for water projects in the poorer countries of the world. These are unexceptional suggestions to operationalize a different approach to solving the world's water problems. What is needed and is yet to be accepted is that be it in crop cultivation, industrial production or residential consumption, only an integrated approach that combines improved efficiency in water use with augmented supplies will tackle deficits and prevent disputes between communities, regions and nations. A pure supply—side approach only aggravates both sets of problems.

WATER AND GENDER

Empowering poor rural women with adequate water rights means strengthening their access to water for both domestic and income-generating uses. Better access to water improves women's health and incomes, and liberates them form the daily drudgery of fetching water. Income-generation through gardening and farming, livestock, aquaculture, forestry, and other water-based enterprises constitutes the mainstay of rural livelihoods, but women's opportunities for watershed income generation are still too limited.

Women's improved access to water can be negotiated relatively easily when new water resources are developed. Worldwide, growing populations with higher living standards are making ever greater claims on finite fresh water resources for agricultural, domestic, industrial, and other uses. In an increasing number of river basins, the physical limits of available water resources are being reached. And pollution deteriorates water quality. Marginalised water users, especially poor women, risk losing even the limited access to water that they now have. Growing absolute water scarcity and pollution hit poor women hardest.

Water Use for Household Consumption

Water use for drinking, cooking, washing, and cleaning contributes to the welfare of all household members. Like other responsibilities for household welfare, the responsibility to provide water for consumption is gendered. In rural communities men may be responsible for the construction and maintenance of domestic water infrastructure, while women ensure that water is permanently available in the home, although other patterns have been reported as well. When the government or private vendors sell domestic water for cash, either men or women or both may be responsible for that household expenditure.

The gender issue at stake is the intra-household division of responsibilities for household welfare. A common long-term vision is that spouses should equally share these unpaid domestic chores. In the short run, however, in poor urban and rural areas, the main issue is that poor men, especially poor women, need to be liberated from the often exorbitantly high cash or labor costs for mediocre water service and low-quality sanitation. What is widely recognized as a basic human right, namely access to safe drinking water and sanitation, needs to be concretized. Whether the private sector or public sector is most effective in realizing this right is an open issue, but public financial support remains necessary. In any case, water supply projects become considerably more effective if men's and women's complementary intra-household roles and perspectives are incorporated in project design and implementation, up to decision making level.

Water use for Income-Generation

Water as input in farming and other productive activities generates an income that is mainly controlled by the manager of the enterprise. The manager is usually also the one who makes investments in infrastructure construction and operation to harness water for her or his enterprise. In order to increase women's incomes from water-based enterprises, access to water and water infrastructure are important factors, but access to land, markets, skills, credits, and so on also critically determines the ultimate profitability of women's enterprises. But even more fundamentally, in areas where most women are excluded from owning economic opportunities, water alone as one input cannot contribute much. A comprehensive perspective is indispensable.

Depending upon the gender of the majority of the farm decision makers in a specific site—for example, in an irrigation scheme-male-managed, dual managed, or female—managed farming systems can be distinguished. In sites with dual—managed or female—managed farming systems, where women are about half or the majority of farm decision makers, water for irrigation directly contributes to women's farming enterprises. Therefore, irrigation agencies that supply water, and in some cases also reallocate the land that is newly brought under irrigation, need to target both men and women farmers on the same footing. However, ample evidence has shown that irrigation agencies, whether state or non-governmental organizations, have ignored the gendered organization of agriculture and interacted exclusively with male farmers, mostly with the local administrative and political elite. Women farmers (and poor male farmers) were typically excluded from these negotiations in which the rights to newly developed water and irrigated land were set. This resulted in the erosion of women's pre-project status and resource rights. Conversely, agencies that took the gendered organization of local farming as their starting point and included both women and men in irrigation institutions and land and water rights, empowered both women and men farmers, tapping the productivity of both groups. Men, women and the project all gained.

The need for gender-balanced allocation of rights to water and irrigated land is most obvious where female-and dual-farming systems already prevail. Examples can be seen in southern and eastern Africa, West African wetlands, and often, on homesteads. Recent interventions that have attended to gender have proved to be smooth and effective. Not only in African countries, but also in India, small schemes that irrigate women-cultivated homestead lands have been successfully constructed for and managed by women's groups. A major challenge today is to develop and widely disseminate affordable, small-scale land—and water—management technologies, such as treadle pumps, bucket drips, and water harvesting techniques. With these technologies more poor rural women and men can obtain access to larger quantities of water. Synergy with marketing, input provision, and other rural development efforts should foster more productive use of the water.

At the other end of the spectrum are the farming systems, such as the large-scale canal irrigation schemes in Pakistan and India. In these system, women are primarily unpaid family labourers who perform transplanting, weeding, and harvesting. At best, when husbands are absent, wives can help to irrigate the land or fulfill canal maintenance obligations. Thus, the local farming system excludes the majority of women from decision making positions in the household enterprise. Women's categorical exclusion from community—level, male dominated irrigation institutions is virtually uncontested.

Even where male farming systems dominate, a minority of women farm is there in their won right. They belong not only to de jure and de facto female-headed households, but also to male-headed households, especially if land is in married women's names and farm sizes are small. These women often have no formal water rights and may face strong taboos against farming, irrigating, and interacting with "strange me" for the women do the work themselves, contesting social norms. This dispersed minority of women farm managers and irrigators in male farming systems need support for achieving stronger water rights and gaining membership and leadership positions in water users associations. The economic empowerment of the majority of women through water-based farming is only feasible if the range of conditions for profitable, women-controlled enterprises is addressed in a holistic way. Women's access to land, and forward and backward linkages are as important as women's access to water.

Water Rights under Growing Water Scarcity

In rural areas in developing countries, a single source is typically used for multiple purposes, such as irrigation, livestock, fisheries, washing and often, for drinking as well. Communal rules for water access are set and implemented through highly complex interactions among the various user groups and local authorities. Technical, institutional, and legal interventions by outsiders may strongly impinge upon local arrangements. Upstream water takers and polluters who share the same river basin further affect local water availability, use and regulation. From local to basin level, large-scale water users and administrative, political or economic elite tend to dominate these interactions, excluding the weaker segments of society. For women's multiple water needs for their own welfare and that of their families often blatantly get overlooked.

New opportunities have emerged to redress this neglect. The 90s decade has been substantive efforts worldwide toward integrated water governance in order to address growing competition for adequate clean water. This has been accompanied by new formal legal frameworks and new water management institutions at basin level that call for genuine representation of all water users' interests. Community—based integrated water management institutions and local water tenure arrangements in rural areas are to be linked with these new basin-level institutions. Poor women's water should get utmost priority from local to basin level.

In sum, the agenda for gender-inclusive and pro-poor water governance in rural areas entails:

- Abolishing the excessive labour and cash costs incurred by poor women and men in meeting basic domestic water needs, while fostering equitable sharing of costs between men and women;
- Developing and disseminating affordable technologies to more poor rural women and men so that they have access to more water collectively and individually;
- Collaborating with other gender and rural—development initiatives to foster women's independent entrepreneurship and make the range of water-based, income generating activities more profitable;
- Informing poor women and men about new water-governance initiatives from local to basin level and including them effectively in the design of new institutions; and
- Ensuring, under competition for water, that is reallocated from large-scale users to small users, to meet both women's and men's water-related basic needs for health and income.

Since the early 1980s more and more attention has been devoted to the relationship between women and the environment, and in particular in relation to water, during international conferences. Agenda 21, the main outcome of the Rio Earth Summit held in 1992, included a chapter on Global Action for Women towards Sustainable and Equitable Development. The same year, the Dublin Statement on Water and Sustainable Development included women in one of its four principles. During the international Conference on Freshwater held in Bonn, Germany, in December 2001, gender was given a prominent place in the Ministerial Declaration Reference[15].

CONCLUSION

As the demand for fresh water is increasing among the competing users, there has been very serious and intensive thinking at various levels all over the world on its limited availability, possibilities of exploitation and the effective management. The world needs sustainable water management. A sustainable water supply which also satisfies people's needs for high quality is an enormous challenge to the governments of many countries, especially in the developing countries, because in many countries the existing water shortage is worsened by a large number of problems or these problems themselves cause or have caused the shortage situation in the first place. Continued rapid growth of domestic and industrial uses, growing recognition of environmental demand for water, and the high cost of developing new water resources threaten the availability of irrigation water to meet growing food demands. A crucial and urgent question for the world therefore, is whether water availability for irrigation—together with feasible production growth in rainfed areas—will provide food to meet growing demand and ultimately improve national and global food security.

The problems to be dealt with make it necessary to develop strategies for combating the current and looming situation of shortage. This is where water supply management and demand management present themselves. Strategies of sustainable water management can only be carried out successfully and sustainably if the framework conditions for efficient use of water are given. Reform of water policies is thus the first and most important step in many countries towards creating conditions which make sustainable use of water possible.

Finally, a vital part of a long-term solution is worldwide recognition of the links between rapidly growing populations and shrinking freshwater supplieds, Recognition, knowledge, and concern can help build the political will to avert a crisis and develop the commitment needed to assure that humanity's apparently unquenchable thrust for freshwater does not exhaust the world's finite water supply.

REFERENCES

1. Wolff, Peter: On the sustainability of Water Use, published in National Resources and Development, Vol. 49/50, Institute for Scientific Co-operation, Tubingen.

2. Ibid
3. Clerk, R: Water The International Crisis—The MIT Press, Cambridge Massachusetts, 1993.
4. Robert Engleman and Pamela LeRoy, Sustaining Water, Population Services International, Washington, DC, 1993.
5. 5. R.A.J Arthur, The Price to pay for scarcity, Water & Environment International.
6. Population Reports, series M, Number 14, February 1999, The Johns Hopkins School of Public Health, Baltimore, USA.
7. The report was presented by World Bank vice-president Ismail Serageldin in the 20th session of the international seminar on planetary emergencies in Erice, Sicilia, in Southern Italy, Economic Times, August 23, 1995.
8. Ingenious Local Solution to Drinking Water Problem, Dr. Sudhirender Sharma, Director, Energy/Environment Group, in Water for Life, F.A.O 1994, p. 63.
9. Postel, S. 1984. Water: Rethinking Management in an Age of Scarcity. World Watch Paper 62 Washington, D.C: World Watch Institute.
10. Sloggett, G., and C. Dickson. 1986. Ground-water mining in the United States. Washington, D.C.: U.S. Department of Agriculture.
11. Ibid
12. Smil, V 1993. China's Environmental Crisis: An Inquiry into the Limits of National Development. Armonk, N.Y. U.S.A. M.E. Sharp.
13. USDA (U.S.Department of Agriculture) 1993a. Production, Supply, and Demand View. Washington, D.C. Computor Disk.
14. Rosegrant, M.W.: Water Resources in the 21st century. Increasing scarcity declining quality, and implication for action—Paper presented at the Conference on the Sustainable Future of the Global System, Tokyo, Organised by the United Nations University and the National Institute for Environmental Studies, Japan, 1995.
15. For further reading see Douglas Merrey and Shirish Baviskar, eds., Gender Analysis and Reform of Irrigation Management: Concepts, Cases and Gaps in Knowledge, workshop proceedings (Colombo, Sri Lanka: International Water management Institute, 1998); Barbaravan Koppen, more jobs per Drop: Targeting Irrigation to Poor Women and men, (Amsterdam: Royal Tropical Institute, 1998): and R. Prdhan *et al.* (eds.), Water, Land, and Law: Changing Rights to Land and Water in Nepal, workshop proceedings (Kathmandu, Nepal: FREEDEAL, Wageningen Agricultural University, and Erasmus University Rotterdam, 2000).

3

Indian Scene

Give us a share of your most beneficent sapidity, like mothers longing with love.

—Rig Veda

Introduction

India, with a population of over a billion is the world's largest democracy, Historically, civilization in India as around the world, have largely evolved and developed around water bodies as most human activities, including agriculture and industry depend on water. Needles to say, access to enough water of sufficient quality is fundamental for all human, animal, and plant life as well as for most economic activity. Throughout the Indian history, farmers have depended on irrigation to produce sufficient, stable food supplies. India's irrigated agriculture sector has been fundamental in its economic development and poverty alleviation. In fact, from 1951 to 1997, gross irrigated areas expanded four fold from 23 million hectares to over 90 million hectares and irrigation continues to be the single largest use of freshwater. Fed by the seasonal monsoon rains and rivers that flow from the Himalayas, India has potentially large water resources. However, the distribution and availability of water is not uniform across the country or through the year. Surface sources often run dry in the summer and groundwater availability varies from the rich aquifers of the Indo-Gangetic-Brahmaputra plains to the comparatively low yielding hard regions of peninsular India.

The water situation in India seems to be going from bad to worse. Not only is there a growing scarcity of water in the country, the agriculturally important states like Punjab, Haryana, Tamil Nadu and

Rajasthan are facing a steady fall in their ground water levels. While the per capita availability of utilizable water in India in 1951 was 3,450 cubic meters in 1999 it came down to 1,250 cubic metres. This according to the ministry of water resources, is expected to decrease to 662 cubic metre per person in 2050. In 1993, only 78 per cent of rural and 85 percent of urban Indians had access to potable water.

An important fact to be noted is that of the total rain water received in the country, only approximately one-third is recharged into the aquifers while the two-third is lost due to run-off and evapo-transpiration. There has been under-utilization of created irrigation potential. High conveyance loss of water, lack of equity, adequacy and reliability of water supply, wastage of water by over-irrigation by the farmers, excessive exploitation of ground water in fresh and marginal zones, increase in operational maintenance (O & M) expenditure coupled with poor maintenance because of lack of funds with the agencies, limited increase in irrigation efficiency, excessive weed growth and silting in the carrier channels, traditional cropping patterns are some of the hallmarks of our water irrigation sector.

Rapid depletion of the ground water level is a cause of great concern for India as its agriculture depends overwhelmingly on this source of water. One estimate has it that groundwater sources account for as much as 70 to 80 per cent of the value of agricultural produce attributable to irrigation. Rainfed farming also has a distinct place in the country's agriculture, occupying 67 per cent of the cultivated area, contributing 44 per cent of the foodgrains and supporting 40 percent population. In fact even after realizing the complete irrigation potential of the country, 50 per cent of the cultivated area will continue to depend on the rainfall.

Concern about quality also needs to be expressed as increasing industrialization and urbanization is taking its toll on the water quality. Safeguarding water quality, particularly in the rural areas, is too inadequate. A primary source of drinking water in India even without treatment is from rivers, lakes, ponds and wells and water-related diseases have become human tragedy, killing thousands of people each year, preventing thousands more from leading healthy lives and undermining development efforts. Thanks to rising population, urbanization and industrialization, water thefts and water conflicts are the realities of the day and water riots and water robberies may be the spectacle of tomorrow. No wonder if someone predicts that the next world war will be fought over water. This chapter highlights in details water scenario of the country, the problems of groundwater depletion, demand for fresh water and conflicts arising out of it and other related issues.

Water Availability in India

- Occupies—2 per cent of World's Land Area
  4 per cent of World's Fresh Water Resources
- Supports 16 per cent of World's Population
- Even the availability of fresh water is highly unevenly distributed over space and time

SCARCE WATER RESOURCES

India—which has 16 percent of the world's population, 2.0 per cent of the world's land area and 4 percent of the world's water resources—is already heading towards a state of water crisis. A late monsoon in 2002 resulted drought in 13 of the 29 states in the country. But drought apart, otherwise too there is a huge gap between the quantity of water that is made available to us by Mother Nature and the quantity that we are utilizing.

Following facts will indicate the water poverty that India faces. And the situation, the experts warn, is only going to get worse. The reservoir of underground water, estimated at 432 billion cubic metres (BCMs), has been declining at a rapid rate of 20 cm annually in as many as 15 states. As a result, it is feared that major metropolitan centers may go dry as early as 2015 on account of over-exploitation and misuse of water resources. Studies conducted by the UN indicate that the rate at which the underground reservoirs are being emptied all over the world, including India, is at least 10 times faster than it can be naturally recharged.

In Tamil Nadu, the watertable has dropped by 30 m in 30 years and many aquifers have run dry. According to a study done by the New Delhi based Central Ground Water Board, it will take just 2,600 additional large tubewells, those of the kind being installed by large housing societies, industries, educational institutions and hotels, running at an average of 10 hours per day, to exhaust the entire reserve of underground water in Delhi. The rate of water withdrawal by existing tubewells in Delhi as it is exceeds the amount replenished by rain to the extent of a good 13 per cent. The annual rains add approximately 7 per cent to the reserves every year. The rate of withdrawal by existing tubewells, however, amounts to nearly 20 per cent. This kind of over pumping of ground water, can magnify the concentration of pollutants in the water that remains, and in many cases polluted surface water or salty sea water pours into the aquifer to replace the ground water, making it impossible to farm. "Over 200 million people are at risk as eight river basins are water-deficit. Another 100 million in 12 cities also thirst for water".

The ecological consequences of drawing heavily on ground water too are grave. Deep aquifers are a vital link in the hydrological cycle because they release water slowly into rivers, lakes and wetlands in the dry seasons and soak up water to prevent flooding in the wet times. The only reason that many of the world's great rivers such as the Niger and the Nile flow all year round is because of ground water release. Taken too much and the result is dried—up wetlands and river beds, environmentalists warn.

The problem of urban water scarcity is further expected to intensify in the near future. Studies indicate that by 2050, more than 50 per cent of the Indian population may shift to cities, making water scarcity an acute problem. The water conflicts between states as well as nations too are expected to intensify on this account. The states that figure high on the list of potential water scarcity are Punjab, Haryana, Bihar, Andhra Pradesh, Gujarat, Karnataka, Madhya Pradesh, Maharashtra and Orissa.

Rainfall

India receives an average annual rainfall equivalent of about 4,000 billion cubic metres (BCM). This source of water is unevenly distributed both spatially as well as temporally. Since the lay of mountain ranges affects significantly the distribution of rainfall, windward slopes receive heavy rainfall while the areas situated on the leeward are very dry. For example, south of 18°N parallel, the west coastal lowlands and the Western Ghats at most of the places, have a rainfall over 250 cm., but a narrow strip of land east of the Western Ghats receives only 38 to 65 cm of rainfall in monsoon season. Southern

Tamil Nadu which is situated to the east of the Cardamom Hills also receives very little rainfall. Pudukottai, for example, receives 5.5cm of rainfall during months from June to September. Excluding the Western Ghats and West coast lowlands, rainfall decreases, in general, from east to west.

Table 3.1: Rainfall Pattern in India

Town	*Highest Rainfall (Annual) (c.m.)*	*Lowest Rainfall (Annual) (c.m.)*
Shillong	241.5	158.9
Kolkata	158.9	115.6
Allahabad	102.7	90.5
New Delhi	71.4	56.2
Srinagar	66.4	19.1
Bikaner	30.5	26.3
Jabalpur	144.8	128.9
Nagpur	112.7	98.3
Malegaon	58.0	45.7
Ahmadabad	82.3	79.2
Nagappattinam	133.7	20.0
Mangalore	346.7	288.4

The number of rainy months and the amount of rainfall are of considerable importance, because they determine the agricultural possibilities of a particular area.

Salient Feature of Monsoon Rainy Season

June to September is the main rainy season of India. As high as four-fifths of her annual rainfall is received during this season. These rains are of great importance for the agricultural prosperity of India as the successful cultivation of about three-fourths of the cultivated land depends on the adequacy of these rains. Hence, this is the period of major agricultural activity. Rice, jowar, bajra, jute, sugarcane, cotton, ground-nut, etc. are raised in this season. Even tea and coffee cultivation is possible largely due to these rains. However, the dates of the onset and the withdrawal of the monsoon in rainy season depart from the normal occasionally. Even when the date of onset of monsoon is normal, monthly and seasonal amount of rainfall deviated widely from the normal. Some months receive exceptionally heavy rainfall in certain years while the others are abnormally dry (Table 3.2).

Table 3.2: Variation in Actual and Mean Rainfall

Town	*Month*	*Mean rainfall in the month*	*Actual rainfall in the month with year Wettest)*	*Actual rainfall in the month with Year Wettest)*	*Mean Annual rainfall*
Hissar (Hissar)	August	11.39.cm.	53.85 cm. in 1926	1.97 cm. in 1957	44.60 cm.
Calcutta	September	28.97 cm.	115.70 cm. in 1900	5.61 cm. in 1938	158.18 cm.
Bombay	July	70.95 cm.	149.98 cm in 1907	10.79 cm. in 1918	209.92 cm.
Nagpur	September	18.05 cm.	62.71 cm. in 1891	5.18 cm. in 1899	112.73 cm.
Allahabad	August	33.31.cm.	91.47 cm. in 1953	5.33 cm in 1883	102.73 cm.

Source: A Geography of India by Gopal Singh, pp. 29-30

The rainfall in most of the areas except North-Eastern and coastal regions is erratic with occasional long dry spells turning agriculture into a gamble. Supplementary irrigation is therefore essential even for kharif while for rabi a substantial supply is needed for assured and higher production. Use of ground water can be helpful in areas where it is available, hence the need for an assessment of water resources. Normally there is a five year rainfall cycle. It is, however, interesting to note that India has witnessed normal monsoon since 1988 (Table 3.3)

Water Availability

- Average annual precipitation 4000 BCM
- Average annual water res. potential 1869 BCM
- Utilisable Surface water 690 BCM
- Replenishable Ground water 432 BCM

As mentioned earlier, India receives about 4,000 billion cubic metres (BCM) of rainfall every year of which 1869 bcm appears as average annual run-off in the various rivers of the country. Due to topographical and geographical limitations, only 690 bcm of surface water can be utilized in addition to 432 bcm of replenshible ground water. Over—exploitation of ground water in some parts of the country is leading to a situation of droughts while mismanagement is resulting in floods in other parts. The country's requirement upto 2025 can be met through optimal utilization of the renewable water resources. The Government is promoting rainwater harvesting through watershed management programmes and other water management schemes.

Table 3.3: Monsoon Track Record

Year	*Rainfall as % of long period average**		*Arrival date in Delhi*
	Actual	Forecast	
1988	119	113	30 June
1989	101	102	1 July
1990	106	101	1 July
1991	91	94	14 July
1992	93	92	11 July
1993	100	103	27 June
1994	110	92	25 June
1995	100	97	11 July
1996	103	96	23 June
1997	102	92	9 July
1998	105	99	17 June
1999	96	108	28 June
2000**	-	99	29 June

* average annual rainfall over 70 years
** anticipated

Water Needs

Some 200 million Indians do not have access to safe and clean water. An estimated 90 per cent of the country's water sources are polluted with untreated industrial and domestic waste, pesticides and fertilizers run-off from fields. According to a publication of the Rajiv Gandhi Drinking Water Mission-'Introduction of Rural Water Supply and Sanitation Programme in India, published in 1993'—about 1.5 million children under five die each year from water born diseases. The country also loses over 200 million workdays annually due to these diseases. According to the World Resources Institute, agriculture accounts for 69 per cent of fresh water withdrawals on a global plane, industry 23 per cent and domestic use 8 per cent. There are regional variations with industry as the largest user in developed countries. In 1995, according to experts, domestic use accounted for 350 cubic kilometers of fresh water. In Africa; the per capita withdrawals for personal use were only 47 litres per day. This figure was 85 for Asia, 334 for the UK and 578 for the US.

Indiscriminate and unregulated extraction of groundwater has led to fall in the groundwater table by about 25 metres in the last 25 years in some of the dark/brown blocks of the country. In the coastal areas it has led to the incursion of sea water into the aquifers (rock of soil layers holding water).

Water use efficiency is low in India, being no more than 30 to 40 per cent. The International Water Management Institute (Colombo) estimates that even if an equitable water use efficiency of 70 per cent were to be attained by 2025, there would be a 17 per cent increase in irrigation water demand. All this implies a need for additional storage through water harvesting, recharge of groundwater, integrated watershed development, participatory management and protection of water sources from getting polluted. Interestingly India has 300 mm more rainfall than the world average of 800 mm though a large amount of it is allowed to run into the river and sea and not collected for proper utilization.

The National commission for Integrated Water Resources Development Plan (NCIWRDP) in the report of September, 1999 adopted the norms for water supply as 220 litres per capita per day (LPCD) for urban areas and 150 LPCD for the rural areas. According to these norms, the national water requirements for domestic and municipal use for the years 2010, 2025 and 2050 were calculated as 43 BCM, 62 BCM, and 111 BCM respectively. Of these, the ground water will be required to contribute 19, 26 and 46 BCM respectively for these years. The water requirements for bovine population is included in the total water requirements for domestic and municipal use. Apart from the above purposes water is also required for industries, power generation, inland navigation and ecological purposes. However, the demand from ground water is mainly for irrigation, domestic and municipal use, industries and power only. According to the estimates made by the NCIWRDP, the total requirements from ground water are 252 BCM (year 2010), 282 BCM (year 2025) and 428 BCM (year 2050).

Peter Gleick has placed a range of 20 to 40 litres per day as a minimum need for drinking and sanitation. Including bathing and cooking, this figure varies from 27 to 200. Gleick, however, proposes an international standard of "an overall basic requirement of 50 litres per person per day" as a basic need for drinking, bathing, sanitation and cooking. This is the norm which is required for rural India. According to him, in 1990, 1 billion people in 55 countries did not meet this standard. Malin Falkenmark suggests a standard of 100 litres per day for personal use in developing countries. However, there is little unanimity in these estimates. It has been shown by Rosegrant and Ringler of the International Food Policy Research Institute that the minimum water requirement for basic needs ranges from 400 to 2,000 cubic metres. This is obviously an area for further work.

Water Resources

Although India possesses about 16 per cent of global population, its water resources are just 4 per cent of the total average annual runoff in the rivers of the world. The distribution of rainfall in the country is also highly uneven with 100 mm in Western Rajasthan to over 11,000 mm at Cherrapunji in Meghalaya. Further, variability of rainfall from season to season is also very high. The compounding effect of such factors is the uneven water availability from basin to basin. India has been divided into 20 river basins and the availability varies from 18,417 cubic metres in the Brahmaputra valley to as low as 180 cubic metres in the Sabarmati basin, Rajasthan, for instance, with 8 per cent of the country's population has only 1 per cent of the country's water resources while Bihar with 10 per cent of population has just 5 per cent of the water resources. Thus, while India is considered rich in terms of annual rainfall and total water resources, its uneven geographical distribution causes severe regional and temporal shortages[1]. According to the Central Water Commission's estimate, total average annual run-off of the river system of the country is around 1,869 cubic kilometers.

Water Use

Directly or indirectly, quantum of consumption of water is related to population, demand for food, production of non-food agricultural and industrial items, consumption of energy, level of quality of life and ecological need of a nation. Availability of water, in turn, determines suitable technology for adoption in these sectors. Water requirement for agricultural and domestic needs including drinking, cooking, sanitation, etc. together constitutes bulk of total water consumption. However, due to competing sectoral water demands, particularly from energy and industrial sectors in next 5 decades, the share of irrigation in total water use would come down from 83.3 per cent in the year 1990 to 72.8 per cent in 2025 and further to 64.7 per cent by the year 2050. (*Table* 3.4)

Table 3.4: Water Use in the Year

Purpose	*1990*		*2000*		*2025*		*2050*	
	BCM	*% of Total*	*BCM*	*% of Total*	*BCM*	*% of Total*	*BCM*	*% of Total*
Domestic	25	4.5	44	6.6	77	6.2	93	5.6
Irrigation	460	83.3	520	78.5	909	72.8	1,072	64.7
Energy	19	3.5	27	4.1	70	5.6	212	12.8
Industries	15	2.7	30	4.5	120	9.6	199	12.0
Others	33	6.0	41	6.2	72	5.8	80	4.8
TOTAL	552	100.0	662	100.0	1,248	100.0	1,656	100.0

Source: 1, Draft of Basic Document on the Assessment of Availability and Requirement of Water for Diverse Users in the Country; Ministry of Water Resources, GOI, New Delhi, 1997-Revised in October 1998.

Further, India is the second largest consumer of water in the world, after China. India's water consumption is approximately 20.1 per cent of the world consumption. Per capita water consumption in India, at 297.7 cubic meter, is more than the world average per capita consumption of 287.3 cubic mt. Following Table 3.5 indicates a comparative picture of usage of water in a few major countries of the world.

Table 3.5: Usage of Water

Country	*Total Consumption (Cubic Kms)*	*Per Capita Consumption (Cubic metres)*
Australia	12.6	648.3
U.S.A.	166.3	598.1
Japan	45.1	355.4
China	391.7	307.7
Germany	18.8	226.7
World*	1,524.4	287.3

* The World figure includes the top 90 countries according to their 9 Bp
Source:-The Economic Times, New Delhi, October 14, 2002.

Over use and Deterioration in Quality

The over use of water in India has been the cause of many serious problems. For instance, the environmental problems which include water quality degradation from agro-chemicals, industrial and domestic pollution, groundwater depletion, water logging, soil salinization, siltation, degradation of wetlands, and ecosystem impacts are of great concern. Further, water—born diseases following increasing water pollution aggravating health-related problems. Mismanagement of water at the farm level is rampant. Where water is available in abundance it is greatly over-used leading to water logging and salinization of once fertile agricultural lands.

Alongside, there is increasing extraction of groundwater to meet the demand of agriculture, especially for the cultivation of water intensive crops, like sugarcane. It is estimated that nearly 90 per cent of groundwater extracted is used for irrigation. Presently, more than 8.5 million electric and diesel pumps are used to withdraw groundwater leading to falling water tables in most states. Farmers are forced to dig ever deeper and in many cases have begun to mine geologically ancient aquifers.

Further, the indiscriminate dumping of untreated industrial and domestic waste into rivers and lakes further exacerbates an already stretched water situation. The pollution reduces the amount of water available for use, creating scarcity in regions which otherwise have abundant water resources. The domestic sector produces about 75 percent of the wastewater and sewage treatment facilities are inadequate in most cities and almost absent in rural India. The Central Pollution Control Board (CPCB) has identified 1,551 large and medium industries in 17 categories of highly polluting industries. They have been given time schedule to install necessary pollution control equipments to comply with the prescribed standards. As of June 30, 2002, 1,349 industrial units had installed the requisite pollution control equipment, 179 units had closed down and 23 units were yet to install the necessary pollution control facilities. Legal action has however been taken under the Environment (Protection) Act, 1986 against all the defaulting units[2].

Out of 726 industrial units along the major rivers and lakes in the country, which have been found discharging their untreated or partially treated effluents into the freshwater bodies including rivers and lakes, as on September 30, 2002, it was found that 409 firms complied with the prescribed standards, 178 had Effluent Treatment Plant (ETP) facilities but these were not operating satisfactorily, three industrial units were getting their ETP installed and 130 industrial were closed down. Only six units had not installed the requisite pollution control facilities. Legal actions have been initiated against all the defaulting units[3].

Of the over two million small-scale industrial units, a number of which-like tanneries for instance, are extremely polluting, very few have any treatment facilities whatsoever and their untreated waste invariably find their way into country's water system. In the North Arcot district of Tamil Nadu, for instance, some 80,000 litres of effluent containing toxic chemicals like mercury and chromium are discharged untreated each day from the tanneries. It is reported that absent 1.5 million children under five die each year from water borne diseases. The country also loses over 200 million workdays annually due to these diseases[4].

Groundwater pollution is particularly serious as some percent of the country's domestic water needs are met from this source. Disease like hepatitis and diarrhoea, roundworm, hookworm and guneaworm are common place and, as reported by the World Bank, some 21 per cent of all communicable diseases in India are water borne.

India to Become Water-Stressed by 2025

India's total annual renewable fresh water resources are estimated at 2,085 BCM. However, the annual average water availability (AWA) in terms of utilizable water resources within the limitations of physiographic conditions, socio-political environment, legal and constitutional constraints and the technology of development as available is estimated at 1,086 BCM, comprising 690 BCM of surface water and 395.6 BCM of ground water. Water scarcity has been defined according to "supply-side" approach in the pioneering study by Falkenmark, Lundqvist and Widstrand. For this purpose, ranking has been done according to per capita amount of "Annual Water Resources" (AWR). They define 1,700 cu.m. per capita per year as the level of water supply above which shortage will be local and rare. Below 1,000 cu.m. per capita per year, water supply will hamper health, economic development and human well-being. At less than 500 cu.m. per capita per year, water availability is a primary constraint to life.

India will be on the list of water-stressed countries by the year 2025 when nearly half the world population will experience water shortage. Today, nearly 1.2 billion people do not have access to safe drinking water. The stress on water resources is the result of a multitude of factors such as rapidly rising population and changing lifestyles that have increased the need for freshwater and intense competition among agriculture, industry and the domestic sector that is pushing the groundwater table deeper.

Nearly 40 per cent of India's urban population, which is below the poverty line, has no access to water. The scenario in rural India is no better. In 1985, there were 750 villages without water resources, with the number increasing to 65,000 villages in 1996. But these figures of fast depleting water table have another story to tell. It has changed the lifestyle of the villages. Researchers at The Tata Energy Resources Institute, with funding from the United Nations Population Fund, tried to understand the linkages between population and water by carrying out a survey at the national and village levels to ascertain the impact of population growth on water resources and that of availability of water and water quality on the life of the people, particularly women and children.

The study conducted in 350 households in 20 villages in the districts of Thiruvananthapuram, Solan, Raichur and Bikaner threw up some interesting facts. In all the villages surveyed, water scarcity was a common problem but the perceptions were different. For some areas in Bikaner and Raichur, water shortage meant shortage even for drinking and in Solan, the problem connoted shortage for irrigation purposes. For instance, of the villages surveyed in Bikaner, there were cases of households with 5-6 members managing with just 10 litres of water a day while in the villages of Solan, drinking water did not appear to be a problem, but water for irrigation was an immediate requirement.

On the other hand, the villages surveyed in Thiruvananthapuram did not face water scarcity except in summer months when few households in coastal villages experienced the problem of salt-water mixing with the water in wells. In all the villages surveyed, it appeared that there was considerable effort for collecting water for drinking and other household uses. The stress varied from region to region depending on water resources, waiting time and number of trips required. In all of the villages surveyed in Raichur, Solan and Bikaner, the burden of water collection was on women and children (mostly girls), who spent several hours every day on the activity. Most women respondents complained of frequent backaches as they had to carry heavy pots of water. In raichur, the villagers also felt they were unable to send their children to school as they were required for household chores, including collecting water.

In the villages of Bikaner, a few relatively better-off households and several of those belonging to the Rajput community (where women were not allowed to go out) spent as much as one-fourth of their income on hiring carts for carrying water and investing close to six-months' income in maintaining storage facilities. Diarrhoea was a frequent problem in all villages surveyed in Raichur, Solan and Bikaner though the villagers did not perceive it as primarily a water-related problem. Group discussion also indicated that there were conflicts over water in some of the villages in Solan and Bikaner. Findings from the field surveys indicated the vicious circular linkages between poverty and environmental degradation.

Management of Water and Mitigation of Scarcity

It has been mentioned that availability of water varies very widely from one region to another. While eastern and north-eastern regions have, by and large, adequate availability of water both for domestic and agricultural uses, the states in western and southern regions suffer from acute scarcity of water, particularly during the summer months. The situation becomes very critical particularly the year the rainfall is scanty. Mismanagement and misuse of water further aggravates the situation. For instance, what governs the volume and depth of groundwater extraction is the cropping pattern. Administered prices play a big role in this regard. The crops suited to the dry lands of India have suffered neglect not merely in poor state-sponsored research & development, they have also been discriminated against in the procurement and price-support systems. Farmers have tended to grow water-intensive crops, as these have been the ones favoured by state procurement and pricing policies. Over-extraction of groundwater for irrigation purpose and its unsustainable use are the major factors intensifying droughts suffered by the hard rock regions of the country. In these regions, comprising nearly 70 percent of India's landmass, natural rates of groundwater recharge are low. The situation demands exercising great care for preserving the balance between rates of groundwater extraction and recharge. Unfortunately, this balance has been repeatedly violated.

However, our tribal farmers show an acute perception of the intricate inter-relationships between surface and groundwater.

Groundwater is not used for kharif irrigation, the crop being protected through farm ponds or nulla bunds made as part of the watershed project, Groundwater is carefully preserved first for drinking water, for livestock and humans and then used, if possible for rabi irrigation. Tubewell, if any, are set aside only for drinking water.

It may be reiterated that there is a complete lack of effective water management and planning which is primarily responsible for the severe water crisis in many parts of the country. In fact, the present institutional arrangements in India including central, state and local institutions do not enable comprehensive water allocation, planning and management. Further, non-implementation of the policy pronouncements is also, to a great extent, responsible for the water crisis situation in India. The top-down approach of water management also results in major economic, social and environmental cost. Moreover, actual utilization of water for different purposes has never been assessed. However, the Central Water Commission projected a demand for water for various uses in the country in 2000 at 750 bcm and assessed utilizable surface water by conventional methods as 690 bcm. To harness available water resources and augment water supplies about 3,600 large dams and reservoirs and 250 barrages have been constructed and a total live storage capacity of 177 bcm was provided by 1995.

Needless to say, water is as essential for human existence as for the propagation of wildlife and fisheries. Water resources management is getting integrated, where there is more focus on storage, supply, consumption and distribution to different segments of the society. In India, the Ministry of Water Resources lays down policies and programmes for the development and regulation of the country's water rescurces. The Central Water Commission is engaged in flood forecasting on inter-state rivers. The Central Groundwater Board (CGWB) is responsible for assessing the groundwater potential in India through hydrogeological surveys, exploration, evaluation and monitoring of groundwater regime. The Central pollution Control Board (CPCB) has under its purview the promotion of cleanliness of streams and wells in different areas of states by prevention, control and abatement of water pollution. Despite having these institutions, the water crisis and non-availability of quality water in most parts of the country are persisting and becoming more serious. However, many non-governmental organizations and panchayat raj institutions have been endeavouring to mitigate the water crisis in various states.

NEGLECT OF RIVERS

The recent release of waters from the Narmada into the Sabarmati must have given rise to much salutation of the river as a life-giver, but we have as nation never done enough to take our lifelines. Several perennial flows like the Ganga-Bramaputra-Barak are becoming seasonal. Rivers are drying up or declining in volume. The Yamuna supplying water to Delhi and Agra is highly polluted, mainly because waste water that is discharged from Delhi, Mathura, Agra and Etawa into the river is untreated and the river does not flow fast enough to dilute the load of pollutants. The marauding of the Ganga is now an oft discussed issue. There are vast stretches of Indian rivers that are unable to support fish on account of the levels of pollution caused by discharge of untreated effluents. Of the total water that we pollute, only one-tenth of the waste water from cities is treated. As a result, of the 85 per cent of the country's urban population that has access to drinking water only 20 per cent receives drinking water that meets the health and safety standards.

Acute scarcity and growing pollution have prompted the Government of India to set for itself the target of cleaning all major rivers of India by 2007. But keeping in mind the government's track record of meeting deadlines, we should not be too optimistic about the issue. Measures being considered by the government to tackle the water scarcity problem are harvesting of rainwater, spreading awareness, intelligent damming of river waters and encouraging traditional water—harvesting techniques. What we all need to understand is that the ecological and environmental crises that we all have been talking and hearing about so far, are no more problems of the future. The calamity is staring us right in the face. If we don't wake up to the facts now, it might just be too late.

Himalayan Waters

Himalayas are water towers for humankind. Uttaranchal mountains are especially rich in their water resources. This area is home to dozens of perennial streams and numerous other rainfed rivers along with innumerable rivulets, waterfalls, and ponds, etc. India's largest river system—the Ganga River System-also takes its origin in the Uttaranchal mountains. In Garhwal Himalayas, the largest glacier Gangotri, the origin of the river Ganga, has an estimated volume of well over 20km^3 of ice, which is much more than even the maximum capacity of storage created by the Bhakra Dam on Sutlej, which is less than eight km^3. This glacier, unfortunately, now continues to recede. Apart from these streams and rivers, there are several lakes in the area as also the perpetual ice fields and glaciers.

However, due to maldevelopment, mismanagement, and continuous neglect, the Himalayas' aquatic resources are in hot waters. Dry spell can be witnessed all over especially in the summer season. In the land of plentiful water, most of the villages are suffering from the lack of drinking water. Majority of the people do not have easy access to potable water. Only some 10 per cent of the arable land is irrigated, and the rest 90 per cent depends on rain.

Uttaranchal's glacier-fed rivers are being seen only as a potential source of hydropower. Ruthless mismanagement of Himalayan waters might trigger potentially negative consequences not only in Uttaranchal, but also in the entire Indo-Gangetic Valley. Aquatic ecology in the mountains on the whole, is in a bad state, affecting the integrity of the entire food production system.

All the water bodies in the mountains can be managed as a potential source of food production. The mountain water sources harbour a variety of food sources; fisheries and aquatic food plants being the most important. Water also has tremendous bearing on all components of the food production system. Water, in fact, is the most important element in the ecological integrity of a farming system. Its very wise use and conservation-based management is vital for sustainable food security in the mountains.

WATER SCARCE BASINS

It will be pertinent to describe the two important river basins in the Indian subcontinent which have their ramifications on relations with the neighbouring countries viz. Pakistan, Nepal and Bangladesh. These are (i) Indus basin and (ii) Ganga basin.

Indus Basin

The Indus basin begins in Tibet and flows through India and Pakistan, about 60 million people reside in the basin area, which covers the northern and northwestern states. The drainage area of 321,289 square kilometers of the Indus River basin encompasses nearly 10 per cent of the total geographical area of India. Crop land in the basin is about 9.6 million hectares, nearly 5 per cent of the total crop land of the country, of which about 30 percent is irrigated. The fight for water has been ongoing in the Indus basin, with water table dropping because of groundwater over-pumping and basins running dry during the summer season. Water scarcity in this basin is an international issue; after the independence, India and Pakistan nearly went to war over this basin. However, a water treaty was established in 1960 and that has proved resilient.

Ganga Basin

The Ganga, the subcontinent's largest and most important river, rises in Nepal and flows 1,400 miles through three densely populated Indian States—Uttar Pradesh, Bihar, and West Bengal—before entering Bangladesh and flowing into the Bay of Bengal. The Ganga River basin within India encompasses nearly 26 percent of the total geographical area of the country, and is inhabited by 323 million people. Physical cropland in the Indian part of the basin is estimated to be 58 million hectares or 30 per cent of the total crop land of the country, of which 20 percent is irrigated. Even though the Ganga does not seem water-scarce based on total annual flows, it often experiences severe water stress from January to April, and floods during other months[6].

There was perpetual serious dispute over sharing of Ganga water, with the then East Pakistan. The conflict persisted even after creation of Bangladesh. However, a new chapter in the Indo-Bangladesh relations opened up with the signing of the Indo-Bangladesh Treaty by the Prime Ministers of India and Bangladesh on 12th December, 1996 on sharing of Ganga Waters for a period of thirty years. The Treaty provides Bangladesh with a guaranteed minimum amount of water sharing during the dry season, especially the three driest months of March, April and May. In order to ensure implementation of the agreement, a team of inspectors from the two countries monitors the flow rate of the Farakka Barrage during the dry months.

Critics however argue that, if the agreement is to work over the long term, India must begin to manage the Ganga watershed much better than it does now. Deforestation in Nepal and northern India has greatly increased the amount of sediment washed from the hills into the river during the monsoon season, clogging waterways and increasing the incidence of damaging floods. Unless ways can be found to capture more stable runoff during the wet season for use during the dry season, Indian farmers might be tempted to take all the water they can get from the river during the driest months, putting the agreement in jeopardy. Despite such caveats, the fact that two neighboring countries have successfully negotiated and reached a comprehensive agreement over such a contentious issue is a positive sign. It promises to permit downstream Bangladesh a more equitable supply of water from the Ganga and to foster better water management practices in upstream India.

Projects for Cleaning Rivers face Funds Crunch

Unable to take up new projects for cleaning rivers due to resource constraint, the Ministry of Environment and Forests has decided to prioritize the approved works under the National River Conservation Plan (NRCP). Priority will be accorded to the projects under way. The Ministry has expressed inability to take up the remaining pollution abatement works of important rivers such as the Ganga and the Yamuna, for which there has been a major demand from a cross section of the Society. But, considering the importance of these two rivers and the fact that a significant investment has already gone into their cleaning, preference would be given to them to consolidate the gains of the work already done.

The Ministry has sought a review of the Tenth Plan, if pollution control at these two rivers is to be given higher priority and works are to be completed during the current Plan period. The shortage of funds was reported only after a year of the Tenth Plan elapsed. The total fund requirement for the remaining works of Ganga and Yamuna alone is estimated at Rs. 2,000 crores and the Ministry is exploring the possibility of getting external assistance. The target is to clean all major rivers by 2007 and others by 2012, though matching outlay is not available in the Tenth Plan to meet the target. A gross budgetary support of Rs. 1,417 crores was initially approved under the Tenth Plan which the Ministry has agreed to raise to Rs. 1,825 crores at the time of mid-term review. The approved works under the NRCP now extend to 157 towns along 31 stretches of polluted rivers in 18 States, involving Rs. 4,064 crores. As many as 763 projects, worth Rs. 2,460 crores, have been sanctioned under the Plan. This is in addition to 261 projects sanctioned under the Ganga Action Plan (GAP) phase-I which has now been merged with the NRCP.

The main objective of the NRCP is to maintain the quality of water of the major rivers through the implementation of various pollution abatement schemes. The river-cleaning proramme was started with the Ganga Action Plan in 1985 and extended to GAP-II in 1995 before being merged into the NRCP in 1996. The NRCP, governed by the National River Conservation Authority, promotes and approves appropriate policies to achieve the objectives, and mobilize resources.

When started, the GAP was a 100 per cent Centrally-sponsored scheme which was then made to share cost on a 50:50 basis between the Centre and State. In 2001, it was decided that the Centre would bear 70 percent of the cost of the schemes while the remaining would be taken care of by the States, of which the share of the public would be 10 per cent.

Other River Basins

Apart from these two important river basins, there are other river basins and a number of states have become embroiled in disputes over water rights and over dams that might provide more water for one state but at the expense of another. "Water disputes, if not attended to, will become a major headache for the stability of Indian Society", says Mohan Katarki, a lawyer representing Karnataka state in a water dispute with Andhra Pradesh[7]. These two states are arguing the court over the height of a dam on Krishna River, which could affect the amount of water available for use by both states.

Similarly, conflict over sharing Cauvery waters among the riparian states of Kerala, Karnataka, Tamil Nadu and Pondicherry has taken a serious dimension. In view of the uncompromising attitude adopted by the concerned states and the direction given by the Supreme Court, the Central Government came to the conclusion that there was no possibility of resolving the dispute by continuing negotiations. In accordance with section 4 of the Inter state Water Disputes Act, 1956, the Central Government Constituted the Cauvery Water Disputes Tribunal for adjudication for the Cauvery Water Dispute and to refer the complaint of Government of Tamil Nadu to it for adjudication. However, the Tribunal has not been able to solve the dispute and later in pursuance of the order given by the Supreme Court the Government of India notified a scheme called " Cauvery Water Scheme" under 6 A of the Inter-State Water Dispute Act, 1956 on 11.08.1998 to give effect to the Interim Order and all subsequent orders. The Authority has been working out on acceptable and equitable procedure for pro-rata sharing of the distress[8].

Thus, water is becoming a contentious issue in India, as the dispute over the sharing of Cauvery waters has clearly shown. This has sharply brought home the urgency of cooperative and integrated management of water resources.

A PANOROMIC VIEW OF WATER SITUATION IN A FEW STATES

Rajasthan

The State of Rajasthan is often affected by droughts during summers. The summer of 2000 had been the worst ever affecting 26 out of 32 districts. This grim picture was worsened by the fact that out of 34,609 villages, 23,500 were reeling under drought and at least 35 million people and 37.5 million cattle were affected.

Marwar region's seven districts account for 12.2 million population and 1,35,519 sq. km geographical area. As per the latest census the growth rate in population here ranged from 47.45 per cent in Jaisalmer district to 22.39 per cent in Pali district. The most populated of all deserts in the world the density in the Thar varied from 13 persons per sq. km in Jaisalmer to 166 persons per sq km in Sirohi. The ground water availability in the region is poor with only 56.83 per cent of the total area holding any potential. Out of the potential areas too, in 45.21 per cent the ground water sources have been over-exploited, in 18.60 per cent it is critical and in 2 per cent it is semi critical. Only 34.20 per cent of the available landmass is under safe category.

Forced Bachelors Blame it on Water

The severity of water crisis in some parts of the state is revealed from the fact a village in Rajasthan is turning age-old social practices on their heads. Here prospective grooms are willing to shell out dowry and are ready to marry widows. This revolution has been forced on grooms of Saderi village, which is facing a severe water crisis. Men here aren't finding brides and the ones who are married cannot hold on to their wives. There are no water sources in Saderi. Women here trudge two kilometers down the hillock to fetch water from the hand pumps and two wells. It is the journey back that is grueling, especially in the killing heat and with pots delicately balanced on their heads.

And now, no man wants his daughter to do the drudgery in Saderi. Which explains why 80 to 90 bachelors are waiting for marriage proposals to trickle in. Worse, over 10 married women fled the village, unable to bear the daily burden.

BAIF Experiment

BAIF—a non-governmental organization headquartered in Pune, with financial assistance from India Canada Environment Facility, New Delhi, has been implementing the watershed programme in 2 villages of Bundi district in Rajasthan since 1997. Thanks to this programme, these villages and 4-5 nearby hamlets today have sufficient water for drinking and have also harvested good crops in both the seasons—Kharif and Rabi.

The Programme

The activities are being implemented through People's Institutions developed in the villages (7 Village Watershed Committees. 1 Charagah Samitee *i.e.* Silvipasture Management Committee and 13 Women's Self Help Groups). The work done includes field bunding on 176 ha. Silvipasture on 45 ha (on Panchayat land), afforestation on 30 ha. and horticulture on 30 ha under the drainage line treatment, 1,300 gully plugs, 25 stone bunds, 13 permanent checkdams/annicuts, 3 underground bandharas (sub-surface dyke) and construction of 11 gabion structures. The indigenous knowledge of the people and scientific methods of geo-hydrological and engineering investigations blended well for initial assessment, planning and implementation of project activities. Innovative measures like subsurface barriers using the clay soil in the drainage, gabion structures and an eco-friendly and low cost structure such as gabion with locally made net using the grass, have proved very successful and effective structures of water harvesting. Improved cooking devices were popularized among the women in order to save on fuel wood. Use of increased quantity of farmyard manure and bio-fertilizers is being popularized to ensure the sustainability of crop production.

Programme Outcome

A mid-term survey and analysis (end of 3rd year, which was one of the drought years) to find out the impact the project revealed very encouraging facts.

- The area under kharif crops in 1999 increased 30 per cent over the baseline year when it was 268 ha.
- Assured irrigation from the wells helped the rabi crop area to increase from 272 ha. before the project intervention to 450 ha in 1999, a 65 per cent increase. There was also increase in the summer crop area by 7 per cent over the baseline (which was negligible). Due to sufficient water availability and improved agricultural practices the yield of maize witnessed a three fold increase (from 1 to 3 t/ha.) and that of wheat doubled (from 2.7 to 5 t/ha.).
- Over 60 per cent of the women are now using improved cooking devices and reducing their cooking by 15 per cent. The firewood quantity saved is about 25 per cent.
- The drinking water sources, mainly the hand-pumps, are now free of microbial counts due to the close supervision of the villagers. Women have to spend only 60 minutes (maximum time) for fetching the drinking water against 2 hrs. before the project.
- Some of the nallahs have been rejuvenated and in a couple of water harvesting structures, water remains in summer as well.
- Milk yield of cows has increased as a result of better feed and purchase of cows of improved breed.

The combined effect has resulted in ground water recharge and increased water table, which benefits most of the wells. This has resulted in increased crop area and higher productivity. The rough estimate based on the additional area brought under cultivation and the increase in yield, indicates a net benefit of 14 million rupees per annum to the villagers. This profit equals to double the investment made so far in the watershed development in these villages.

Plan for Groundwater Recharge in Thar

After the grand plan of linking the rivers countrywide, now there is a mega-plan for groundwater recharge in the Thar desert. The Union Ministry of Water Resources has put forth a proposal to fill the aquifers in 13 districts falling under the Thar desert in Rajasthan with the surplus water from the Indira Gandhi Nahar Project (IGNP) and the floodwaters from the Ghaggar river. The proposal to recharge depleted aquifers of the Thar, which might involve a few hundred crores, is at the feasibility stage, but the Rajasthan Government, quick not to let go any opportunity to enhance water availability in the state, has expressed its keenness to join as a party.

According to estimates, groundwater resources of Thar desert amount to 92,000 MCM (million cubic metres) and the annual replenishable groundwater resources are about 3,613 MCM. The assessment is that a total of 71,000 MCM of water can be recharged to the groundwater in the Thar desert, while in the Lathi basin alone it could be 14,700 MCM. Experts say that they have tried it out on a smaller scale at Aligarh in Uttar Pradesh with considerable success. The idea is to create a water sanctuary from where water could be taken out as and when required. The underground storage also would replenish the wells and other water bodies in the region and also help the re-generation of vegetation. The concept hinges on the observation that the depleted aquifers like Lathi formations, Nagaur-Palana formations and alluvial formations in 13 districts-Barmer, Bikaner, Churu, Sriganganagar, Hanumangarh, Jaisalmer, Jalore, Jhunjhunu, Jodhpur, Nagaur, Pali, Sirohi and Sikar-can hold the excess water available during the monsoon.

What is required is injecting water into abandoned dug-wells and tube-wells after necessary filtration. The bowels of the earth can carry water in alluvial aquifers-Borunda limestone. Lathi formation and Nagur Palana Sandstone. It is estimated that IGNP would have 1,300 MCM of surplus water while the Ghaggar floodwaters would account for another 148 MCM. It is a scheme for which there is no alternative. The desert's survival and prosperity are linked to its implementation. At certain places, the groundwater level in Rajasthan-which has 10.5 per cent of the nations' land area but only 1.15 per cent of the water resources—has gone down by 50 metres in recent times. As the water level goes down deeper, the quality of the water too deteriorates with the presence of fluorides, nitrates and other minerals.

The water level in the Thar region has gone done drastically in recent times. At some places the exploitation of groundwater had been 400 per cent. In many areas they are indulging in water mining and if this continues unchecked the whole region would have to be provided water from outside in a few years' time. The new project would both help solve the water problem in the Thar and at the same time reduce the hazards of waterlogging along the IGNP and flooding in the Ghaggar region. The experts also say that 1-8 MAF (million acre feet) water flows down to Pakistan every year during the rains-which is not welcomed by that country as well—could also be put into use this way. While the details are yet to be worked out, apart from the existing lift canals on the IGNP, two more lift canals—Churu-Ladnulift canal and Barmer lift canal-would have to be constructed for the project.

The whole point is to ensure that not a single drop is allowed to flow more than 10 metres above the surface. The paradoxical scenario will be of dry streams and unfilled tanks in the rainy season. Instead, the water that seeps in, reappears in wells, tubewells, ponds and stop-dams later when and where wanted.

This strategy is the result of a calculated comparison of costs. An assessment was made of the amount of water saved and the expense incurred in each of the approaches available in the village of Dantha in Khandwa district. The cost of water secured through contour trenches, earthen checks etc. worked out to just 4 paise per litre. The system is so water-tight that even if it rains up to 10 centimetres a day-double the average daily downpour at Cherapunji, one of the heaviest rain receivers in the world—not a drop of water will flow down the drain. The result is a water surplus even in conditions of low rainfall-which was the case during the last three years.

Water University

Realizing the importance of water, a "water university", the first of its kind in the country, is now being set up in Alwar district of Rajasthan. An insitution which would impart training to youth from all over the country in water conservation, rainwater harvesting and on ways to involve the local communities in water management, it is being set up by Tarun Bharat Sangh (TBS), the NGO run by the much talked about "Waterman," Rajendra Singh. The Swedish International Development Agency (SIDA) will fund the proposed university. SIDA is planning to sign a five-year agreement with TBS to work jointly in the water sector. In all probability the water university will be located near Bhikampura, the headquarters of TBS in Thana Gazi, Tehsil of Alwar. The place is not far from the Arvari river, rejuvenated to life by the systematic building of a series of *johads* (checkdams) by the local people with the support of TBS.

The university would impart training to youth to work on water conservation in all sorts of agro-climatic zones of the country. The zones have been identified during the country-wide "padayatra" Mr. Singh had led recently. To start with, the water university would require Rs. 2 crores, though total investment would come Rs. 9-10 crores. This would include the water harvesting work proposed to take up along with the varsity activities.

Punjab

Punjab had 34.8 million acres feet (MAF) of water in 1955. However due to water sharing and consumption, the state is left with only a reserve of 23.8 MAF now which is pathetically low. This is relevant in the context that the state needs 36.5 MAF now. Water levels have fallen alarmingly in Malenkotla, Sangrur and areas around the Ghaggar river. While water tables are rapidly falling in the centre, the levels in the waterlogged areas on the corners of the State are rising which is a threatening situation to the State agriculture. One big problem that is imminent in the wake of falling water tables is availability of contaminated water for daily usage.

From October to June for a period of nine months all the rivers in Punjab are just like dry beds of sand which do not help in recharging the underground water-table. The waters generated by melting snows and rainfall in the hilly areas of Himachal Pradesh are dammed up in Bhakra and Pong reservoirs and are conveyed to the arid areas of Punjab, Rajasthan and Haryana. Water from the hill areas is utilized to the optimum extent, but no effort is being made to store Punjab's own annual rainfall of 25 inches plus.

Lack of recharging potential due to negligible flow and no storage in the rivers for nine months in a year is the principal cause of the problem of falling underground water-table. What would have happened slowly anyway has been accelerated by the cultivation of rice. The State's water-table has dropped more significantly during the past two decades. Where the water-table was 50 feet deep in 1980, it was 250 feet deep in 2002. Such a significant drop in the level of water should have rung alarm bells long ago, but our nature is to keep on ignoring our problems in the hope of getting some miracle to happen. Scientific thinking suggests that such a miracle has neither happened nor will happen. The solutions, at this late stage, are difficult but not impossible. Under the Indus Water Treaty between India and Pakistan, out of six western rivers called waters of the Indus basin, India has been given full rights to utilize the waters of three rivers, the Sutlej, the Beas and the Ravi. Pakistan gets the remaining three rivers namely the Chenab, the Jhelum and the Indus. What has been given to India in a treaty, the country, has miserably failed to make use of.

Until 2001, India by not damming the Ravi was letting its entire water to flow into Pakistan. The story is not as dismal in the case of the Sutlej and the Beas. The discharge contributed by its catchment area in the hills is dammed up and released into canals for the use of three states. Only the inflow contributed by the drainage area in Punjab goes into Pakistan. Pakistan on its own part has constructed several barrages, especially in its Sind province to irrigate the formerly desert like areas of that province. Farmers from West Punjab have settled in big numbers in these newly irrigated but otherwise virgin tracts of land. The area is blooming with all sorts of plantations.

At the level of the state government and other agencies a lot of advice is being given to the cultivators in Indian Punjab to abandon the practice of cultivation or rice and adopt other low water requirement crops instead. Can the people issuing these sermons guarantee that the water-table will come back after the plantation of rice is discontinued? The answer to this question is a big no. By stopping the cultivation of rice, we can certainly arrest the further downward slide in the water-table, but recharging will happen only if lakes are developed for constantly seeping the surface waters into the sub-oil.

At this stage we need to develop a series of cascade lakes by damming the Sutlej, the Beas and the Ghaggar at intervals of three to four miles each. First two such lakes should come up near Beas town on the Beas and close to Phillaur on the Sutlej. The Ghaggar dam should logically come up somewhere between the crossing of the Chandigarh-Ambala highway and the Grand Trunk Road. The primary purpose of these lakes will be to recharge the underground water-table. The secondary purpose can be the production of fish creation of recreational opportunities and augmentation or rural and urban supplies of potable water for the cities and districts of Amritsar, Jalandhar, Ludhiana Chandigarh and Patiala. At this time all the big cities of Punjab are suffering terrible shortage of potable as well irrigation quality water.

Punjab's water problem is very serious. It deserves serious action.

The solutions are not cheap, but they are a dire necessity, no matter what the price tag is. Once these dams are in place, the rise in the underground water-table will be noticed in the very first year in the areas located up to two miles from the river banks. Within a few years the impact on the water-table will travel to more than 10 miles on each side. Ultimately the entire state and some parts in Pakistan and Haryana will also benefit from these dams. In America wherever a dam has been built across a stream its impact on the underground water-table has been positive and permanent. The excessive consumption of water during the peak years of industrial revolution of the nineteenth century resulted in deepening of water-table all over the industrial America. With the construction of hundreds of small and medium sized dams the imbalance was rectified in the early part of the twentieth century. A lot of dams are still coming up in the area of water shortage.

Gujarat

Nature may be often harsh on Gujarat, but the parched state speaks of a situation which has been pushed to the brink of disaster due to man-made blunders more than the scanty rains. The huge pile of records at the Central Ground Water Board (CGWB) headquarters in the Capital has reports that warn of alarming depletion of the ground water level in Gujarat. The CGWB discovered a huge reservoir of groundwater spread over 1,500 kms in Kutch about three decades back. Billions of litres of precious drinking water was pumped out of the soil unmindfully. It was used for growing sugarcane, sold by private operators and even used for water sports. Today the region does not even have enough water to drink. Gujarat is an arid state and is basically not suitable for growing water intensive crops like sugarcane. Same has happened in places like Rajkot and Ahmedabad. Even in the heart of Rajkot city, residents get barely 20 minutes supply a day, and in the more remote areas, they have to buy every drop of water. The crisis has been aggravated as the water mafias have taken over the private tubewells and have started selling water. The most dismal part is while extraction is overdone, the recharge activities have been negligible. As a result in places like Mehsana, Ahmedabad, Bhuj and Kutch, the water table has fallen by over 4 metres in last two decades.

In April 2003, the situation was so grave in Saurashtra region that the Government contemplated running water trains during the peak of summer. The frequency of water supply in a few towns (all district headquarters) in Saurashtra region in May, 2003 was as follows:

Surendra Nagar	:	Once in a week
Wadhwan	:	Once in five days
Amreli	:	Once in five days
Porbander	:	Once in 2-3 days
Junagadh	:	Once in 3-4 days
Jamnagar	:	3-4 days in week
Bhavnagar	:	Five days a week

The situation is not any better in rural areas. For instance, in Surendranagar district, out of 657 villages, nearly 104 rely on tanker supplies and by mid-May the numbers swells up to 150. During peak summer, the main water sources for the district, Dholidaja and Nayka dams are dry. The administration banks on the deep bores, many of which however fail. Amreli district too faces the same fate. The main water source, Khodiyar dam is drying up and another source Thebi has touched rock bottom. Junagadh city gets water once in three days. Junagadh's main source, Anandpur dam, can fetch 1.5 crore litre water every day. But only 25 per cent of it can be procured because of the worn-out pipelines. The district authority declared 368 villages, out of 860 villages, as scarcity hit in May 2003, and nearly 290 villages got water through tankers. In order to recharge water, as early as in 1977, under an UNDP project, the CGWB had successfully demonstrated the technology for artificial recharge of deep tubewells under pressure and recommendations were made accordingly to the state Government. But no action was taken thereon.

Water Cooperative Society

The Mohini Water Cooperative Society in the Kakrapar system Gujarat, is one of the most successful examples of improved water management and more efficient cost recovery through farmer's participation. In Surat district, in the village Mohini of Taluka Chorasi,, the society was started under the State Cooperative Societies Act of 1961 that has been involved in irrigation management since March 1979. The water Cooperative undertook effective implementation of revised rotational schedules and charged the farmers government-fixed water rates. It has been fairly successful in almost eliminating the arrears in payment and also in minimizing the wastage of water. The members of this society now receive dividends, and sizable reserves have also been built up. There has been considerable improvement in water-use efficiency and there appears to be no complaints about inadequate or non delivery of water. This has been accomplished through cooperative efforts, with minimum interference from either the revenue department or the irrigation department. Another factor that has contributed to the financial success of this society is that it buys water at a bulk volumetric rate from the government and sells it to the farmers at the crop-area rate. Since there is a substantial difference in the design and actual crop area under sugarcane, the difference between the two rates yields almost Rs. 270 per hectare as profit to the society. The gross irrigated area has also almost doubled as a result of water savings and acquiescence of farmers in receiving a measured volume of water supplies from the authorities.

Narmada Water for Kutch[9]

By the end of 2003, the drought prone Kutch district may get assured drinking water from Narmada through pipe line project from Malia canal. The 231 crore rupee 56 km long pipe line will provide 10 crore litres of water per day.

The special Malia branch of the canal has already reached the door step of the district. Also known as Saurashtra branch which supplies drinking water to Rajkot, Jamnagar and Kutch. Work of construction is over at Kherot village of Maliya taluka. Its entry point of Kutch is just 14 kms away from Suribari bridge which links Kutch with Saurashtra. After completion the pipeline project and other internal lines, 949 villages and ten urban centers in the district will get assured water. Rural areas will get 70 litres of water per head from the present 25 to 30 litres per head and urban areas with drainage system will get 140 litres of water.

The canal design has been prepared taking into consideration the future requirement of increasing population of Kutch which is estimated to increase from 18 lakh in 2001 to 22 lakh in 2021. The project will also provide 450 lakh litres of water for industrial purpose of the district. The district will get assured supply of sweet water for drinking purpose for the first time after independence which is a victim of frequent droughts as it receives scanty and irregular rainfall severely depleting its under ground water resources. The pipeline scheme could become possible with the Supreme Court ruling for the five metre increase in the height of the Narmada Dam. With renovation and strengthening of existing internal network, entire Kutch will get pipe line water within next two years. The pipeline will supply water upto 20 crore litres per day as the requirement of the district increases. With the gradual increase in the height of dam as per the directive of the Supreme Court, the water from Malia canal will start flowing by gravitational force and water will not have to be pumped which will save lot of electric power.

The canal based pipe line master plan was sanctioned in 1996 to provide drinking water to 12 districts in the State. This was as per the Narmada Tribunal award under which 1.06 million acre feet water was earmarked for drinking and industrial purposes. The Saurashtra branch canal covered 1,200 villages and 21 towns in Kutch and Saurashtra. Kheroi canal which is to receive 50 crore litres comes from Kadi near Viramgam 150 km from Malia. The Narmada water upto this point will flow by gravitation.

The Gujarat Government has also prepared detailed plans to repair water supply schemes damaged in the killer quake which hit the state in 2001, with the financial aid of World Bank and Asian Development Bank. The Rs. 498 crore package includes water supply schemes of Kutch. Jamnagar, Rajkot, Patan, Surender Nagar and Banaskantha districts.

Water Crisis—Position in a few States

Thirsty Orissa

At Jamut, Balangi village, a water harvesting plant was constructed at a cost of Rs. 16 lakh under employment assurance scheme by the Soil Conservation Department in 1996-97. Now it is defunct with no water. Since 2000, six watershed projects were taken up in Birmaharajpur block of Sonepur district at a cost of Rs. 1.50 crore under the Dorught Prone Area Programme. The funds have been used up, but the projects have not contributed in any way towards making Birmaharajpur, one of the backward blocks of the country, drought-proof. Various anti-poverty programmes that are being implemented for more than a decade in eight districts of KBK (Kalahandi-Balangir-Koraput) region have failed to deliver. The state government in consultation with the Centre, had formulated a long-term plan of the KBK. An outlay of Rs. 6,251.08 crore as planned for 1998-99 to 2006-07. However, up to 2002, the

Centre released only 22 per cent of this amount. But the state government could not utilize even that. It has used only 74 per cent of that amount, and has failed to fulfill the primary objective checking migration of labour to the cities, making the area drought-proof, and providing irrigation.

A survey by Nabakrusna Choudhury Centre for Development Studies, Bhubaneswar, revealed that a poor family in the KBK region might have earned only Rs. 375 at the minimum and Rs. 725 at the maximum per annum. No wonder, more than 50,000 people migrated from these areas to other states in 2002. Even the latest Orissa Development Report 2002, prepared by the Planning Commission along with the Orissa government admits that funds spent on anti-poverty programmes have not been utilized properly.

Karnataka

Nearly 72 taluks in Karnataka state are facing perennial drinking water crisis and Kolar among them is the worst affected. According to a report submitted by experts, groundwater has dried up in the district, Fluoride and nitrate contents are higher in groundwater in the district. In fact, excessive exploitation of groundwater is a major reason for the acute drinking water shortage in many districts. The state headquarters Bangalore imports most of its water from almost 95 kms away at Rs. 12 per kilolitre. This elixir of life may be more than three times dearer at Rs. 45 by the time Cauvery IV stage water supply scheme gets under way. There are 10 lakh borewells in the state. Over exploitation of groundwater, poor management of irrigation projects and wastage of rainwater are among the major reasons for water scarcity in the state. The united Nations had adopted the slogan, "water is future". Harvesting of rainwater is the only solution to overcome the shortage.

The state Government has called upon the people to use water judiciously and adopt harvesting and conservation measures in their houses. The state budget for 2003-04 has earmarked more than 50 per cent of funds for water projects. As much as 31 per cent of the funds under the sectoral Outlay Plan has been set aside for irrigation and flood control, seven per cent for drinking water supply, and 12 per cent for other water—related works. The state Government has also been implementing several World Bank aided watershed projects, drinking water supply schemes and hydrology projects at a cost of Rs. 2,800 crore.

The State Government is now armed with a comprehensive policy for rainwater harvesting that can be implemented almost anywhere, and the State Department of Forest, Ecology and Environment ready to share the know how with interested parties, Rs. 300 crores over 10 years to implement the policy could well help cover the deficit of 209 mld anticipated by the year 2011. The three existing and proposed new sources of water supply—Cauvery river, Hesarghatta and Thippagondanahalli are projected to tocuh 1,494 million litres per day (mld) 11 years from now, short of the required 1,703 mld by 209 mld. The Department of Forest, Ecology and Environment had commissioned a study by the Center for Symbiosis of Technology, Environment and Management (STEM), which had been accepted in principle, and the work of disseminating information would begin soon. The alarming rate of ground water depletion, escalating costs of Cauvery IV Stage, which have already touched Rs. 3,000 crores, and over exploitation that could bring down the water supply drastically were the provocation for commissioning the study.

Himachal Pradesh

The water crisis in Himachal Pradesh is going to reach some unprecedented proportions and magnitude. A number of districts like Kangra, Chamba, Billaspur and Hamirpur along with the famous tourists destinations of Shimla, Palampur, Kullu and Manali are reeling under a severe water crisis. Due to unplanned growth of towns in the State and especially, places like Shimla and Manali which have now virtually turned into concrete jungles, all the available water sources have dried up and the forest catchment area has been drastically reduced. There are also reports of alleged misuse of water by some big hoteliers in the town. Some of the standard hotels are buying tankers from private water supplying agencies of which majority of residents are unaware.

The water level has gone to an all time low in river Beas. The tourists towns of Manali, Kullu and Mandi are situated on the banks of Beas. Places like Palampur, Pragpur, Ramchowk, Naura, Banoori, Kotla, Lahru and Sambha in Kangra District are also facing water crisis. In Mandi district Bargaon, Koti and Sarkaghat and in Bilaspur district Nayndevi, Ghumarwin and Jhanduta are facing water shortage.

The Sikh pilgrim center of Ponta Sahib and Shillai in Sirmaur and Nalagarh, Kandaghat, Salogra, Arki and Barotiwala in Solan district are also reeling under cute water shortage. Similar reports have been received from Chamba, Una and Hamirpur districts of the State. The environmentalists and experts feel that there is a need of massive mobilization of the harvesting techniques, focus on maintaining traditional water source and recharging of water sources are required as immediate and sustainable steps to tackle the crisis in the state.

The Himachal Pradesh EPH and Parliamentary Affairs Minister, Thakur Kaul Singh, recently (May 4, 2003) cautioned the people of the state to be mentally prepared to face a drinking water crisis due to successive droughts which had dried up perennial water sources as early as April and May. Referring to the acute scarcity of drinking water in Mandi town, the minister lamented that the Rs. 9.41 crore drinking water augmentation project taken up earlier had come to a halt in the state as no funds were provided for the project during the five year tenure of previous government. Work on the project had been taken up on a warfooting now. Two other proposals were also under consideration—to harness either the Uhl water or that of Jiumi Khud. Value—added execution of the drinking water supply schemes is being taken up by tapping and sealing the water sources, and water supply sources and tanks are being treated with chlorine to avoid the outbreak of water-borne diseases.

Hamirpur district (Himachal Pradesh) is facing a water crisis due to drought conditions, burning of forests and non-arrival of timely and heavy rains last winter (2002). The district has received no rains during the 2003 summer season. The situation is so bad that the people are forced to walk 5 km to 10 km daily to fetch water for their needs. Traditional water sources like khads, bowlis, wells and khatris have dried up. Nearly one-third population of the district living in around 277 villages is the worst affected. Traditional water sources in these areas have dried up and the government water supply system has been paralysed almost completely, Sixtynine water supply schemes of the district have been affected in varying degrees. Eight water supply schemes have almost no water and people living there are forced to share water from the ponds and khads.

Jammu & Kashmir

During the summer months, Jammu city and its suburbs with a population of about 12.60 lakh plung in a scarcity of drinking water.

Tanker-owner make fast buck as residents of most of the new colonies skirting the main town are dependent on them for their requirement of drinking water. A tanker full of water is normally sold for anything around Rs. 450 in the town and Rs. 500 outside. The tankers or the PHE Department undertake as many as 102 trips a day, which the private operators make slightly more number of trips to supply water to consumers. The position is worst in the Kandi areas of Samba, Kathna, Udhampur and around Akhnoor in Jammu Division as the water bodies dry up during summer season. It is equally bad in the old city of jammu where water is supplied from Tawi which flow right in the middle of the town. Water is supplied once for about 30 minutes a day in most of the areas.

Uttar Pradesh

And now, it is the turn of Uttar Pradesh to reel under a water crisis. The state Irrigation Minister, Mr. Om Prakash Singh, has acknowledged that the state is facing a serious water crisis. The situation in more than 100 blocks in the state is grim with 22 blocks being included in the 'dark' zone and another 80 blocks comprising the 'grey' zone. The water table in these blocks is fast receding. About 150 check dams are being constructed at various places to stem the tide. There is a proposal to restore all water bodies to the pristine position they enjoyed in the pre-Independence era. Efforts were also afoot to enhance their water-holding capacity. Work on the Yamuna basin had been started. The second phase of the work was already inaugurated. In this phase, work on 29 places would be completed between Muzzafarnagar and Aligargh at a cost of Rs. 66 crore. From Javer to Tappal, work on the 27 km-long dam, which would cost about Rs. 24 crore, had also been taken up. Work on the Tapti river basin would be strengthened at a cost of Rs. 45 crore. This would include work on 63 projects. All these projects would be completed by 2005.

OVERCOMING CRISIS

Dewas—A Success Story

Here is a story to lap up for all those thirsting for good developmental news. It is a report from Dewas, a low-profile town near the capital of the hardly high-tech Madhya Pradesh, about a wonder of a water conservation mission, and it makes two noteworthy revelations. It shows what an individual agent of development can do about the distress of a drought-prone area and what India can do about one of its perennial problems. Collector M. Mohan Rao is no mere rain-maker, his has been the rarer miracle of meeting the problem, to which a merciful monsoon was supposed to be the only solution, with a simple but scientific solution backed by popular initiative and mobilization. Dewas is today a picture of water-harvesting structures on building roofs, all raised during a one-year drive, which testify really to the success of a groundwater-tapping and conservation programme. Funded largely by public contributions and carried forward with near-complete public participation, the pilot scheme under a groundwater recharge mission has proved a pioneering experiment, inviting similar other efforts in the neighbourhood. Clearly, the example cries out for emulation elsewhere in the state and the country as well. The objective of the programme-effective management and utilization of groundwater resources—is to tackle an intractable issue which has impeded development. Other similar issues that escape solution range from recurring droughts over large parts of the country to inter-state river disputes. There are many a dust-gathering expert study urging adoption and expansion of dry farming as the ideal remedy to both these afflictions.

It is not as if science alone can achieve the objective or the community as a whole can be counted upon to cooperate in every case. The task has called for a tough will even in the MP town, and it has been displayed by Rao, determined to regulate use of groundwater resources and their reckless exploitation by the richer sections of the community. The task enlarged manifold for the country as a whole, will call for a proportionately tougher political will. Easy optimism on this score is not encouraged by the country's experience of the political clout of the affluent farmers lobby prevailing over public interests, including those of agriculture.

Illustratively, the National Bank of Agricultural Research and Development has, for long years, been trying to persuade states, particularly the drought-prone ones, to legislate for some regulation of groundwater tapping—and trying in vain. It is common knowledge that free or heavily subsidized power for farmers has only increased the extent of the scarce water resource inexcusably inefficient use, but the consideration has not compelled any serious contemplation of even steps short of the NABARD's suggestion. Attempt at introduction of drip and sprinkler farming, water saving techniques known to have helped a desert bloom in Israel, have never taken off, Industries have been extremely reluctant to encourage such initiatives and innovations in sectors of farming that feed them, with the sugar industry and sugar cane farning providing a striking instance. If the right lessons are drawn from it, Dewas can lead to a much bigger success story.

Majhgawan (Satna, M.P.) Watershed Scheme

The Deendayal Research Institute, in collaboration with the District Rural Development Agency, Majhgawan (Satna), took up an area of 12,536 hectares and formed 17 micro-watershed schemes to tackle low productivity, lack of irrigation, a receding water table, shortage of drinking water, high soil erosion etc. The 'watershed area'—the area over which rainwater collects—was taken as a unit. Contour trenching and bunding check the flow of water. Trenches are excavated and bunds erected (with the soil taken out) to check the water's free flow. This prevents soil erosion while plenty of percolated water raises its sub-soil level.

The vision of water management programmes, thus, is as simple as they can be. Instead of tanks or dams, water is kept where nature keeps it-underground, Thus, there is no loss due to evaporation. The old philosophy of 'water should not be allowed to run, it should walk' has been replaced by the new one which states that 'water should creep and, ultimately, come to a halt'.

Gurgaon Project

As a fire-fighting measure to arrest the depleting ground water resources, the Gurgaon administration is putting in place a mechanism for harvesting rain water. The administration has roped in INTACH, a semi-government body which has considerable expertise in rain water harvesting. The body will work out an integrated approach to make its exercise more effective. A transparent plan for rain water harvesting has become necessary following the amendment in the building bylaws of Haryana Urban Development Authority (Huda). The amendement has made it obligatory on certain categories and sizes of residential buildings to incorporate a mechanism for rain water harvesting in order to procure a "completion certificate". Although the amendement in the bylaws would be effective post facto, the authorities are considering working out a methodology to enlist the owners of old houses, to put in place a system for harvesting of rain water in their respective premises.

The administration has also approached the Centre for Science and Enviornment for vital inputs, so that its remedial measures for recharging the sub-soil water become more holistic. Incidentally, the Central Gorund Water Board has notified Gurgaon city and its surrounding areas as possessing a fast depleting underground water table. The notification enjoins regulatory measures, such as installing new tubewells etc. One of the elements in the proposed integrated approach is to refill the pond in village Sukhrali near the main bypass. This is one of the largest ponds in the district, about three acres in size and running about 30 ft deep. While the refill has the potential or recharging village Sukhrali and its surrounding areas, it can also be used as a place to absorb the run off water from houses in the areas.

Ranney Wells in Mewat

A Rs. 303—crore scheme for setting up ranney wells in the Mewat area of Haryana would be introduced to enhance the availability of drinking water. A decision to this effect was taken at a meeting in which the Chief Minster was shown a presentation on this project following the recommendations of a committee headed by the Financial Commissioner and Principal Secretary, Irrigation. Eight ranney wells would be set up in a phased manner starting from Nuh followed by Hathin and Punhana. These would benefit 242 villages by making available 120 millions litres of water per day. Out of a total of 497 villages in the area 255 were already being fed by the tubewells.

Dialogue—The Pune Experiment

The Kolwan Valley farmer knows a thing or two about water how to conserve and harvest water in a region where rain plays elusive and hides in the shadows. The water committees in these few villages in Mulshi taluka, a few kilometers away from Pune, have set an example on how to ensure water security as well as equitable distribution of the precious commodity to its entire population. To support such initiatives, the Dialogue, a global consortium set up by 10 organisations including the World Health Organisation and the United Nations Environment Programme (UNEP), came into being. In February, 2003, a Dialogue team visited the Kolwan Valley to jot down a few lessons on water harvesting and traditional farming.

Today, as the Kolwan farmer manages the 'village watershed committees', after watershed development was initiated by Pune-based NGO, Gomukh, he is ensured of two crops in place of one. Citing this as an example of "community-based movements that has a great scope in India owing to the social fabric that supports them," the Dialogue team observed that local action was of great significance to precipitate change and ensure food and water security in India. The 'Dialogue on Water, Food and Environment', as it is known, aims to bridge the gap between the food and environment sectors. Formed after the World Water Forum 2000, which witnessed an intense debate between the agriculture and environment sectors and their views on managing water, the Dialogue works in more than 20 countries to foster understanding between these two sectors.

Of greater significance is the fact that Dialogue will help local initiatives like the one at Kolwan valley, find sponsor groups and help initiate a national-level dialogue to suggest practical solutions to policy makers. Dialogue could be the facilitator between initiatives like the Kolwan Valley and similar others across the country as well as abroad. Says Robert Bos, senior scientist with WHO, who was in Pune as part of the Dialogue team: "Although we realize the local action cannot be applied as a generic term to a large country like India, we do believe that the knowledge here can provide water options while contributing to a global body of knowledge".

With little encouragement from the government to harness the traditional knowledge of managing water and the environment, "even to initiate a dialogue between government departments and rural communities we will need some consensus from the government", he says. Another drawback is the gender inequality, where men and women limit themselves to specific jobs, and illiteracy, which creates a communication problem for the exchange of ideas". In Bhalgudi village of Kolwan, the NGOs and villagers have overcome official hitches and trap spring water. In the absence of a sanction for natural spring development, they called it excavation. Now, the water belongs to the village, who pooled in the money and the farmer on whose land it was developed, is not complaining. He like the others, grows two crops.

Hariyali—A New Initiative

The Prime Minister, Shri Atal Bihari Vajpayee underlined the need to make water conservation a mass movement. He said, " our country is facing a severe water problem not because we do not have adequate resources of water but because of our inability to conserve all the rainwater". He emphasized upon the start a big drive for the revival, renovation and maintenance of all available resources of water. Launching a new initiative named 'Hariyali' the Prime Minister said that under watershed management, villagers could take up many small works to conserve water for drinking, irrigation, fisheries and afforestation. This would not only add "Hariyali" to the rural landscape but also create new employment opportunities for the rural poor. This programme is going to be implemented by Panchayati Raj Institution (PRIs) with peoples' contribution and participation.

While declaring water conservation a national mission, the Prime Minister, Atal Behari Vajpayee, has suggested that all industrial units, hotels, Government offices and commercial establishments conduct a water audit to see how significant and targeted saving of water could be achieved. It is suggested that July should be observed as 'water month' and the period may be used for conducting awareness activities shramdaan in schools, colleges and workplaces. The Prime Minister urged, "Let every household, every housing society and colony, every panchayat and municipal body consider how it can save water and implement rainwater harvesting, water recycling and other measures before the end of this rainy season. Our scientists should take up development of low-cost water saving technologies as a major challenge. Let us collectively address the problem of water shortage, which is assuming crisis proportions with every passing year, by conserving every drop of water. This problem can be dealt with effectively only through a sustained partnership between citizens, people's organizations various water users, and the Government at all levels. There is the need to call for short-term as well as long-term measures, all of which have to be taken up with utmost urgency and a sense of national mission".

Conclusion

It is axiomatic to state that access to enough water of sufficient quality is fundamental for all human, animal, and plant life as well as for most of the human activity. Though, India has 14 major rivers and receives about 4,000 BCM of water annually from precipitation, water shortage has become an incessant problem across various states, in the country. In fact, it is becoming increasingly clear that water is the most serious constraint for the well being of the India's population. Though traditionally, water management has been the Government's responsibility, private sector participation in water projects now recognized as a growing imperative. However, in view of extremely uneven geographical distribution of water resources causing severe regional and temporal shortages, and growing demand for water, there is every likelihood of expansion of water markets, water lords and water conflicts including growing inter-state rivalry in sharing the river water. In fact, it is time we move towards the direction of rainwater harvesting, watershed development and the conjunctive use of rain, river, ground, treatment of sewage and sea water on the one hand, and augmentation of utilizable water availability through inter-basin transfer of surplus waters on the other. Keeping such a scenario in view, the National Perspective for Water Resources Development envisaging inter-basin transfer of water assumes a great significance. Long distance inter-basin transfer of water is not a new concept but has been in practice in India for over five centuries. The Western Yamuna canal and the Agra canal built in Mughal times are examples, water was carried from the Himalayas to the district of Punjab, Uttar Pradesh and Rajasthan. The Kurnol-Cuddappah canal and the Periyar—Vaigai canal are other examples of inter basin water transfers executed in India in the 19th century. In the last century. Indira Gandhi Canal Project is another example of inter basin transfer of water form the Himalayas to the desert of Rajasthan.

Mention may be made in this context that the revised National Water Policy (April 2002) of the Government of India states:

> "Non-conventional methods for utilization of water such as through inter-basin transfer, artificial recharge of groundwater and desalinization of brackish or sea water, as well as traditional water conservation practices like rainwater harvesting, need to be practiced to further increase the utilizable resource. Water should be made available to water short areas by transfer from other areas including transfers from one river basin to another, based on a national perspective after taking into account the requirement of the area/basins"

Surprisingly, while we are endeavouring to generate additional water resources at a prohibitive cost, hardly we express any concern about the conservation of existing water resources and necessity of their efficient utilization at the farm level. Hence, the first line on attack should be to create a huge mass movement on "water literacy" by increasing awareness of the people, imparting education/training/ skills to farmers, improving irrigation—related understanding of the state (and even private) development agencies and motivating local action about improving region/crop/site-specific on-farm irrigation efficiency. We must prioratise research towards water (conservation, efficient irrigation)—related technology generation.

Finally, Water (fresh harvested and brackish) and its efficient management should be an important subject even in schools. Students can act as technology-transfer agents. Farmers/people should have free access to user-oriented water-related information. Enactment of legislation (if needed), mobilization of farmers/people to promote efficient water management and establishment of participatory linkages between farmers/people, scientists, NGOs and different State development departments will effectively trigger synergistic push to the whole farmer/people's participatory "water literacy" and efficient water management" programs. In fact, ensuring efficient on-farm utilization of "whatever water we already have" is as important as generating additional water resources.

REFERENCES

1. India Assessment 2002, Water Supply and Sanitation A WHO—UNICEF sponsored Study, Planning Commission, Government of India.
2. Government of India, Economic Survey 2002-03.
3. *Ibid.*
4. Introduction to Rural Water Supply and Sanitation Programme in India; Rajiv Gandhi Drinking Water Mission.
5. Sehgal, K.L. 1995. Aquatic Resources of North Western Himalaya. In Singh V, *Ecological Carnage in the Himalaya*, 35-47, Dehradun; IBD.
6. Rosegrant, Mark, W., Cai Ximing, Cline Sarah A; World Water and Food to 2025—Dealing with Scarcity; IFPRI, Washington, D.C.
7. Patel; T: India faces chaos over water rights. *New Scientist*, April 26,1997. p.12.
8. Government of India, Ministry of Water Resources, Annual Report 2001-02.
9. Jaswanie, RKA: Narmada Water for Kutch; *Yojana*, April 2003.

4

Water Situation in Delhi

"Nostradamus may not have predicted it, but Delhi is heading for a catastrohe. Delhi will be a dust bowl by 2010"

—World Bank Report

Introduction

Delhi's antiquity, in fact, goes far beyond the ken of history. The capital city has now grown beyond manageable capacity. It has engulfed all the ancient city sites and overflows across the Yamuna and the Northern Ridge. The earliest known capital city of Indraprastha was founded by the Pandavas in 14 OO BC. Maya, the Danava architect built a splendid palace and court for them. Recent finds support the view that it was located in the Purana Qila area. Delhi of today has spread into the old cities of Suraj Kund, Qila Lal Kot, Qila Rai Pithora, Siri, Tughlakabad, Adilabad, Firuzabad, Shajahanabad and Qutab Minar area and Lal Qila area. Ala-ud-Din Khilji who built a new capital for himself at Siri (1,302 AD) with a palace of thousand pillars which saw many triumphs and tragedies and which has since completely gone to ruins, had also built the magnificent tank called Hauz Khas. At that time it had enough water to meet the needs of the city.

Civic administration in Delhi was established by the British rulers in 1,862 with the setting up of Delhi City Commission. At that time Delhi occupied an area of only 3 sq. km. with a population of only 1.21 lakh. Presently the city has an area of nearly 1,500 sq. km. and a population of more than 14 million. The city of Delhi is almost perpetually in the grip of water crisis, more so during the dry season when serious water shortages afflict the city. Apart from 1,600 unauthorized colonies and 1,100 slums, there is a large chunk of the

> *"Here we stand in Delhi City, symbol of old India and the new. It is not the narrow lanes of houses of old Delhi or wide spaces and rather pretentious buildings of New Delhi that count, but the spirit of this ancient city. For Delhi has been an epitome of India's history with its succession of glory and disaster; and with its great capacity to absorb many cultures and yet remain itself. It is gem with many facts, some bright and some darkened by age, presenting the course of India's life and thought during the ages. Even the stones here whisper to our ears of the ages of long ago and the air we breathe is full of the dust and fragrance of the past as also of the fresh and piercing wind of the present. We force the good and bad of India in Delhi city which has been the grave of many empires and the nursery of a republic. What a tremendous story is here, the tradition of millennia of our history surrounds us at every step, and the procession of innumerable generations passes by before our eyes...."*
>
> **—Jawaharlal Nehru**

city's rural belt which falls in the water deficit zone. At the end of nineteenth century water supply to Delhi was from the percolating wells constructed in two lots in 1892 and 1894.

The yield at that time was only 6,10,000 gallons per day. Around 1900 A.D., settling tanks and filters were provided to increase the availability of potable water—about 17,00,000 gallons per day for a population of two lakh population. As the capital of India was shifted from Calcutta to Delhi in 1912, a scheme was implemented for an intake well near Wazirabad, which was completed in 1926. With the passage of time local bodies were created for various functions and for providing basic amenities of water and treatment of sewage, Delhi Joint & Sewerage Board was constituted in 1926(1). Presently, the water requirement in Delhi is about 800 million gallons per day (MGD) whereas the supply from available sources is about 650 MGD. This really is the gross supply. A big chunk of this is being lost of the 9,000 km of the existing pipelines, some 1,700 km. have no doubt, been replaced but loss of water on leaking pipes, overhead tanks and taps is still substantial.

For instance, about 31.5 MGD is estimated to be lost in illegal tapping at ten places of the raised mains coming from Bhagirathi Water Treatment Plant to Okhla. There is large-scale tapping of water at Sarai Kale Khan and Preet Vihar. It is for this reason that DJB has put security guards for protecting these mains. "But not much has been done as the local politicians have been opposing it. Further, despite a number of proclaimed measures to reduce transmission and commercial

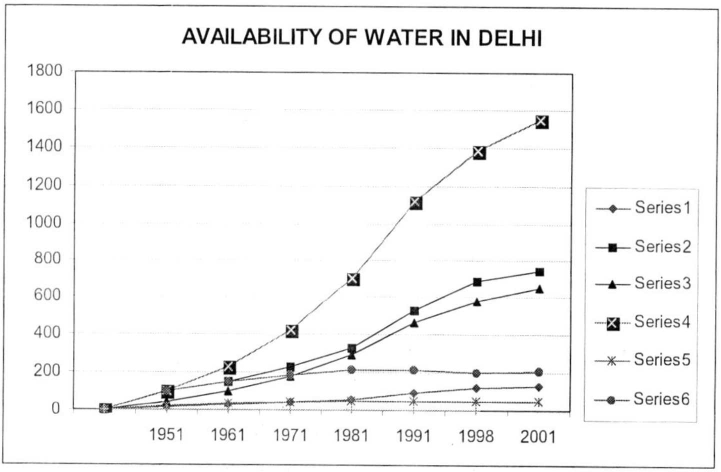

losses, it remains around 48 per cent. This means that of the 650 MGD of water treated by the DJB, just 312 MGD of water reaches the people. The DJB's Leak Detection Cell has not been able to do much in this area. Senior DJB officials also put the blame for shortage of water on reduction in supply of raw water from Haryana. The water treatment plant at Nangloi has been producing 20 MGD less than its capacity as water is not received from Haryana. If it is supplied, at least several colonies would not experience any shortage. The gap between demand and supply is partly met by extraction of groundwater through wells, tube wells, deep bore hand pumps, etc. However, the wells and bore wells are drying up and the vegetation has also been affected severely. Faced with such alarming situation, today the only solution seems to be massive rain water harvesting. Position regarding population growth and water availability in Delhi is given in Table 4.1.

Table 4.1: Availability of Water in Delhi

	1951	*1961*	*1971*	*1981*	*1991*	*1998*	*2001*
Population (lacs)	17.45	26.50	40.65	57.68	93.07	120.00	130.00
Index	100.00	152.00	233.00	331.00	533.00	688.00	745.00
Water Produced (MGD)	42.00	97.00	177.00	296.00	469.00	580.00	650.00
Index	100.00	231.00	421.00	705.00	1,117.00	1,381.00	1,548.00
Availability of water	24.00	37.00	44.00	51.00	51.00	48.00	50.00
Index	100.00	154.00	183.00	213.00	213.00	200.00	208.00

Freshwater Consumption in Delhi

Although the average water consumption in Delhi is estimated at 240 litres per capita per day (lpcd), the highest in India, the figure is not indicative of an adequate supply to every resident since the water supply in Delhi is far from being uniformly distributed. The NDMC and Delhi Cantonment area (about 5 per cent of population) get an average supply of above 450 lpcd, while areas in Narela and Mehrauli zones get less than 35 lpcd. The gap between the supply and demand of water has resulted in over-extraction of groundwater which has led to serious problems with both quantity and quality of groundwater. Residents of European cities like Copenhagen consume only 111 litres a day. Further, Delhites end up wasting more water than other cities.

At least 15 per cent of the water is lost during distribution and about 40 per cent is stolen. Of the water that reaches people, only 20 per cent is used for cooking and drinking. The rest goes towards washing and other purposes. That is not the end of the matter. Beside illegal tapping of water in certain colonies affecting supply, Delhi also faces the problem of keeping reserve a certain volume of water for industrial use and fire-fighting, besides stocking potable water. Going by figures available with the Central Ground Water Authority, the fall in groundwater level varies between 4 and 20 m across the city. In certain areas in South Delhi and the Capital's fringes, however, the dip is as much as 40m. Admittedly, Delhi is no stranger to paucity of water. But what has happened this time round, is that many households equipped with tubewells have become dependent on DJB supply because of the rapid fall in groundwater levels.

Source of Delhi's Water Supply

While 90 per cent of drinking water in the country comes from groundwater, nearly 89 per cent of the Delhi's water supply is from surface water and most of which is from Bhakra-Beas Management Board and regenerated Yamuna. Nearly 17 per cent of water is from Haryana and 17 per cent from Uttar Pradesh. The denizens of Delhi are a major groundwater-dependent supplied by the DJB, Approximately 11 per cent comes from groundwater reserves. However, it is difficult to establish the total groundwater extracted, because a large number of private tubewells dug by households and the industrial sector for their own supply are unaccounted for in the official estimates. Many water tankers and bottled water companies are drawing and selling groundwater. Presently DJB is supplying 650 million gallons of water per day. It is expected that by the end of 10th Five Year Plan *i.e.* 2007, it will be increased to 950 MGD.

Delhi banks on 'padosi pani' to quench its thirst. While 100 MGD of Delhi water supply stems from UP's Ganga overflow, 480 MGD comes from Haryana. On its part, Delhi generate 80 MGD by tapping underground water through Ranee wells and deep-bore wells. The Sonia Vihar plant, which will augment the supply by 140 MGD, is expected to be operational in December, 2003. Besides, Delhi is ready to bear the construction costs of the Rs. 12,000 crore Renuka dam in Himachal Pradesh with a view to benefiting from its water supply.

While the 14 million population of Delhi needs about 3,768 million litres a day (MLD), the DJB produces only 2,950 MLD. Of this, about 30 per cent is lost from leaking pipes and pilferage. Effectively, people get just about 2,065 MLD. With the demand steadily on the rise, additional sources of water are very limited. The current shortage is as much as 45 per cent, which means if a person needs 100 litres, he has to make do with 55 litres. While VIPs wash their cars with piped water everyday, many people hardly have enough to drink. The total shortfall of about 1,700 MLD is partially met by drawing groundwater. And the massive exploitation of groundwater has led to drastic fall in the water table all over the city. There's hardly any fresh water left in the city except in the river bed. As a result, the city homes have taps that spring to life for hardly an hour a day and tubewells that are running dry.

There are more factors contributing to water scarcity during dry summer. The DJB gets raw water from the Yamuna, western Yamuna canal from Haryana, upper Ganga canal and groundwater. Almost 80 per cent of the Yamuna water flows in the months of July to October, For the remaining months, the river has scant water. Haryana often releases polluted water. This forces the Government to close the Nangloi treatment plant frequently. While Haryana was supposed to release 40 million gallons a day for the Nangloi plant, it released just 25 MGD during February, 2003.

The Delhi Jal Board (DJB) claims that the city enjoys one of the highest per capita consumption of water in the country. Water consumption here is much higher than that of people living in France, Germany and United Kingdom. But for scores of Delhiites who spend their entire day waiting for a drop of water, these figures are just a cruel joke. Only few areas in New Delhi and Cantonment get close to the average supply. While in New Delhi and cantonment people get over 325 LPCD, in Mehrauli, a densely populated locality in South Delhi, the average figure is just about 30 LPCD. There is an urgent need to bridge this gap. Losing 40 per cent of the 650 million gallons a day produced adds to the gap between the demand of 830 mgd and supply of 630. Despite this, the DJB claims that a Delhiite gets 215 litre per person, per day (LPCD), while in Kolkata and Mumbai, the figure is 202 and 178 LPCD, respectively. In France, Germany and England, the figure is 150, 135 and 175 LPCD, respectively.

Then what explains the crisis? The drastic fall in water table has led to the problem. Earlier, there used to be lot of groundwater and people would just drill and get as much water as they wanted to, in addition to the piped supply. Now the groundwater has fallen to almost inaccessible depths. The worst affected is south Delhi where the water table is falling by almost two metres every year. In south west Delhi, there is hardly any sweet water left. If the exploitation continues at the same rate without making attempts to harvest the rain, the entire city's sweet water will be exhausted in a decade's time.

Water Exploitation

The exploitation of groundwater over the years has led to a sharp drop in both quantity and quality of the sub-soil reserve in the city. In a vicious circle, as the water table drops, the dangerous pollutants like nitrates, fluorides, pesticides, heavy metals and salinity keep getting concentrated in the remaining water. The water at lower levels in the city is saline. Very little fresh water is left. The drought—like conditions in 2002 left a thin layer of sweet water in southwest Delhi that housed densely populated colonies like Janakpuri and Vikaspuri. In Dwarka, the Central Ground Water Board (CGWB) says, the sweet groundwater is not sufficient to meet even a year's demand of the sub-city. In south Delhi, the quality is still good but the levels of water have plummeted to as low as 44 metres. North and west Delhi's groundwater is a cocktail of various contaminants at lower levels. In east Delhi, only the water close to the surface was potable.

According to the CGWB, water quality is fast declining. Bacteriological contamination levels are rising. There is also a high content of cadmium, lead and chromium in the water. The iron levels have increased. Most factories which were shifted outside Delhi because they were polluting the environment, bury their wastes. These wastes, sometimes poisonous, are getting into the aquifers. Thus, the water that comes to each of us is full of these pollutants. This can lead to health problems.

In parts of northeast, west and south west districts of Delhi, one has to dig deep to get fresh water. Otherwise, the water is brackish. Parts of north and central Delhi have the same problem. In other parts, like the ridge and parts of central, south and south west Delhi, the water quality is better. Fresh water aquifers at the west of the ridge, the water reserves are not plentiful at shallow depths and the water in the Nizamuddin and Connaught Place areas is marginally brackish to saline.

DJB Chief P.K. Tripathi had recently conceded that the water crisis in the city was due to the limited access of people to the groundwater. Ironically, this is happening at a time when the city's dependence on the sub-soil water is growing. As stated earlier, the 14 million population of Delhi requires 830 mgd. The DJB supplies 630 mgd, of which nearly 40 per cent is lost in pilferage and losses. Different agencies offer different figures on how much groundwater is extracted to particularly meet the gap of nearly 450 mgd. DJB member (Water) G C Nandiwani said: We just take 78 to 80 mgd. A CGWB source said both individuals and DJB take nearly 300 mgd. And a publication of the French Research Institute in India, Centre De Science Humanies, estimates the DJB takes 120 mgd and individuals extract 274 mgd—a total of 394. Though agencies may differ on the figures, the fact remains that the ground water in Delhi is fast depleting.

Dual Supply System

About 40 percent of the 630 million gallons of water supplied by the Delhi Jal Board daily is never used for drinking. Instead, it is used for domestic chores, including flushing of toilets. The wastage has continued for years even as the city's demand for drinking water is rising by 20 MGD a year. However, in a much belated move, city planners and developers have chalked out plans to supply less-expensively treated water for washing and cleaning.

The parched city is being forced to consider a dual water supply system—one for drinking and the other for washing. A more timely switchover, would have saved colonies like Vasant Kunj from their water woes. The underground water in these colonies is not fit for drinking and the DJB doesn't have enough treated water to supply there. There is no reason why water meant for drinking should be used for washing or flushing. It is an international practice to use recycled or less-treated water for purposes other than drinking. At Frankfurt airport in Germany, almost 90 per cent of the requirement is met with recycled water.

In Delhi, it seems nobody ever thought of a separate supply of drinking water. We continue to wash our cars and clean our floors with prime quality chlorinated water. Chennai and Chandigarh are some success models which Delhi can emulate. It is time to implement a dual system than just sit over planning. When it is difficult to increase the quantity of drinking water, using different qualities of water for different purpose can ease shortage. The first dual supply system in the city was introduced by the DDA in Dwarka. Such a system has existed in some parts of Lutyens' Delhi for years but it was never meant to be a solution for water shortage. We have taken an initiative to substitute chlorinated water with brackish ground water in Dwarka for purposes that don't require high purity levels. From now on, all houses being built in Dwarka will have a dual piping system. DJB officials have welcomed DDA's initiative. In areas like Dwarka and Najafgarh, where the subsoil reserves are largely saline, it is better to draw groundwater and create space for recharge with sweet rainwater.

Two separate water supply lines are provided to each dwelling unit. One is for potable water and leads to the kitchen for drinking and cooking. The other is the domestic water supply for the toilets, bath rooms etc. The domestic water does not need extensive treatment as it is not meant for drinking. Even tubewells with water unfit for consumption can be used to augment supply to a colony. In the dual supply system, less potable water is used and civic agencies can divert the saved amount where needed.

WATER WATCH

Precautions for a dual supply system:

- Over 40 per cent of treated water is used for purposes other than drinking.
- A dual supply system with treated and untreated water for drinking and other purposes, respectively, can save 80 per cent treated water.
- The system is only marginally costly and is best suited for upcoming colonies.
- Domestic water should not contain any banned substance.
- The domestic supply have to be fully purified.
- Water lines carrying domestic water have to be clearly marked.
- The internal supply line can be painted red.

The domestic water is pumped from underground sources and a dose of disinfectants is added. It is stored in ground-level tanks for complete disinfection. And pumped into overhead tanks for distribution to homes. In more advanced dual systems, even rainwater is recycled and stored in ground level tanks. This helps to conserve underground water reserves. Dual water supply systems require a higher investment in infrastructure like two parallel water lines and additional overhead tanks. This is the prime reason why civic agencies are reluctant to introduce it. In countries like Saudi Arabia where water is scarce but funds are not, the system has been in existence for decades.

Experts say that once the infrastructure is in place, vast amounts will be saved on treatment costs. For example, the DJB in Delhi spends around Rs. 5 to treat one kilolitre of potable water. To treat kilolitre of domestic water might cost Rs. 3 or 4. It would be difficult to introduce the dual system in existing colonies due to high costs and also the increased danger of the two varieties of water getting mixed. The system is best introduced in new colonies where the two-pipe system can be included in the development plans.

While water is one of the most scarce and precious resources in the parched city, just 44 per cent subscribers of the 1.4 million water connections pay on the basis of actual consumption. This is because only 77 per cent DJB connections are metered and of them 43 per cent meters do not work at all. A whopping 23 per cent or 322,000 connections remain unmetered. Majority of them are very old connections where the metres were not replaced after they stopped working long ago. By DJB's own admission, 260 MGD of its 630 MGD supply flows free everyday.

With meters not working people either pay the minimum of the average charge. The minimum charges for a domestic connection is Rs. 30; Rs. 150 for commercial and Rs. 450 for industrial connections. The meters are designed to work at a certain pressure and if there is a variation, it starts malfunctioning. In fact, the DJB gets large number of requests for replacing meters but there are no right meters with them. So even those who wish to pay for the amount they use, can not do so. Due to these factors, the board earns about Rs. 225 crore annually, which is just enough to meet the power bill of Rs. 200 crore. The white elephant that employs about 29,000 people meets its total revenue expenditure of Rs. 480 crore that includes paying salaries from loans from the government. It has no money to plug the estimated 40 per cent leakage in transmission and make more water available to Delhiites.

Treated Waste Water

Instead of the treated water, CPWD sends raw river water to New Delhi areas. Since this water stinks, mostly it is not used. Even for gardening, people use either groundwater or the DJB supply. This adds to the decline in the water table and further limits the availability of potable water. The only way to meet the ever-increasing demand for water is by ensuring that pure water is used only for drinking, cooking and bathing. Treated waste water can be used for other purposes like gardening. About 80 per cent of the 380 million gallons of waste water treated every day is released into the river, as none of the city's civic agencies has the infrastructure to utilize it. Being one of the greenest cities in the country, Delhi requires a lot of water for gardening and treated waste water suits this purpose perfectly. Instead of laying pipelines for drawing the treated water to their areas, civic agencies find it easier to drill tubewells. The result is a massive fall in the water table. DJB was willing to provide the treated water free of cost to the agencies if they just lay the pipelines. This water is full of nutrients and if used, would even help cut down on fertilizers. Apart from gardening, this treated waste water can also be used for flushing and washing pavements. At present, a huge quantity of potable water is being used for these purposes every day.

The Najafgarh drain in South-west Delhi alone has the potential of meeting the secondary requirements of a huge sub-city like Dwaraka if its waste is treated and used. The pressure on groundwater will also reduce. The city has no option but to start reusing treated waste water. Several government agencies are now thinking of laying pipelines for using treated waste water. But these plans are yet to take shape. One of these plans is to supply treated waste water to the Garden of Five Senses near Mehrauli, which is currently facing water shortage. Also, the water tank at Hauz Khas will be revived using treated waste water. The Pragati Power station has already started getting 2.2 mgd of treated waste water. According to the agreement, they will pay for the treatment plant's maintenance and power bill. DJB will not charge anything for supplying the water. In Bangalore and Chennai, the industrial units are supplied with only treated waste water. If there is assured demand from the industry in Delhi, DJB can perhaps set up the infrastructure and supply good quality treated water.

Delhi May Face Water Rationing

Delhi may finally force a water usage quota on its residents and hike prices to boot if pressure from neighbouring states like UP and Haryana persists. Upset over the capital receiving a huge amount of water and still needing more, Haryana has already joined battle in the apex court. Meanwhile, agitations over Tehri water for the much hyped Sonia Vihar plant in Delhi continue. It is acknowledged that the neighbouring states are irked by the over-use and even wastage of water in Delhi.

Despite political compulsions, it is felt that DJB simply cannot continue with its present price slab of a mere 35 paise per kilo-litres (up to 10 kl.) and Rs. 100 for 1,000 litres from producers of bottled water as the cost of production is Rs. 5 to 6. And this arithmetic has become a sore point as the Delhi government subsidises the DJB to the tune of Rs. 260 crore out of its annual expenditure of Rs. 480 crore. In Bangalore, the cost of production per Kl is Rs. 16 against Rs. 5 to 6 in our case. But how long we will keep getting water so cheaply? If we get water from Himachal Pradesh's Renuka dam, as is being proposed, then the cost of production will go up enormously, leaving the authorities with no other alternative but to hike prices accordingly.

Indiscriminate Groundwater Extraction

A survey of the trends in water levels between 1960 and 2001 showed a constant decline. During the 60's, ground water levels were approximately four to five meters and in several parts, waterlogged conditions existed. Between 1960-2001, water levels declined by two to six metres in most parts of the alluvial areas. As further decline of 8-20 metres has been recorded in south-west district and in the southern district, the decline ranged between 8-30 metres. In Mehrauli, for instance, water plummeted to a depth of 30 to 35 metres below the ground. The other areas with a depleted water table are Nizamuddin, Lajpat Nagar, Greater Kailash, Chittaranjan Park, Kalkaji, Tughlakabad, Saket, Malviya Nagar, Panchsheel, Vasant Vihar and Vasant Kunj. The prime instrument in this case has been the tube-wells. While there are court restrictions on the boring of new tube wells, those that are already operating are wreaking havoc on the precious little underground water reserve, which has dropped by about 20 metres in various places within the past two decades. The ideal level is 5 metres below ground level. Furthermore, at least 15 per cent of water is lost during distribution and 40 per cent of the total supply never reaches its actual destination as it is stolen.

In 2002, the water table in South and Southwest Delhi fell by two to five metres. Many Delhi Jal Board (DJB) tube wells in these areas dried up and this was effecting the supply of water to the city. The latest survey carried out by the CGWB shows that the maximum decrease in the water table has been recorded in Lado Sarai adjoining Tughlaqabad and Saket. The water table here has gone down by 10 m and is currently at 51 m.In Aya Nagar near Gurgaon, the table has fallen to 30 to 35 m. A steep decline has also been noted in R.K. Puram, Vasant Vihar and Vasant Kunj. In R.K. Puram, the Water table has gone down to 31 m. Delhi Jal Board officials conceded that the continuous decrease in ground water was drying up their tube wells used for augmenting muncipal water supply. Of the 1,200 tube wells being operated by the DJB in South and Southwest Delhi, about 100 had to be re-bored in the recent past to access ground water.

WAVES: A NEW INITIATIVE

It is proposed to launch The Citizens Water Alliance for Voluntary Efficiency in Saving (WAVES) Program. An important initiative towards bridging demand and supply gap in the water situation.

Goals of the Waves Program

Imparting timely information regarding leakages, bursting of pipelines to Delhi Jal Board and eliciting quick response to ensure effective and efficient functioning of the water distribution system at all levels. Educating all citizens of Delhi regarding water conservation methods and techniques to bring about greater water efficiency and thus bring about an effective increase in supply of filtered water.

Inculcating an awareness of the rapid depletion of ground water level and the need to harvest rainwater through simple Rain Water Harvesting Structures.

Components of the Waves Program

Water Wardens and Asstt. Water Wardens: These 2000 wardens shall be imparted training to make them conversant with DJB water supply systems and method of conservation of water including saving that can be affected in day to day household uses. The selection of the Water Wardens will be made from the RWAs, whereas the Asstt. Water Wardens will be nominated from the ECO CLUBS.

In order to achieve short term/long term objectives, training will be imparted by means of workshops/seminars/site visits etc. on a regular and sustained basis. A monthly meeting will be held under the chairmanship of CEO, where mutual exchange of information will take place. Suggestions/ideas on dissemination of information/Rain Water Harvesting would be welcomed and suitably acknowledged.

Responsibilities of Water Wardens/Asst. Water Wardens

Watch for leakages, bursting of water lines, faulty water devices etc., passing on of information regarding leakages etc. to DJB control rooms/concerned officials at the earliest. To ensure rectification of the reported matters.

Bring about awareness of water conservation measures in all spheres of activity.

Disseminate information regarding depletion of ground water level and its consequences.

Assist DJB in providing technical assistance to citizens desirous of implementing simpler and effective Rain Water Harvesting techniques/methods.

Tanker to Mouth

Just like the dismal food situation during mid sixties of 'Ship to Mouth, Delhi is in a worse position regarding water. Water Tankers, called the lifeline of the Delhi Jal Board, are unable to keep up with the demand for water in south Delhi colonies hit by a severe shortage. According to one source, the DJB owns 597 tankers, while private operators run 400 which supply water to Mehrauli, Najafgarh, Bawana, Nangloi, South Delhi and Shahdara. The tankers together supply 22 million litres of water a day, only a fraction of the total DJB production of 2,610 mld. Only 8 mld of water is supplied by private operators. The DJB spends Rs. 49 per thousand litres for water supplied by tankers, against Rs. 1.50 for piped water.

Another source says that almost 1,200 private tankers operate in the city, which is more than the 1,000 odd tankers run by the DJB. Since they charge a steep price, their service is more efficient and often people have no choice but to take water from them, instead of the unreliable DJB tankers. A large number of hotels, hospitals and residents depend on the private tankers that sell groundwater.

Residents of localities like Greater Kailash, Kalkaji, Munirka DDA flats and Vasant Kunj complain of receiving water for just half-an-hour to an hour a day. The evening supply had stopped completely. Many households spending amounts ranging from Rs. 20 to Rs. 40 every day to buy drinking water. The worst hit area in south Delhi is Chittaranjan Park, which is at the end of the south Delhi distribution system. The underground tank with a pumping station and other modern gadgets in Chittaranjan Park though commissioned three years ago, but no water is being pumped.

Augmenting Supplies

In an ambitious move to augment water supply to the national Capital, foundation stone of a carrier line canal was laid in May 2003. The ambitious venture, projected to cost Rs. 314,75 crore, would provide additional 360 million gallons of water per day to Delhi on its completion after three years. In view of the severe water crisis in Delhi during summers, project will be completed within three years. The new canal, with a length of 102.43 km will have a carrying capacity of 707 cubic feet per second and will be fully dedicated to the Delhi Jal Board. The concrete lining of the canal will limit the water losses due to seepage to only 4-5 per cent.

The annual water woes of Delhiites may be solved, when the Sonia Vihar Water Treatment Plant in East Delhi becomes functional in December 2003. The new plant will have the capacity to supply 140 MGD while the existing gap in water supply was 160 MGD. As for the quality of water, the water which is being drawn from the river turned out to be very good after treatment, but this can not be guaranteed as ground water forms 10 per cent of the DJB supply.

Problem of Suppliers

Four decades ago, Delhi had three million people. The water of the Yammuna was adequate to feed them. Today NCT of Delhi imports water. The population has risen to 14 million. The Yamuna water is not available for most part of the year. The level of groundwater has retreated from about 2-5 metres to 8-12 metres. Delegates to a seminar held recently in the Capital believe that the NCT of Delhi could be usurping other's rights over drinking water to serve its burgeoning population.

The Capital imports 250 million cubic metres (MCM) of water from the Ganga, Yamuna and Beas catchments. The import of 250 MCM of water from the catchment of these rivers is causing ecological imbalance at the cost of the inhabitants. The Tajewala barrage, about 220 km north of Delhi (near Yamuna Nagar), is the seat of Western Jamuna Canal from where water is fed to NCT of Delhi through the Haiderpur water works. The total Yamuna water availability annually is about 376 MCM per year. Of this, there are large evaporation and other minor losses that may be estimated to be at least 30 per cent. So the actual water use from Yamuna is 264 MCM per year. Water from the Upper Ganga Canal is brought through a pipeline from Murad Nagar. This amounts to 200 MCM and another 290 MCM from the Beas in an open canal that joins the Western Yamuna Canal. After the 30 per cent evaporation, the water available from this source is about 190 MCM per year. The total water use from all these sources is about 630 MCM per year, a large fraction of which is not local but imported from other river basins.

The population density and high irrigation and fertility of hinterland around NCT of Delhi does not permit the import of water to service the Capital. In the NCR, the water requirement for agricultural use is roughly 200 MCM per year or 0.2 MCM per sq, km per year. But the hinterland catchment have higher fertility and more intensive cropping, so their water use may be higher, say, 0.25 MCM sq. km a year. The carrying capacity from the total river water resource for NCT of Delhi works out to about 6.3 million people. The carrying capacity from the total catchment augmented reserve works out to about 1.5 million people, a total carrying capacity of about 7.8 million people, which is about half of the total population of NCT of Delhi. Any excess of population beyond a city's carrying limits can be to the detriment of the people who live there and also to those whose catchment areas are being usurped to serve the city's excess population. As degradation of their life sustaining sources due to excessive transfer of their water to serve other regions, people from these catchment areas may soon be agitating to hold their rights on water.

NEED FOR CONSERVATION OF WATER

Though most of the surface water comes from Yamuna, nearly 282 mcm of allocated water allegedly flows unutilized. The unutilized rainwater runoff, from a rainfall of 611.8 mm in 27 days is estimated to be 192 mcm. Increasing domestic and industrial activity have led to greater generation of wastewater. There also exists a wide gap between 4,225 million litres wastewater discharged per day and treatment capacity of 1,332 million litres per day.

As already stated, the growing population of Delhi has put a tremendous pressure on the existing civic services including the supply of water. Due to rapid urbanization there has been over-exploitation of groundwater resources. It may be reiterated that unplanned and uncontrolled extraction of groundwater has disturbed the hydrological balance, leading to decline in productivity of wells, rise in energy requirement and deterioration in quality of water with increase in salinity that has made it unfit for human consumption. In addition to overexploitation of groundwater, the uncontrolled disposal of effluents and sewage in the city has contaminated the groundwater levels. Studies conducted by Central Ground Water Board in Delhi revealed that groundwater in most parts of Delhi is contaminated with fluoride and nitrate and is unfit for drinking without treatment. This precarious situation urgently calls for conservation of the limited freshwater which is available in Delhi. In this grave situation water harvesting seems to be only solution.

When we mention about water harvesting, it refers to collection and storage of rainwater and also other activities aimed at harvesting surface and groundwater, prevention of losses through evaporation and seepage and all other hydrological studies and engineering interventions with the intention of conservation and efficient utilization of the limited water endowment of physiographic unit such as watershed. However, in common parlance, water harvesting is the activity of direct collection of rainwater. The rainwater collected can be stored for direct use or can be recharged into the groundwater.

In this context, it may be stated that rain is the primary source of water for us. Rivers, lakes and groundwater are all secondary sources. Presently, we however depend almost entirely on such secondary sources of water. In the process, we forget that rain is the ultimate source that feeds all these secondary sources and remain ignorant of its value. There is now a need to make optimum use of rainwater at the place where it falls.

In point of fact, in view of impending water crisis in Delhi, each individual will have to start water conservation so that he may not have to keep on re-boring his well and lengthening his pipe. If the people in Rajasthan can conserve rainwater, even Delheites can do so. In Rajasthan, even poor households conserve water without Government's intervention. In Delhi, unnecessary cementing and tiling of pavements, courtyards and storm water drains has precipitated the water crisis. It is necessary that Delhi residents keep their front and back courtyards kachha. Cemented courtyards donot allow water to percolate into the ground. The other benefits of keeping kachha spaces means there is no water logging. Even the storm water drains should not be cemented.

Recharging the groundwater through rainwater harvesting is not doubt one aspect of the initiative for water conservation. We also need to use the available water supplies wisely. If our water demands keep on increasing at the present rates, no amount of rainwater harvesting or any other form of water augmentation can help. In fact, there is an urgent need for judicious use of water and prevention of wastage.

Some Tips for Water Conservation

- You need not be an environmental activist to contribute your bit to guard the blue gold. Here are a few small tips that can help:
- Close the tap well after use.
- While brushing your teeth do not leave the tap running, open it only when you require it.

- See that there are no leaking taps. Leaking of one trop by another will fill one bucket during the night.
- Fix all the leaking taps at once and close any tap you find open, even though it might be a roadside one.
- Do not leave the taps running while washing dishes and clothes.
- Water in which the vegetables and fruits have been washed can be used to water the flowers and ornamental potted plants.
- At the end of the day if you have water left in your water bottle do not throw it away, pour it over some plants.
- When washing dishes by hand, do not let the water run whilerinsing. Fill one sink with wash water and the other with rinse water.
- Never put water down the drain when there may be another use for it such as watering a plant or garden or cleaning.
- Run your washing machine and dishwasher only when they are full and you could save 4500 litres a month.
- Wash your produce in the sink or a pan that is partially filled with ware instead of running water from a tap.
- When you clean your fish tank, use the water you have drained on your plants. The water is rich in nitrogen and phosphorus, providing you with a free and effective fertilizer.
- Collect and use rainwater for watering your garden.
- Don't let water run while shaving or washing youre face.
- Brush your teeth first while waiting for water to get hot, then wash or shave after filling the basin.
- You waste up to 7 litres of water when you leave the tap open to brush your teeth. Use a mug or a tumbler to shave, brush or wash your face.
- You use up to 50-100 litres of water every time you wash your car. Avoid doing so everyday. Give your car a sponge bath and if possible use a bucket instead of a hose pipe to wash the vehicle.
- Teach your children to turn the faucets off tightly after each use.
- Do not water your lawns on windy days.
- Use a watering can to water the plants. Reuse the water you use to wash the vegetables in.
- Bathe your young children together.
- Direct downspouts or gutters towards shrubs or trees.

- Create an awareness of the need for water conservation among your children.
- Toilets require 7 to 10 litres of water for flushing every time. Use flushing systems that allow you to control the amount of water you flush.
- Take a bucket bath instead of a shower bath. As much as 25 litres of water, the amount equivalent to letting a shower run for only 2.5 minutes, assuming a flow rate of 10 litres a minute, might be all the water that a whole family may need in a day.
- Avoid the purchase of recreational water toys, which require a constant stream of water.
- There are a number of ways to save water, and they all start and end with you. Water-Use it Wisely.
- Plant a tree. Trees help conserve water.
- Use recycled stationery. Consume less paper. Paper is made from felled trees.
- Be aware and care for the environment. The problem is ours and not somebody else's.

Table 4.1: Unintentional Wastage of Water

Activity	*Method Adopted*	*Qty. used Ltr.*	*Method to be Adopted*	*Qty. Required Ltr.*	*Qty. Saved Ltr.*
Brushing Teeth	Running top for 5 min.	45	Tumbler or Glass	0.5	44.5
Washing Hands	Running tap for 2 min.	18	Half filled wash basin	2.0	16.0
Shaving	Running tap for 2 min.	18	Shaving Mug	0.25	17.75
Shower	Letting shower run while soaping staying under shower too long	90	Wet down, tap off, soap up, rinse off	20.00	70.00
Flushing Toilet	Using old fashioned large capacity cistern	13.5 or more	Dual System short flush liquid waste, full flush solid waste	4.5 9.0	4.5 or more
Watering Plants	Running hose for 5 minutes	120	Water can	5.0	115.00
Washing floor	Running hose for 5 minutes	200	Mop and bucket	18.0	182.00
Washing Car	Running hose for 10 minutes	400	Buckets (Two)	18.0	382.00
					831.75

Water Harvesting in Delhi

Nature has its own methods of rainwater harvesting. In open spaces, the water seeps through and keeps recharging the groundwater resources. But in places like Delhi, there are hardly any open spaces left. Most of the land is covered with concrete and since only surfaces covered with vegetation are capable of carrying the recharge process, nature fails. Thus, rainwater harvesting takes the form of a necessity.

Rainwater harvesting has a huge potential in Delhi. Currently, 60 per cent of rain water flows down the drain. INTACH says storing only one per cent of the total rainfall in India is enough to meet the country's domestic requirement of water. Delhi has a total land area of 1,486 sq. km. The annual average rainfall is 60 cm. which is received over a period of 27 days, eighty per cent of which falls in the period between July to September. The rain water therefore has to be harvested during this short period. Rainwater harvesting, even at 50 per cent efficiency, means 450 billion litres annually-equal to about 35 per cent of the total demand of the city.

Water can be collected in rooftops or compounds. The amount of water harvested depends on the frequency and intensity of rainfall, catchment characteristics, water demands and how much runoff occurs. It depends on how quickly water can get down to the sub-soil and recharge the aquifers. In urban areas especially, adequate space for surface storage is not available and roof-top rain water harvesting is an ideal solution to solve water supply problems. An individual can do it very easily.

The Chief Minister of Delhi stated that the rain water harvesting schemes of the Delhi Government had been a big success and as many as 163 rain water harvesting structures were recharging the ground water aquifers successfully by harvesting 4.50 lakh cubic meters of rain water from around 8 lakh sq. meters of rooftop and compound area. In order to promote the concept of rainwater harvesting and water conservation financial incentive to the extent of 50 per cent of the total cost of the rain water harvesting structure or Rs. 50,000, which ever is less, has been announced by the Delhi Government for the Resident Welfare Associations and Cooperative Group Housing Societies of the worst affected areas of South and South—West Delhi.

The issue is how much of this rainwater is possible to harvest when we assume its great potentiality. The total amount of water that is received in the form of rainfall over an area is called the rainwater endowment of that area. Out of this, the amount that can be effectively harvested is called the water harvesting potential. The Centre for Science and Environment (CSE), New Delhi estimates that with an area of 1,486 sq. km., the rainwater harvesting potential of Delhi comes to about 907 billion litres annually. This is equal to about 270 days of water requirement for the entire city.

It may be relevant to point out that rainwater harvesting is a single, economical and eco-friendly method of water conservation. The rainwater that falls on the surface/roof tops is guided to bore wells or pits through pipes to recharge the under groundwater which can be used later whenever required. Rainwater can be harvested to the extent of 55,000 litres per 100 sq. metres per year from rooftops alone. This amount is sufficient to meet the demands of a family of five members for four months. For the entire city, the recharging potential is about 76,500 million litres per annum. In fact, rooftop harvesting has been practiced since ages and even today it is practiced in many places throughout the world. It may be stated that domestic roof water harvesting (DRWH) is not a new concept in India and has been a part of traditional wisdom. However, before practicing this in the modern context, a number of additional issues have to be looked into because of changes from the traditional (essentially rural) to modern (urban life) styles. In doing so we should draw upon our rich traditional experiences.

Advantages of Rainwater Harvesting

- It is an ideal way to solve the water problem
- The ground water levels will rise.
- It reduces the runoff which chokes the storm water drains
- It reduces flooding of roads, also reduces soil erosion
- The quality of water improves
- Power consumption is reduced as an one metre rise in water level results in saving 0.4 kwh of electricity.

Of late, the Delhi government has also decided to make water harvesting compulsory in all new residential buildings, group housing societies and plotted residential areas of over 500 square metres. As per the Government order, all building plans submitted to local bodies will have to indicate the system of storm water drains. The plan will also show points of collection of rain water in surface reservoirs or recharge wells. The government has also instructed that a system of dual piping be installed in all new buildings so that water stored in overhead rainwater for water coolers and garden taps, the second will supply treated Jal Board Water to wash basins, both rooms and kitchen outlets. All non-residential buildings having a discharge of over 10,000 litres a day, will have to incorporate a waste water re-cycling system. The re-cycled water will be used for horticulture purposes. The city government has also been preparing special ponds in several parts of south and outer Delhi to collect rain water. The water collected in these ponds and wells eventually percolates into the ground and helps replenish the water table.

Finally, following actions are also needed in order to make water harvesting a more successful venture.

1. The basic components of a DRWH in both traditional and modern systems are the roof, a storage and a gutter to channelise the flow into storage. Noting the increased pollution in the atmospheric and on the surface catchment area (especially in metropolis) and increasing in sanitary problems, special attention is needed on the quality of water from the roof.
2. Also in the modern context the water quality standards are better defined. Especially the drinking water quality standards, whether national or international are quite stringent.
3. Given the decreasing per capita space available, innovative measures are needed for water storage. Material of storage may prove quite costly and this may be a deterrent.
4. For propagating this technology in the domestic sector, provided sufficient number of customers are available, there should be adequate number of architects and technologists to make the units readily available on a commercial scale.
5. Suitable policy support would also be needed for the above. Independent initiatives for awareness generation will of course be of great value.

Water Bodies

Initially, there were lakes of natural water catchments which acted as sponges. Also, there were open spaces which helped in the recharging of ground water. Now, there are buildings on these open spaces. The entire city has virtually developed into a concrete jungle and we have forgotten about the dead and decaying bodies and ponds strewn across the city. For example, there is a huge lake in Munirka where, it is said, the famous Khazan Singh used to swim. Today the lake has totally dried up and there are plans to convert it into a swimming pool. This lake could easily be utilized to conserve and hold water. A majority of the 508 water bodies in Delhi are in the south and southwest district, the places worst affected by exploitation of groundwater. While we all agree that these hold the key to our survival, the big question is who will revive them. It would be of interest to know about the fate of 800 years old reservoir in Hauz Khas. It is a 50-hectare expanse, green and full of birds. The park is nearly 800 years old, but what has disappeared is the lake. Once, this lake supplied water to the people who lived in the Hauz Khas area. Today, it is a dry bed. The southern part of Delhi has always faced water shortages. Even during the reign of Alauddin Khilji, the residents faced extrme water shortages. So, the Khilji king built a tank near the city. The tank was so large that historian Sharfuddin Yazdi said that an arrow shot from one end would not reach the other. This tank stored rain water. The deep reservoir flanked by steps was an example for the ingenious water storage system during that time. This tank was so impressive that Timur who came to invade Delhi, couldn't stop singing its praise.

Yet, over a period of time it got silted and the tank dried up. For many years, it lay almost buried till Firoz Shah Tughlak chanced upon it one day and got it excavated. Shah was justified in feeling proud of this achievement. At that time, people used to sell water. The more enterprising ones would dig wells near the lake area and draw water. So, assessing the needs of the people, he got it excavated and cleaned. At that time, Hauz Khas was known as Hauzi Alai. But when Firoz Shah built a magnificent college at one end, it became known as Kauz Khas. So, this tank of the third city of Delhi was resurrected by the third Tughlaq king.

The tank, however, again did not survive for long. Today, while there is little water, the lake's catchment area is still very evident. It is not really barren. The catchment region has been replaced by the kikar and other trees. What is disheartening is the fact that a sizeable part of the area has been paved, thus making it impossible for the rain water to seep through the strata to refill the ground water.

There is then the huge and heavily silted Yamuna basin. Sanitary landfills and the discharge of huge amounts of untreated sewage in the river are polluting the aquifers. The landfills must have an impermeable lining so that the leaching of toxic material into the ground does not happen in future for **saving the flood plain**: The river bed and flood plain of the Yamuna are the lifelines of Delhi. In the peak summer months, the Delhi Jal Board draws groundwater from the flood plains to supply it to the worst affected areas. During rains, when the river is in a spate, the emptied aquifers get refilled. The encroachment on the flood plain in the name of development must be stopped at once. Already stadia, temples, and the Metro Rail headquarters are built on reclaimed flood plain. Nobody should encroach on the river's carriageway. It is one of the biggest sources for recharging the aquifers and a large reservoir of biodiversity. The place should be left alone. Reviving water bodies can be possible only if the people themselves take the initiative. It has to become a peoples' movement.

Role of Women

In a unique scheme for drinking water management, the Delhi Government has decided to seek the help of women to help tide over the summer crisis. It has decided to set up "Pani Panchayats" of women across the Capital, mainly in rural areas, to store, manage and distribute water at the local level. Announcing this at a press conference to unveil the Summer Action Water Plan, the Chief Minister, Shiela Dikshit, said that the five-member committees would be formed as part of Pani Panchayats' which would manage drinking water. In addition, it has been decided to install plastic water tanks of 1,000 to 5,000 litres capacity in rural areas, slum clusters, unauthorized colonies and other water scarcity areas with the help of the local people and MLAs.

The "Pani Panchayats" could also have some other eminent members of the area but the authorities would prefer to have an all-woman show as females have a better knowledge and capacity to handle water-related crises. These women would be provided with the keys to the water tanks to ensure that neither the water is stolen nor contaminated by some miscreants. It is hoped that substantial relief would come to areas which are covered under this scheme which is being adopted in Delhi for the first time.

Tata Energy Research Institute's Findings

The report claims that the water table in most parts of the metropolis is falling rapidly and in many places in South Delhi is almost bone-dry. The crumbling of over half-a-century-old supply infrastructure and the burgeoning population have compounded the problem. Delhi "attracts more than 4 lakh people each year. Even as it accepts, within its fold, all these new citizens, the city's infrastructure is on the verge of a breakdown". The study suggests that the groundwater level is receding steadily, and at places, rather rapidly. Due to unbridled urbanization, the scarce rainwater runs off without being tapped, which is a complete waste of resources. Delhi receives its water supply from surface flow and groundwater. The availability from surface sources, that is, Yamuna, Ganga, and Bhakra is approximately 1,150 MCM (Million cubic metres) and, of this, 60 per cent is available from Yamuna.

Delhi gets an average rainfall of 600 mm per year, of which 80 per cent is received in three months (July—September). More than half the rainwater runs off into the Yamuna. It is ironical that while the Yamuna is frequently in spate during mansoon, it reports 'zero flow', and that too in peak summer. The pressure on ground water is already beyond capacity and recent reports suggest that people are increasingly tapping this water resource. Even the DJB depends on groundwater.

Though the courts have put curbs on new tubewells, those already installed are gnawing at the dwindling reserves. The latter operate without any restrictions, some even for 24 hours. According to the study, the groundwater table in the Delhi region, at varying depths of 20 to 50 metres below ground level, is saline. In some areas near the Yamuna these saline aquifers occur at a depth of about 65m. This effectively makes the water unfit for consumption within that level. Electrical conductivity is also reported to be high, making the groundwater further unfit for consumption. In few areas, comprising blocks of Najafgarh, Alipur and Mehrauli, the salinity is as high as 3,000 to 5,000 micro siemens/cm. Nitrate concentrations have been found to be higher than the permissible limit of 100 mg/l in IIT Delhi, NCERT campus, Mehrauli and a few Shahdara and Nazafgarh blocks. The status is similar with respect to fluoride levels in certain areas. Already, the average level of water table for blocks away from Yamuna is between 15 to 20 m. The water table has receded by 10 m in many areas during the last two decades. Nazafgarh and Mehrauli blocks have shown a decline of more than 10 m. South and south west districts have shown a fall of 10 to 30 m in the last four decades.

As if the scarcity is not enough, the municipal infrastructure is 50 years old and ill-maintained. This is evident from the fact that nearly 15 percent of the water supply is lost in transit and a whopping 40 per cent remains unaccounted for. The gap between waste water discharges and treatment capacities is seriously worsening the quality of the existing water sources and, if not managed properly, it will increase considerably in the future.

The decline in the water table would also mean that the households, especially in South and Central Delhi, that have their own tube wells will have to dig deeper to access the 500 to 600 litres of water per day to meet their needs. The 11-meter deep tube wells in Lajpat Nagar have gone dry and these will now have to be re—bored to a depth of 13 m. Though there are no exact figures available with the CGWB about the number of private tube wells, it is estimated that there are around 2 lakh illegal tube wells in the city in addition to the 1 lakh registered with the CGWB.

The crisis has bred an unhealthy practice—consumers are being forced to tap groundwater reservoirs in violation of norms laid down by the CGWB. The ban on tapping groundwater, however, exists for two reasons. First indiscriminate boring of tube wells by residents has sharply depleted the groundwater as stated earlier. In effect, the practice has seriously jeopardized the availability of an additional 11-12 per cent water supply, which could otherwise have been tapped systematically. Further, in many cases, the boring is not done to the correct depth, the water extracted is often extremely dangerous for human consumption.

In the recent past, as a part of the Fluorosis Mitigation Programme of Delhi Government, over 3,000 water samples taken from MCD run schools and houses in Delhi were tested for fluoride in the first phase of a school based survey and it was found that 550 samples contained more than the permissible limit 1.5 ppm (parts per million). Some of the samples tested even had over 4 ppm. It may be pointed out that long-term intake of excess fluoride may lead to severe, permanent bone and joint deformation.

Survey Committee

Joint survey committee appointed by the Delhi High Court to suggest ways of maintaining water bodies in the Capital has recommended that only those water bodies having an area of at least 4,000 square metres should be preserved. It has concluded that maintenance of smaller water bodies than this will be unviable. The committee included representatives of various government agencies. It has suggested that smaller water bodies should make way for green areas and social amenities. The Committee maintains that the loss can be offset by the creation of new water bodies twice the size of the ones 'killed' at a suitable location close to the original ones. These recommendations have been strongly contested by an NGO Tapas which is canvassing for preservation of water bodies. According to V.K. Jain, Chairman of Tapas, who had filed a petition in the High Court two years ago on the issue of neglect of water bodies, the recommendations are based on the presumption that smaller water bodies are difficult to recharge. The logic is flawed, he argues.

Most of the water bodies have a widespread catchment area, which is generally encroached upon by human settlements. This prevents the recharging of the water bodies. The authorities should, therefore clear the catchment areas so as to preserve these water bodies instead of removing them. Tapas contends that local authorities have a tendency to preserve slums at the cost of these water bodies. The joint survey committee has identified 508 water bodies in the capital. While taking note that these are in a state of neglect and disrepair, the Delhi High Court recently observed that the problem of water-shortage in the capital would be solved if these water bodies were maintained by the land-owing agencies. Just like Agrasen Ki baoli (stepwell) near Connaught Place, several other Archaeological Survey of India (ASI) protected water bodies in Delhi may soon start recharging the fast depleting aquifers. And the ancient technological marvels may become functional soon, as that is the time the high court has given to the ASI to make them presentable. Nine of the 10 water bodies on which work has started are baolis and one is a pond. These include Rajon and Gandhak ki baoli in Mehrauli, Arab ki sarai near Humayun's Tomb and the stepwells in Old Fort, Red Fort and Ferozshah Kotla.

The desilting work at Rajon ki baoli is over and the place already has some fresh water. The stepwells in Mehrauli need some restoration. The desilting work in Agrasen ki baoli had to be temporarily suspended because of the rains. Again DJB had about 250 acres of marshy land in Jahangir puri. Of this 60 acres were given to Delhi Police and 40 acres to Public Works Department (PWD) for making houses. PWD filled toxic fly ash in its land. But that can pollute the aquifers. The court has asked PWD to restrain from further dumping of fly ash and to see if it can be removed. DJB has also pointed out to the PWD about the harmful affect of dumping of fly ash on the sub-soil aqua reserves. Since a majority of the water bodies to be restored have a limited catchment areas, from where rainwater will gush into them, a body of technical and legal experts should be set up to examine how the effectiveness of these structures can be optimised. The ancient harvesting structures in places like Hauz Khas and Tughlaqabad have a large catchment area and can be very effective in boosting water tables of these stress zones.

THE ROLE OF VARIOUS AUTONOMOUS BODIES

After Independence, various organizations were set up to provide and maintain the essential requirements of ever growing population of Delhi. In 1957, Parliament passed the Delhi Municipal Corporation Act and the Government of India appointed commissioner of local bodies. The management of the various local bodies was amalgamated in the proposed Delhi Municipal Corporation. However, the New Delhi Municipal Corporation (N.D.M.C.) and Delhi Cantonment Board remained two distinct authorities. The Delhi Water Supply & Sewage Disposal Undertaking was formed in 1958. The Delhi Jal Board (DJB) was constituted on 6th April, 1998 through an Act of the Delhi Legislative Assembly incorporating the previous Delhi Water Supply and Sewage Disposal Undertaking. DJB is responsible for the production and distribution of potable water after treating raw water from various sources and also for arranging treatment and disposal of wastewater.

CGWB

The Central Ground Water Board (CGWB) has created a special water harvesting cell. The cell focuses on artificial recharge and helps in removing difficulties faced by people in catching the rain. With water scarcity looming large over the city, particularly during the summer months, efforts are being made to spread awareness on the need to harvest water. Apart from training a large number of people on the technical aspects of harvesting, the CGWB has also been endeavouring to strengthen the regulatory mechanism to ensure that the directives are properly implemented. The CGWB first checks the places where groundwater is being exploited the most to findout whether harvesting structures have been installed.

The CGWB also proposes to implement the first phase of its 24,500 crore, 10-year Master Plan for water harvesting at the earliest. In the first phase, Rs. 3,000 crore plan will be implemented in constructing harvesting structure in areas like south and southwest Delhi where the groundwater situation is very bleak. The CGWB has also made it mandatory for all to adopt rooftop rainwater harvesting systems.

Delhi Development Authority (DDA)'s Efforts

After taking the blame for constructing massive housing complexes without water and power, the DDA has, of late, realized the necessity to go in for water conservation schemes. It has notched up some success in water conservation in recent years in two different projects in South Delhi. One of the projects uses treated water from its Vasant Kunj sewage treatment plant. The DDA has also constructed five weirs to regulate the flow of the treated water in its 500—acre Sanjay Van reserve forest. The forest also has a check dam to tap rain water which was otherwise going waste. According to the DDA even during poor monsoon season, this check dam collects about 10 feet water over a 200-metre long stretch.

In another project, the DDA has set up two water harvesting sites in Vasant Kunj (Phase II) to tap rain water in the ridge between Vasant Kunj and Vasant Vihar. Still, a joint project between the resident welfare association (RWA) and the DDA, it has unfortunately not taken off due to high cost involvement. The plan developed by INTACH (Indian National Trust for Arts and Cultural Heritage) could not see the light of the day because of a lukewarm response from the RWA.

Buoyed by success in their attempt to conserve and re-use waste water, DDA is now experimenting and incorporating new technologies for tapping water. The organization has selected Dwarka and Vasant Kunj for a project for tapping roof top and surface runoff water. Water from roofs, terraces, roads, parking lots and paved surfaces will be collected in a network of underground storm water pipes. This water will be led to green area, the project indicates.

New Delhi Municipality Council (NDMC)

Though this elite Municipal body has a limited area, it has already taken a lead in conserving water through various means. NDMC is tapping three gardens-Lodhi Garden, Talkatora Garden, Nehru Park— and a mildly-used Kushak nullah for rain water harvesting. Within Lodhi Garden itself the NDMC is tapping rain water at six places. The Talkatro Garden has three pits for water harvesting, while Nehru Park has two pits. The NDMC has dug 12 feet deep pits and then dug a 40-feet bore at the middle of the pits. The pits have been filled with layers of stones, boulders, gravel and sand. Rainwater and surface water from nearby areas flows into the pits and filters down through the various layers and finally recharge the ground water.

Water conservation at Kushak nullah however adopts a different technique. The nulla was seasonal and collected water during the monsoon used to flow away. Now, the NDMC has made a wall of boulders and wires to stop the water from flowing away. As this wall is not cemented, it lets water flow through slowly.

The NDMC has also focused on roof-top water harvesting at the Bapu Dham housing complex. Further, large buildings like the senior secondary school at Mandir Marg, New Delhi have also been selected for roof-top harvesting.

Delhi Jal Board (DJB)

We have mentioned earlier that the Delhi Jal Board, constituted on 6th April, 1998 through an Act of the Delhi Legislative Assembly incorporating the previous Delhi Water Supply and Sewage Disposal Undertaking, is responsible for the production and distribution of potable water after treating raw water from various sources and also provides treatment and disposal of wastewater. DJB is committed for augmentation of water supply and has taken many steps in this direction. One of these steps is 'roof-top rainwater harvesting' for augmenting groundwater storage in National Capital Territory (NCT) of Delhi. DJB has created a special cell for this purpose which is manned by qualified engineers. The cell is disseminating information about rainwater harvesting in NCT of Delhi and nearby states. It is providing technical guidance to schools, institutions, RWAs and individuals by personally visiting their premises.

Since 2000, when rainwater harvesting was taken up by the DJB, the Board has provided rain water harvesting structures in its 103 buildings covering an area of 8 lakh sq. metres, which will harvest about 4.5 lakh cubic metres of rainwater annually for recharging the groundwater. Work is in progress at another 27 sites of DJB (3). The DJB has also been making sustained efforts to spread awareness of the need to harvest rainwater through various media.

In order to minimize wastage of water, DJB has replaced 1,000 km old, damaged and sub-standard pipe lines in the last four years. Though a World Bank report alleged 40 per cent leakage of water, DJB claims to have been able to save 25 mgd water by stopping only leakages.

Delhi Jal Board (DJB) is not only responsible for making adequate arrangements to provide potable/ wholesome water to the citizens of Delhi, but is also responsible for collection, conveying and disposal of sewage and wastewater after its treatment. Presently, the installed capacity for portable water production by DJB is about 660 MGD. It is estimated that the residents of Delhi draw the sub-soil water to the tune of 90-100 MGD from private bore holes. The present sewage treatment capacity of DJB is 512.4 MGD. In addition, DSIDC is constructing 14 CETPs and one more CETP by DDA with a total capacity of 40 MGD. Thus, the waste-water treatment capacity likely to be in position by 2003 is 552.4 MGD against waste-water generation of 612 MGD. The DJB plans to augment the capacity of sewage treatment plants by 143 MGD by March 2005 thereby, raising the wastewater treatment capacity to 695.4 MGD against an estimated wastewater of 751 MGD.

The DJB has raised the sewage treatment capacity from 284 MGD in 1995 to 512.4 MGD as on date. As for as utilization of treatment capacity is concerned at various sewage treatment plants, it has increased the utilization of various sewage treatment plants from 200 MGD in 1995 to 380 MGD. The trunk sewers exist in a length of 130 km in Delhi. Some of them were constructed as early as in 1930s. Most of them have outlived their life, silted up and settled at many places. Therefore, it requires rehabilitation/de-silting.

The rehabilitation/ de-silting of trunk sewers in a length of 22 km has been completed. The work for a further length of 10.5 km is in progress and is likely to be completed by 2004. The sewage network exists in a length of about 6,000 km consisting of branch sewers/

peripheral sewers and trunk sewers. DJB has prepared an Action Plan to replace the 600 km length of peripheral and branch sewers out of which, approximately 160 km of under-size branch sewers have been replaced and the balance are likely to be replaced in a phased manner.

The Center for Sciences and Environment (CSE)

The whole saga of rain water harvesting and conservation of water will remain incomplete unless the role of the CSE, a non-government organization, in this sphere is not narrated. The CSE has been involved in raising awareness about the need of community-based water management. CSE is of the view that unless people are involved from individual households, farmers and industrialists to urban and rural communities, it will be difficult to combat the looming water crisis. A water crisis that has come about because rain, as a source of water has been ignored. As a technology solution, CSE is therefore promoting the concept of community and household based water harvesting as this technology can be adopted by all concerned and also promote a participatory paradigm of water management[4]. The most unique performance of rainwater harvesting through adoption of CSE technology has been in the Rashtrapati Bhavan.

The Presidential Estate covers an area of 133 hectares (1.33 sq. km.). The water requirements of the presidential estate are huge since there are about 7,000 people residing in the estate. Approximately 3,000 people visit the presidential premises everyday. The Mughal Gardens in the estate require a lot of water. The total demand is about 2 million litres a day (730 million litres per year). This demand is met through the NDMC supply and the estate's own borewells.

Since about 35 per cent of the water requirements are being met through groundwater sources, there had been an alarming decline of groundwater levels in the estate. Levels had gone down by 2 to 7 metres in the past decade, with three wells running dry. In 1999, work began on tapping one lakh cubic metres of rain water annually inside the estate. All major drains through which rain water flowed in to the Yamuna were blocked and diverted to the three dried wells dug up during the British era. By May 2000, the entire infrastructure had been put in place at an estimated cost of Rs. 16 lakh.

Other Departments/Organizations

Various other departments and organizations have also taken initiatives to avert Delhi's imminent water crisis. In April, 2003, the Delhi State Chief Minister announced that a Rs. 380 crore Munak-Delhi canal would be constructed to get uninterrupted water supply from Haryana. The State Forest Department, one of the largest owners of forest land in Delhi, has also been harvesting rainwater, although for entirely different reasons. The aim is not to create water bodies, but to replenish groundwater so that it enriches the flora and improves bio-diversity. The Delhi State Industrial Development Corporation (DSIDC) is also planning to incorporate roof top harvesting measures in its industrial projects which are on the anvil. The Irrigation and Flood Control Department tapped water on Najafgarh drain and Old Bawana Escape. The idea behind this is to collect surface and rain water so that water percolates into the ground.

Conclusion

Though water harvesting and water conservation efforts have taken off on a serious note, the government agencies have a long way to go, if they are to stave off the water crisis which the city is otherwise likely to face in the very near feature. A limited water supply from the Yamuna, depleting ground water and a fast increasing population have ensured the supply never catching up with the demand. But the general public is yet to realize the gravity of the situation. It is a common misconception among the people that this temporary crisis will ultimately help in augmenting the supply by building up more and more dams. Even now many feel that we can drill deeper and deeper and get more water. Also quite a few feel that ground water is a community resource and it is not individual's responsibility to recharge. Needless to say that in a situation where the scope of getting more raw water from other sources is remote, there is no alternative to recharging of the groundwater through rainwater harvesting besides re-use of waste water after proper treatment. We also need to use the available water supplies wisely and prudently. We have to remember that if our water demands keep on increasing at the present rates, no amount of rainwater harvesting or any other form of water augmentation can help. Hence, we need to use water most judiciously and prevent wastages.

REFERENCES

1. Singh Kanwarjit: Water Scenario in Delhi—Present and Future, *Shelter*, Vol. 5, No. 1, January.
2. Sobhana K, *The Statesman*, New Delhi Edition, April 3, 2003
3. Brochure released on the occasion of a Seminar on Rain Water Harvesting held at India Habitat Center, New Delhi on 30th December, 2002.
4. A Water Harvesting Manual for urban areas—Case studies from Delhi, Center for Science and Environment.

Part II

CAUSES AND PROBLEMS

5

Irrigation and Changes in Cropping Pattern

"Within Waters-so Soma has told me—are all healing balms, and Agni, benign to all"

Introduction

Worldwide, agriculture sector is the largest consumer of water. During the 1950s to the 1980s, irrigation expanded rapidly and currently accounts for about 72 per cent of global water withdrawals, and about 90 per cent of water use in the lowest-income developing countries. Such a major role for irrigation had been justified by the contribution of irrigation systems to stabilizing, then expanding national and world food supplies during the Green Revolution, especially in Asia. Dramatic increases in yield during and after the green revolution were largely based on the introduction and successful adoption of high-yielding varieties of wheat and rice that depended heavily on timely nutrient and pest control management as wells as irrigation applications to secure and control soil moisture. Increased availability of water through creation of irrigation potential has resulted in major change in cropping system in the various states in India. To date, most increases in food crop yields took place in irrigated land. Over the period the country has become self-sufficient in the production of foodgrains and some other commercial crops, there are also signs of crisis in agrarian economy in many states following switching over to the water consuming crops like paddy, wheat and sugarcane.

State-wise Ultimate Irrigation Potential and Creation till 2001-02

Before analyzing the impact of irrigation on cropping pattern, it is necessary to closely look into the progress made by each state in tapping the estimated irrigation potential. It is estimated that 64 per cent of the ultimate potential has already been explored in the country (Table 5.1).

Table 5.1: Statewise Development of Irrigation Potential till 2001-02

'000' hect.

State	*Ultimate Irrigation potential*	*Potential created till 1996-97*	*% of Ultimate irrigation potential*
Tamil Nadu	5,532	3,672	66.18
Uttar Pradesh	30,499	30,290	100.00
Punjab	5,967	5,970	98.32
Rajasthan	5,128	4,929	91.56
Haryana	4,512	3,730	81.03
Nagaland	85	76	78.82
Jammu & Kashmir	1,358	562	40.43
Gujarat	6,103	3,429	53.82
West Bengal	6,918	5,476	67.37
Andhra Pradesh	11,260	6,322	52.82
Maharashtra	8,952	6,181	55.09
Karnataka	5,974	3,706	53.52
Bihar	13,347	8,339	59.26
Sikkim	70	30	37.14
Manipur	604	231	20.36
Orissa	8,803	3,301	33.12
Kerala	2,679	1,249	40.54
Madhya Pradesh	17,932	5,052	27.75
Himachal Pradesh	353	174	45.32
Tripura	281	115	33.81
Meghalaya	168	81	27.38
Assam	2,870	848	27.53
Arunachal Pradesh	168	100	49.40
Mizoram	70	17	18.57
UTs (Including Goa State)	260	100	63.08
All India	1,39,893	93,979	67.18

Among the states, Punjab leads in exploring cent per cent of the ultimate potential, followed by Uttar Pradesh (99%), Rajasthan (96%) and Haryana (82%). The progress made by all these four states is much above the national average (67%). Even in West Bengal, the progress is higher than the national average being 79 per cent respectively. The rest of the states have a long way to go in tapping the estimated ultimate irrigation potential. The situation in the North-Eastern States, Madhya Pradesh, Kerala, Jammu & Kashmir and

Himachal Pradesh is deplorable as these states have failed to create even fifty per cent of the ultimate potential.

It has also been observed that though nearly 67 per cent of the potential has already been created, the gross irrigated area in the country is only about 40 per cent of the total irrigated area (*Table 5.2*). However, this ratio varies from state to state. For instance, in Punjab almost 91 per cent of the cropped area is irrigated, followed by Haryana (85%), Uttar Pradesh (66%) and Tamil Nadu (55%). In Andhra Pradesh, Bihar and Jammu & Kashmir, about 40 to 48 per cent of the gross cropped area is irrigated. In Gujarat, Manipur and Rajasthan, it is between 35 and 37 per cent. In states like Goa, Karnataka, Madhya Pradesh,

Table 5.2: Proportion of Gross Irrigated are in the Total Cropped Area in 1999-2000

States/UTs	*Total Cropped Area (TCA)*	*Gross Irrigated Area (GIA)*	*GIA as percent of TCA*
Andhra Pradesh	13,023	5,746	44.12
Arunachal Pradesh	264	36	13.64
Assam	4,093	572	13.98
Bihar	9,979	4,808	48.18
Goa	171	34	18.88
Gujarat	10,152	3,840	37.83
Haryana	6,029	5,124	84.85
Himachal Pradesh	957	179	18.70
Jammu & Kashmir	1,078	438	40.63
Karnataka	12,097	3,162	26.14
Kerala	3,002	471	15.69
Madhya Pradesh	26,207	7,091	27.06
Maharashtra	22,351	3,769	16.86
Manipur	199	75	37.69
Meghalaya	266	55	20.68
Mizoram	91	11	12.09
Nagaland	295	73	24.75
Orissa	8,524	2,512	29.47
Punjab	8,240	7,487	90.86
Rajasthan	19,286	6,934	35.95
Sikkim	121	16	13.22
Tamil Nadu	6,519	3,585	54.99
Tripura	420	60	14.29
Uttar Pradesh	26,640	17,676	66.35
West Bengal	9,545	2,491	26.10
UTs Including Delhi	189	92	48.68
All India	1,89,740	76,336	40.23

Meghalaya, Nagaland, Orissa and West Bengal, the ratio varies between 20 and 30 per cent. In other states, it is insignificant being less than 20 per cent. Thus, majority of the states still have to depend on rainfall for raising crops.

Cropping Patterns

Cropping pattern refers to proportion of total cropped area allocated to different crops. In any region, the present cropping pattern is the result of past and present decisions by individual farmers, communities, government or their agencies. These decisions are governed by various factors including productivity, expected profit, resource structure and so on. The trend in cropping patterns points to the likely demand—supply graph of agricultural products in future.

Irrigation and Cropping Pattern

Choice of crops and cropping pattern in a region is governed by, inter alia, extent of available irrigation facilities and the prevalent water management practices. Availability of irrigation water induces the farmers to use more yield increasing inputs particularly fertilizers and high yielding varieties with the result that some crops become more profitable than others.

There is also a fact that not only the availability but also the cost of irrigation water is an important determinant in the choice of crops. It will, therefore, be necessary to study the impact of irrigation on cropping pattern as also the manner in which the cost of irrigation and water management practices can influence cropping pattern.

In most of the irrigated area in India, water supply is erratic and limited, and thus warrants its judicious use for increasing crop production. By and large, there is a tendency to take rice, sugarcane and wheat crops as soon as water is made available. But these crops are very high water consuming, and therefore, productivity per unit volume of water is low as compared to the light irrigated crops such as mustard, groundnut, barley and pulses. Crops/cropping patterns for limited water availability area should be so selected that higher returns are obtained per unit of water applied.

Considerations

The water management problems are very location specific, and thus need a special attention for their solution on individual farm units. Criteria most suitable for scheduling irrigation may vary in different situations, but in all cases, it should permit favourable crop yields, optimum use of water, and proper attention to other factors (irrigation cost, control of ground water level, leaching of salts, accommodation to other farm operations) involved.

Except for flooded rice, quantity of water required for irrigation of a given crop grown in a particular environment is determined by (i) the quantity of useable water stored in the crop root zone and (ii) the rate at which this supply is depleted. Usable soil water storage in this context includes water storage and transmission properties of the soil, the crop's effective root depth, proficiency in water absorption and the ability of the crop to tolerate soil water stress imposed by increasing soil water suction and interacting climatic parameters. These factors independently determine the quantity of water that may be extracted safely before irrigation is needed. Rate of depletion of this supply is determined primarily by climatic parameters with certain limitations imposed by soil and plant factors which interact to control the evapotranspiration and deep percolation rates. In the case of submerged rice culture, water is stored above the soil surface and irrigation water must be supplied to keep the depth of water within certain limits. Different irrigation methods are adopted because of variations in soil types, topography, water supply, crops and other management practices. However, efficient water application method should permit uniform application at rates consistent with intake and water retention characteristics of the soil.

Thus, an efficient use of water warrants the selection of a proper crop/cultivar, and management practices directed for minimizing water loss by crop in transpiration. In addition to crop transpiration, the water may be lost from the system in runoff, through deep percolation to soil depths below root zone, through soil evaporation, and through transpiration by competing weeds. Weed control, tillage mulching and fertilization can all aid in maximizing the proportion of water which passess through the crop.

It is also possible to supply the nutrients to the crops through the pressurized system of irrigation popularly called as fertigation. Supplying liquid soluble fertilizers through drip system (discussed in chapter 13) can lead to savings in fertilizers applied to the extent of 40 to 60 per cent without affecting the yield and a much higher application efficiency compared to the conventional methods. This is in addition to the savings in water supplied which could be anywhere from 50 to 70 per cent.

Enormous research data are available for efficient use of limited irrigation water supply in field crops. In narrow-spaced grain crops like wheat and barley, a substantial amount of water economy and high yield application efficiency can be achieved, if fields are leveled, properly graded and borders of appropriate size are prepared. Greater irrigation depth results in lower grain yield. In case of widely spaced crops like cotton, mustard, gram etc., the depth of irrigation can be further reduced by 30-50 per cent if the paired row planting in ridge and furrow system of seeding varies from 2-4 cm as compared to 5-8 cm in flat sown plots. In case the ground water available in the area is of poor (brackish) quality than surface water, it can be used either mixed with canal water or in alternate irrigation after establishment of semi-tolerant and tolerant crops grown on sandy loam soils.

In the area which are under high water table conditions, wheat is a most promising crop in winter season. Gram suffers from poor aeration, mustard from diseases and excessive vegetative growth and barley from lodging, and thus they are poor performer than wheat. With water table depths around 150 cm., wheat responds to 1-2 post-sowing irrigation as compared to 5-6 irrigation applied to wheat under low ground water table conditions.

Farmers need on the spot advice and demonstration of proper water management practices for increasing crop production under limited water supply conditions. Earlier, farmers used to apply, generally more water in wheat and lesser to mustard than the recommended quantity of water. The mustard and gram on farmer's field were sown in flat border strips. Further water economy in these crops have been achieved by paired row planting in ridge and furrow system. There should also be sincere efforts for further transferring the water management technology to the farmers. However, the general recommendation will not find full solution of location specific problems of water management.

Cropping Plans

When water supply is limited, the proper water conveyance and land development (leveling, grading etc.) are important steps to minimize water losses in channels and fields. Thereafter, proper methods of irrigation have to be followed. As mentioned earlier, the moisture stress at all the critical stages of crop growth is not equally damaging. The most critical stages of a crop are seedling and flowering. Irrigation should be applied at critical stages of crop growth under limited water supply conditions. Selection of proper crops/cultivars, control of weeds, application of fertilizers and plant protection measures all play a great role in attaining high water use efficiency at higher level of productivity. The development of optimum crop plans for maximizing the net profit per unit amount of water use and land area will depend upon the agroclimatic conditions and ability of individual farmers in using the other inputs, besides water. However, it will not be very difficult to advise the farmers, if the information on water supply is known.

Table 5.3 reveals that during the first decade of planning *i.e.* 1950-51 to 1960-61, there was not much change in the cropping pattern in spite of significant increase of about 21 million hectares in cropped area. The proportion of area under rice and wheat together remained around 30 per cent of the total cropped area during this period while there was a marginal increase in proportion of area under sugarcane, groundnut, cotton and pulses.

Introduction of dwarf high yielding varieties for major cereals e.g. rice and wheat and hybrids and composites of millets in mid sixties ushered in the Green Revolution*. Adoption of high yielding varieties on a large scale, increased use of fertilizers, plant protection chemicals and expansion in irrigated area led to a shift in area towards crops in which the impact of technology on yield was apparent. The proportion of wheat was showing a continuous increase over time. It increased from 8.5 per cent in 1960-61 to 11 per cent in 1970-71 and further to 12.8 per cent in 1980-81. During the next decade ending 1990-91, though there was a further increase in area under the crop, as proportion of total cropped area it remained constant. However, the ninties saw a further jump both in area and its share in gross cropped area. On the other hand, the proportion of area under coarse cereals recorded a continuous declining trend. The percentage of area under jowar to total cropped area declined from 12.0 in 1960-61 to 6.8 in 1991-92 and 6.1 during ninties, while that of bajra and small millets decreased from 7.5 and 3.3 to 5.4 and 0.8 respectively. Even though there was some increase in the percentage of area to total cropped area under maize in 1970-71, it stagnated thereafter, and in 80's and 90's decades, there was a marginal decline.

---

* A term coined by Dr. William Gaud of United States in October, 1968.

Table 5.3: All India Cropping Pattern—1950-51 to 1996-97

Crop	1950-51	1960-*61	1970-71	1980-81	1991-92	1999-2000
(1)	(2)	(3)	(4)	(5)	(6)	(7)
Rice	31,056 (23.6)	34,056 (22.3)	37,381 (22.6)	40,281 (23.3)	42,657 (23.0)	45,403 (23.9)
Wheat	10,010 (7.6)	12,931 (8.5)	18,293 (11.0)	22,247 (12.8)	23,357 (12.8)	27,744 (14.6)
Jowar	15,554 (11.8)	18,426 (12.0)	16,871 (10.2)	16,381 (9.5)	12,498 (6.8)	10,256 (5.4)
Bajra	9,744 (7.4)	11,470 (7.5)	13,391 (8.0)	11,833 (6.8)	10,280 (5.6)	8,896 (4.7)
Maize	3,250 (2.4)	4,401 (2.9)	5,856 (3.5)	6,056 (3.5)	5,888 (3.2)	6,509 (3.5)
Ragi	2,254 (1.7)	2,478 (1.6)	2,474 (6.5)	2,512 (1.4)	2,117 (1.2)	1,743 (0.9)
Barley	3,198 (2.4)	3,140 (2.1)	2,556 (1.6)	1,805 (1.0)	966 (0.5)	750 (0.4)
Other Cereals & Millets	5,576 (4.2)	4,997 (3.3)	4,962 (3.0)	4,060 (2.4)	2,168 (1.4)	1,468 (0.8)
Total Cereals	80,642 (61.1)	91,899 (60.2)	1,01,784 (61.4)	1,05,170 (60.7)	99,931 (54.7)	1,02,769 (54.2)
Gram	7,803 (5.9)	9,274 (6.1)	7,820 (4.7)	6,566 (3.8)	5,593 (3.1)	6,329 (3.3)
Tur	2,228 (1.7)	2,429 (1.6)	2,639 (1.6)	2,940 (1.7)	3,652 (2.0)	3,436 (1.8)
Other Pulses	10,523 (8.0)	11,962 (7.8)	12,667 (7.7)	13,290 (7.7)	13,605 (7.4)	12,266 (6.5)
Total Pulses	20,554 (15.6)	23,665 (15.5)	13,125 (14.0)	22,795 (13.2)	22,850 (12.0)	22,031 (11.6)
Total Foodgrains	1,01,196 (76.7)	1,15,564 (75.7)	1,24,910 (75.4)	1,27,955 (73.9)	1,22,781 (67.0)	1,24,800 (65.8)
Sugarcane	1,757 (1.3)	2,417 (1.6)	2,589 (1.6)	2,830 (1.6)	3,970 (2.2)	4,489 (2.4)
Condiments & Spices	1,244 (0.9)	1,569 (1.0)	1,864 (1.1)	2,081 (1.2)	2,384 (1.3)	2,872 (1.5)
Total Fruits	753 (0.6)	1,096 (0.7)	1,450 (0.9)	1,909 (1.1)	2,029 (1.1)	3,360 (1.8)
Potatoes	177 (0.1)	368 (0.2)	482 (0.3)	727 (0.4)	1,021 (0.6)	1,334)0.7)
Onions	78 (0.1)	105 (0.1)	179 (0.1)	254 (0.1)	325 (0.2)	509 (0.3)
Total Vegetables	1,514 (1.2)	1,553 (1.0)	2,233 (1.3)	2,969 (1.7)	4,170 (2.3)	4,859 (2.6)

(Contd.)

Table 5.3 (*Contd.*)

(1)	(2)	(3)	(4)	(5)	(6)	(7)
Groundnut	4,406 (3.3)	6,467 (4.2)	7,552 (4.6)	6,924 (4.0)	8,680 (4.7)	6,957 (3.7)
Rapeseed & Mustard	1,053 (0.8)	1,145 (0.8)	1,415 (0.9)	2,263 (1.3)	633 (3.5)	6,069 (3.2)
Sesamum	1,957 (1.5)	1,575 (1.0)	1,853 (1.1)	1,795 (1.0)	259 (1.0)	1,814 (0.9)
Linseed	1,226 (0.9)	1,267 (0.8)	1,199 (0.7)	1,152 (0.7)	88 (0.1)	698 (0.4)
Other Oilseeds	2,326 (1.8)	2,323 (1.5)	2,700 (1.6)	3,778 (2.2)	884 (4.0)	11,271 (5.9)
Total Oilseeds	10,968 (8.3)	12,777 (8.3)	14,719 (8.9)	15,912 (9.2)	27,337 (14.9)	26,809 (14.1)
Cotton	5,655 (4.3)	7,610 (5.0)	7,830 (4.7)	7,868 (4.5)	773 (4.2)	8,952 (4.7)
Jute	561 (0.4)	633 (0.4)	753 (0.4)	940 (0.5)	88 (0.5)	861 (0.5)
Mesta	65 (0.1)	281 (0.2)	332 (0.2)	360 (0.2)	23 (0.1)	174)0.1)
Total Fibres	6,667 (5.1)	8,762 (5.7)	9,109 (5.5)	9,328 (5.4)	895 (4.8)	10,086 (5.3)
Tobacco	379 (0.3)	401 (0.3)	431 (0.2)	463 (0.3)	48 (0.3)	430 (0.2)
Other Crops	7,433 (5.6)	8,633 (5.7)	8,486 (5.1)	9,533 (5.6)	998 (5.5)	12,035 (6.3)
Total cropped Area	1,31,893 (100.5)	1,52,772 (100.0)	1,65,791 (100.0)	1,73,096 (100.0)	1,82,728 (100.0)	1,89,740 (100.0)
Area Sown more than once	13,147 (10.0)	19,573 (12.8)	25,007 (15.1)	32,797 (18.9)	41,320 (22.6)	48,510 (25.6)
Net Sown area	1,18,746 (90.0)	1,33,199 (87.2)	1,40,784 (84.9)	1,40,299 (81.0)	1,41,410 (77.4)	1,41,230 (74.4)

Area under rice as a proportion of total cropped area remained unchanged at around 23 per cent during the entire period of planned development (1950-51 to 1999-2000). That is to say, growth rate of area was similar to that of growth rate of total cropped area. In respect of pulses, the proportion of area continuously declined from 15.5 per cent in 1960-61 to 14.0 in 1970-71, 13.2 in 1980-81, 12.5 per cent in 1991-92 and further to 11.6 per cent later. Among the non-foodgrain crops, the proportion of area under rapeseed & mustard and potatoes recorded a marginal increase upto 1990-91, but a sharp increase, thereafter. The position in respect of other crops recorded very minor change.

Impact of Irrigation on the Change in Cropping Pattern

Besides the complexity of socio-economic, agro-climatic and physiological factors, change in the cropping pattern in a given area is primarily determined by the opportunity cost of the farmers, which is the rate of return from investment. Availability of high yielding variety seeds, irrigation water, chemical fertilizers, etc. has helped to a large extent to motivate the farmers in going for superior grain food crops. It is therefore not feasible to quantify the impact of irrigation alone on the change in cropping pattern that has taken place over the years. However, expansion in irrigation facilities has played an important role in causing a shift in cropping pattern. It has been observed that over 63 per cent of the additional irrigated area, of 48 million hectares developed between 1960-61 and 1999-2000 was accounted for by wheat and rice. Wheat alone accounted for 41 per cent of the additional area (*Table 5.4*) .

Next to rice, oilseeds crops, particularly rapeseeds and mustard and groundnut, have also benefited from the expansion of irrigation facilities. By 1999-2000, almost about 13 per cent of the additional irrigated area went under oilseeds crop.

Thus, as irrigation facilities are expanded to newer areas, farmer in those areas favour rice crop in kharif and wheat crop in rabi season, as also other assured and remunerative crops depending however upon the agro-climatic conditions and soil type of the regions. Other principal commercial crops which accounted for additional irrigated area of slightly more than two million hectares each were sugarcane, fruits & vegetables and cotton and the share of these crops in the total irrigated area had gone up over time.

Table 5.4: All India Cropping Pattern in Irrigated and Unirrigated Areas

Crops	*1960-61*				*1999-2000*				*Distribution of Additional Irrigated area as compared to 1960-61*	
	Irrigated		*Unirrigated*		Irrigated		*Unirrigated*			
	Area	*Percent*	*Area*	*Percent*	*Area*	*Percent*	*Area*	*Percent*	*Area*	*Percent*
1	2	3	4	5	6	7	8	9	10	11
Rice	12523	44.7	21533	17.3	23359	30.6	22044	19.4	10836	22.4
Jowar	655	2.3	17771	14.2	827	1.1	9429	8.3	172	0.4
Maize	556	2.0	3845	3.1	1465	1.9	5044	4.4	909	1.9
Wheat	4233	15.1	8698	7.0	23911	31.3	3833	3.4	19678	40.7
Barley	1334	4.8	1806	1.4	469	0.6	281	0.2	-865	-1.8
Total Cereals & Millets	20166	72.1	71733	57.5	51024	66.8	51745	45.6	30858	63.8
Gram	1107	4.0	8167	6.5	1774	2.3	4555	4.0	667	1.4
Total Pulses	1899	6.8	21766	17.4	2970	3.9	19061	16.8	1071	2.2
Total Foodgrains	22065	78.9	93499	74.9	53994	70.7	70806	62.4	31929	66.0
Sugarcane	1674	6.0	743	0.6	4132	5.4	357	0.3	2458	5.1
Condiments & Spices	417	1.5	1152	0.9	1540	2.0	1332	1.2	1123	2.3
Total Fruits & Vegetables	477	1.7	2172	1.8	3366	4.4	4854	4.3	2889	6.0
Groundnut	195	0.7	6272	5.0	1335	1.7	5622	5.0	1140	2.4
Rapeseed & Mustard	138	0.5	1007	0.8	3635	4.8	2432	2.1	3499	7.2
Total Oilseeds	421	1.5	12356	9.9	6631	8.7	20178	17.8	6210	12.9
Cotton	967	3.4	6643	5.3	3143	4.1	5809	5.1	2176	4.5
Tobacco	88	0.3	313	0.3	199	0.3	231	0.2	111	0.2
Other Crops	1871	6.7	7914	6.3	3331	4.4	9837	8.7	1460	3.0
Total Area	27980	100.0	124792	100.0	76336	100.0	113404	100.1	48356	100.0

Source: Directorate of Economics & Statistics (LUS)

In this context, it may be stated that in terms of cropped area paddy is the single most important crop of India, accounting for almost one-fourth of the national crop acreage. It is much-favoured crop of the Indian farmer if he has access to enough water, even in low and medium rainfall regions. Nearly 30 per cent of the country's total irrigated acreage is devoted to paddy cultivation. However, irrigated land allocated to paddy varies across regions and by type of irrigation. In the case of canals and tanks where there is no field-to-field irrigation, farmers have little scope for choosing a crop other than paddy which alone can thrive under a moisture regime of ponded fields. In the absence of field-to-field irrigation paddy cultivation is less pronounced.

Paddy is a water-loving plant. For each hectare of paddy yield of 4 to 6 tons in the semi-arid region of North-Western India, as much as 16,000 tons (16 million litres) of irrigation water gets applied at the farm gate level alone. In high rainfall States paddy grown in rabi/summer season too gives rise to 10-15,000 million litres of irrigation application, though requirements are in the range of 5 to 7,000 million litres for kharif paddy. As a result, paddy alone consumes over two-fifths of the irrigation water so far made available through both public and private investments. Hence, methods which boost paddy productivity per unit volume of water in water-short regions of southern, western, central and northern India attract attention of paddy growers as well as policy makers.

In this context, the change in the cropping pattern has been examined in four specific states viz., Haryana and Punjab in the northern region, Tamil Nadu in the southern region, and Maharashtra in the Western region from 1967-68 to 1999-2000. While in Haryana and Punjab states about 85 per cent and 91 per cent respectively of the total cropped area is irrigated, in Tamil Nadu it is about 55 per cent which is highest among the states of the region. On the other hand, in Maharashtra only about 16.9 per cent of the total cropped are is irrigated. (Table 5.2)

It has been observed that with the increased availability of irrigation facilities, the percentage of area under rice crop to total cropped area recorded an increase over the years both in Haryana and Punjab states (Table 5.5). On the contrary, in Tamil Nadu there was a small decline in the proportion of area under rice crop to total cropped area due to water stress and increased salinity of the soil. In Maharashtra also there was a marginal decrease in the proportion of area under rice crop during this period. There was a major shift in area

Table 5.5: State-Wise Cropping Pattern During Triennium Ending 1967-68, 1984-85 and 1999-2000

(per cent of cropped area)

Crops	*Haryana*			*Maharashtra*			*Punjab*			*Tamil Nadu*		
	1967-68	*1984-85*	*1999-2000*	*1967-68*	*1984-85*	*1999-2000*	*1967-68*	*1984-85*	*1999-2000*	*1967-68*	*1984-85*	*1999-2000*
1	2	3	4	5	6	7	8	9	10	11	12	13
Rice	4.41	9.57	16.68	7.03	7.31	6.76	5.75	21.28	30.60	35.98	32.90	34.00
Jowar	5.88	2.55	2.01	32.14	31.90	23.34	0.10	0.01	–	10.39	10.04	5.56
Bajra	18.51	14.39	9.66	9.37	8.02	7.80	3.54	0.69	0.07	6.05	4.86	2.44
Maize	2.11	1.07	0.36	0.20	0.42	1.20	8.42	4.32	2.00	0.07	0.32	1.02
Ragi	–	–	–	1.17	1.10	0.71	–	–	–	4.54	3.34	1.78
Wheat	16.15	31.60	35.50	4.53	5.16	4.24	32.07	44.36	41.44	–	–	–
Barley	4.21	1.36	0.62	0.02	0.04	0.01	2.05	0.92	0.46	–	–	–
Gram	22.53	10.77	4.68	1.86	2.33	3.85	11.60	0.47	0.16	0.03	1.97	0.12
Tur	0.11	0.29	0.37	2.97	3.50	4.61	0.04	0.47	0.12	0.72	1.97	1.06
Other Pulses	1.93	0.75	0.54	7.28	7.69	7.20	0.99	0.89	0.71	4.80	8.66	8.57
Sugarcane	3.69	2.46	2.19	0.92	1.47	2.59	3.10	1.28	1.56	1.49	2.53	4.84
Groundnut	0.30	0.15	0.02	5.61	3.80	2.44	3.49	0.86	0.07	12.80	14.68	12.61
Rapeseed & Mustard	4.34	4.09	8.12	0.02	0.02	0.04	2.21	1.44	0.87	–	0.01	0.02

from coarse grains to rice both in Haryana and Punjab as cultivation of rice crop was more profitable than that of coarse grains though rice was a water consuming crop. However, in Tamil Nadu, though there was a decline in the proportion of area under most of the coarse cereals, as has been stated earlier, there was also decline in the proportion of area under rice crop.

In the changing scenario of cropping pattern, groundnut, an important kharif oilseed crop, received a raw deal both in Haryana and Punjab states and the percentage area under this crop to total cropped area sharply declined over the years. In Maharashtra, it registered initially a sharp and then a full recovery. But in Tamil Nadu, the proportion of area under this crop after a sharp increase reverted to in level of sixties, following shift in area from coarse cereals. Among the kharif crops, cotton crop was also benefited from the increased availability of irrigation water in Haryana state.

In so far as rabi crops are concerned, both in Haryana and Punjab, there was marked increase in the proportion of area under wheat crop which was caused by major shift in area from gram and other pulses crops. In Maharashtra state, while initially the proportion of area under wheat crop recorded an increase in the last decade or so and then there was a decline.

Among the rabi commercial crops, there was sharp expansion of area under rapeseed & mustard crop in Haryana state where the percentage of area under the crop increased from 4.34 in 1967-68 to 8.12 in 1999-2000 though in the neighbouring Punjab state there was decline over the period from 2.21 per cent to 0.87 per cent. The proportion of area under oilseeds as a whole recorded moderate to significant increase in all the states except in Punjab where after recording sharp fall from 6.02 per cent in 1967-68 to 2.5 per cent in 1984-85, increased marginally to 2.75 per cent in during ninety.

Thus, the impact of irrigation on the cropping pattern in Haryana and Punjab was reflected mainly in the coverage of area under rice and wheat crops. Besides, rapeseed and mustard has also been benefited in Haryana state. In Tamil Nadu, area under groundnut also recorded an increase. These increases resulted in improvement in the coverage under total oilseeds in all the states save Punjab. Coarse grains in general had to give way to rice and wheat. In the changed scenario, except in Punjab, the proportion of area under foodgrains as a whole declined in all other states.

A Few Major problems

Though it is admitted that irrigation has always been a central pillar of the national agricultural development strategy, sharing credit with improved seeds and fertilizer for the success of the green revolution in the country, many problems have already surfaced for the long-term sustainability of Indian agriculture and there is an undesirable distoration in the cropping pattern which was being traditionally followed. Intimate knowledge of local soil, climatic and agricultural condition held by farmers is squandered and heirarchial top-down state irrigation departments typically pay little heed to farmers requirements of water for growing a particular crop in a given season and situation.

It is therefore necessary to adopt appropriate cropping pattern in the irrigated areas so that agriculture in these areas become sustainable in the longer run without, however, compromising with the required growth rate in agriculture. Progress in rainfed agriculture associated with new crops, supplemental irrigation, deficit irrigation, rainwater harvesting, and some forms of precision irrigation, could lead to fast yield increases in rainfed lands. This would generate an increase in food production without a concomitant increase in water irrigation. However, the opening of major rainfed areas for more intensive food production may lead to increased deforestation.

Intensive Versus Extensive Irrigation

If water is abundantly available, as in eastern India, its intensive use per unit of arable land area is worthwhile both from private and social angles. But whenever it is scarcer than land resource, as in much of north western, southern and central India, its use per unit of area ought to be on extensive basis. Such an extensive irrigation policy can be justified on grounds of productivity, equity, stability and sustainability. For water-short regions, the issue of intensive versus extensive irrigation has arisen in the shape of heavily-irrigated versus lightly-irrigated crops. Most notably, scholars associated with the Gokhale Institute of Politics and Economics (V.M. Dandekar, Nilakantha Rath and Asok Mitra) forcefully questioned the wisdom of promoting sugarcane cultivation in Maharashtra State through public canals.

As the income elasticity of demand for sugar is presently high in India, the demand for sugar is growing rather rapidly. If Maharashtra, which is now the number one state in the production of mill-made sugar in the country, is to curb sugarcane raising on public canals, one may rightly ask: how do we go about meeting the growing sugar needs of the economy? The author has been pleading for a shift in central sugar location policy away from the Deccan to northern India where water endowment vis-à-vis land is much better, and where sugarcane is indeed a superior crop to wheat-paddy sequence (per unit of irrigation water as well as per unit of crop area). The Gokhale Institute Scholars, however, feel that once extensive canal networks are created for promoting the cultivation of lightly irrigated crops, the seeped in canal waters can sustain sugarcane cultivation when withdrawn through private investments in dugwell irrigation. This, then brings us to the question of development of conjunctive use of waters within the commands of surface irrigation works.

Waterlogging and Drainage

Poor drainage in canal tracts has been the bane of major irrigation works in the Indian sub-continent. Cumulative seepage of canal water tends to raise groundwater to the crop root zone before long. The process is hastened in the absence of investment in canal lining. In flat topography, canals themselves impede surface water drainage after a downpour. Other normal development activities, like building of roads and railways, and flood protection embankments aggravate surface water drainage. In the absence of requisite investments forthcoming for drainage purpose (including cross drainage), waterlogging becomes inevitable. Following two tendencies hasten this process.

One is the emergence of a sharp deviation in irrigated crop pattern in project commands from the one contemplated by the project planners in their project design. A hallmark of this deviation is the farmer's marked preference for heavily-irrigated crops like paddy and sugarcane over lightly-irrigated crops of pulses, oilseeds, coarse cereals, etc., which are, acre for acre, much less remunerative. The deviation has been reported by many experts, especially in the upper reaches of canals of the Deccan. As irrigation application per unit of land area rises above the level visualized by the project planners, field percolation losses shoot up and water table rises faster.

The other tendency on the part of the farmer is to consciously over-water the crop during the course of an irrigation operation. This is essentially his behaviouristic response to the uncertainties of canal supplies, which are partly due to natural diminution in river flows on account of drought and partly due to mismanagement (including unauthorized use of canal water) of the canal distribution system. This is indicative of his desire to store extra water in the crop root zone till his next turn for watering comes. Evidently, it is more prevalent among farmers served by protective canals which are inherently designed to provide far less than the requisite number of waterings at which full yield potential of a crop is realized. Needless to say, when the two tendencies reinforce each other, the field percolation losses shoot up well above the envisaged level.

Outside the engineering circles, it is not duly realized that establishing a network of man-made drains is a costly proposition. An ideal drainage system (that includes field drains) is probably no less costly to construct and maintain than the canal system itself. This raises the issue of viability of investment in drainage network. Evidently, the viability prospects are better under modern, high-yielding agriculture. However, one may raise the question of cost-effectiveness of horizontal versus vertical drainage. Vertical drainage with tubewells is now considered as a more cost-effective option. In fact, vertical drainage as part of conjunctive use of groundwater surface water and surface water is now the watchword of the planners.

Reform in Irrigation Sector in Tamil Nadu

In Tamil Nadu, of the total quantity of water used for irrigation, 70-80 per cent is taken away by rice crop leaving only about 25 per cent for other irrigated dry crops such as sugarcane, banana and cotton. Water use efficiency (WUE) is also as low as 30 to 35 per cent in rice crop whereas average WUE of other crops ranges from 45 to 50 per cent. Hence, it is necessary that attention should be focused on efficient water management practices for rice without compromising with yield levels. There is no fair distribution of irrigation water for different crops. But there is an enormous scope of reducing the quantity of water allocated for rice, while maintaining the same production level which in turn increases the use efficiency of water. As the state has almost reached the ceiling limit in water exploitation, WUE assumes greater importance in the state.

The State Government has to initiate the "reform process" in the irrigation sector. This was one of the key messages delivered by a

World Bank team, on a recent visit (May, 2003) to Tamil Nadu. In other words, in place of the 'monolithic' Public Works department, which also takes care of buildings, the multilateral agency has mooted formation of an exclusive and autonomous organization to deal with water resources with its own dedicated cadre. Another World Bank idea relates to establishment of a water tariff panel, on the lines of the Tamil Nadu Electricity Regulatory Commission.

The reason behind the suggestion for a separate agency for water resources seems to be the "virtual non-functioning" of the Water Resources Control and Review Council, headed by the Chief Minister. Set up in mid-1990s, as part of execution of the Water Resources Consolidation Project (WRCP), the council has met only on a few occasions. The proposed body, possibly headed by a retired IAS official, can address irrigation issues and take decisions. The Government could confine itself to laying down the broad policy framework for the working of the new organization. On the water tariff commission, consultations may be required with the Municipal Administration and Water Supply department, the Tamil Nadu Water Supply and Drainage Board and Metrowater.

In fact, the idea of a separate body for water resources is not new. The World Bank made it a conditionality at the time of giving funds for the WRCP. The Government too created the Water Resources Organization (WRO) within the PWD but, in practice, it never took off, as transfer of officials from the Buildings wing to the WRO or vice versa continued. The rationale behind the establishment of the WRO was not only to avoid transfers but also to have personnel specialized in irrigation. However, the Government agreed to act in one area immediately. Its officials would prepare a model, incorporating a holistic approach to utilization of water resources. The PWD has already identified the Hanumannadhi project in Tirunelveli district for the experiment and if successful, it will be replicated in other areas. (Hanumannadhi is a tributary of the Chittar and forms part of the Tamirabarani basin. The Adaivinainarkoil reservoir has been constructed across the river, having an ayacut of 6,500 acres).

Beneficiary contribution, instilling an ownership concept among ryots and women's participation constitute integral parts of the model project. Besides, social benefits flowing out of irrigation projects and having implications for other fields of activity such as industry and agriculture are taken into account and hence the respective departments will have to be involved.

Crisis in Punjab Agriculture

Punjab is one of the smallest states of India covering only 1.53 per cent of the total geographical area of the country. The State, popularly known as the "food basket" of India is contributing nearly 21 per cent of wheat, 10 per cent rice and 8 per cent cotton of the country's total output, apart from sizable share in the production of other crops. By exploiting high potential agriculture of the State, the country which was in the grip of serious food shortage till sixties has now been able to generate even surpluses apart from meeting the needs of its rapidly growing population. The technology encompassing intensive cultivation, high use of agro-chemicals, greater requirement of water and rapid growth of farm mechanization in Punjab, has resulted in the manifestation of several adverse effects on environment and ecological balance such as fall in water table, developing pest resistance, degrading soil fertility, eroding bio-diversity, etc. The underground water which is a precious resource in the state, is still being used indiscriminately.

The state, until recently the most prosperous in the country, is thus in a deep economic crisis. After more than three decades of the adoption of new agricultural strategy with its obsession with grains to the exclusion of other food and cash crops, Punjab's agrarian economy with a highly skewed agricultural base is in deep crisis. The production pattern in the state has become predominantly a monoculture of rice-wheat rotation because of higher profitability of these crops resulting from faster increase in productivity and effective price support in comparison to other competing crops. The economic considerations in the choice of crops have thus overruled the natural resource endowments particularly water. In fact, merely due to shift in crop pattern, the water requirement, according to the agricultural scientists of the state, has gone up by about 76 per cent during the past four decades.

It has been stated at the very outset that the food scenario at the national level has completely changed from the food shortages of mid-1960s to the burgeoning surpluses of late 1990s and early years of the 21st century. In this changed context Punjab's wheat-paddy dominated agrarian economy is becoming economically as well as ecologically non-sustainable and the state is being treated as if it has become a liability for the nation. Falling water table in the state is not only threatening to seriously disturb the ecological balance leading to make much of Punjab land barren, but will also effectively exclude small and marginal farmers from making use of their common resource leading to tension and social strife.

Steps Needed to Meet Challenge

There is an urgent need to diversify the state's predominantly agrarian economy. Diversification in the narrow sense of the term within the crop husbandry sector would mean going away from wheat-rice combination to production of other crops such as sugarcane, cotton, gram, pulses, oilseeds, coarse grains, soyabean, etc. However, in a broader sense, diversification would also mean moving away from crop husbandry to other allied activities such as dairying, poultry, fishery, piggery, bee-keeping, horticulture, sericulture or flouriculture, etc. In a still broader sense, it means moving away from agriculture to the other rural non-farm activities and agro-processing. However, we may confine our attention to diversification in the crop husbandry sector as the prevalent production pattern does not seem to be anymore sustainable.

Punjab Agriculture University, Ludhiana has recently suggested that at least 20 per cent area under paddy must be diverted to other crops if the deterioration in the underground water balance of the state is to be checked and the state's land is to be prevented from becoming barren. But a mere advice to the farmers to go for crops other than wheat and/or rice may not work because given the input-output price structure and yield levels still wheat-rice is the most profitable and least risky combination for the farmers in the state.

For diversification to became successful, Punjab should be divided into a number of agro-climatic zones. Each zone should consist of a cluster of villages which are homogeneous in terms of soil type, weather, quantity, quality and type of irrigation and suitability for a particular type of cropping pattern. Each zone should be designated as most suitable for a particular cropping pattern. For instance, we can have wheat-basmati rice zone; wheat-parmal rice zone; wheat, gram sarson, cotton zone; wheat, maize, oilseed, pulses zone; potato, tomato, vegetables zone and so on. Farmers in each zone should be allowed to go in for only those specific crops and specific varieties of those crops for which a zone has been designated as 'most suitable' by experts on the basis of soil, irrigation, quantum of underground water available, etc. There should be synchronization of micro level (*i.e.* zone level) and macro level (state level) plan and the state may decide how much area should be allowed to go under a particular crop. Small and marginal farmers may be persuaded to go in for labour intensive crops such as ginger, turmeric, menthe, tomato, vegetables, chillies, etc. which give much better returns with less consumption of water.

While the central districts of Punjab have contributed substantially to the central pool of foodgrains, the south-western zone comprising Bhathinda, Ferozepur, Faridkot, Moga and Mansa districts produce cotton mostly. The central part is fast emptying its sweet water reserves due to the water-intensive paddy-wheat rotation. The south-western zone is facing the problem of rising watertable which is brackish/alkaline/saline, neither fit for drinking nor for irrigation. There is an apprehension that the unwanted salinity affected/brackish water with its rising trend may penetrate into the low sweet water central zone. This is likely to plunge Punjab into a still deeper crisis in the years to come. To check this, farmers in the central part should stop growing paddy and switch over to crops like maize, oilseeds, pulses, fruits, vegetables, groundnut and fodder. Besides, soil and water conservation measures need to be taken to enhance the recharging of the sub-surface watertable along with the propagation of furrow, sprinkler & drip irrigation.

An altogether different approach will have to be adopted for arresting the rising trend of brackish/salinity affected watertable in the south-western zone by taking recourse to paddy, sugarcane and wheat and trees like eucalyptus and poplars. The cultivation of horticulture crops, which are not only eco-friendly but also remunerative, need to be encouraged. Similarly, fisheries can also work wonders in the area as there are species of fish which can better flourish in the saline/brackish waters. Besides, other occupations like dairy, poultry, beekeeping, flouriculture etc can also pay rich dividends to farmers of this area, provided their marketing is ensured.

Summing up, it may be reiterated that Punjab's agrarian economy is at the crossroads as agricultural production and crop yields have nearly stagnated. Land and water, the two most critical resources on which Punjab's rural economy is built, are under severe stress because of excessive use of chemical fertilizers and growing the same crops over and over again. The future of Punjab agriculture will depend upon their conservation, sustainable use and efficient management.

Conclusion

Globally, of the three standard categories of freshwater use-for agriculture, industry and domestic (personal, household, and municipal)—agriculture dominates. While in the developed countries the use of fresh water for agriculture is less than 70 per cent of its availability, in India, it is nearly 90 per cent. On the other hand, the uses of water for industry and domestic purposes are only 7 and 3 per cent respectively. In fact, a country's level of freshwater use reflects and, no doubt, is one of the key measures, of its level of economic development.

The population in the country is growing rapidly, especially in urban areas and may reach the level of 1.3 billion by 2025 and 1.6 billion by 2050. This would further stimulate demand for food, and hence for agriculture. At the same time, the demand for water by households and industries further stimulated by economic growth is rising. These factors imply escalating competition for irrigated agriculture which, as mentioned, takes a hefty share of available water. Surely, agriculture will continue to be the main claimant of water far outstripping other uses. More and better irrigation is imperative to meet food and fibre needs of a growing and more prosperous population.

A substantial share of the area now being irrigated is by the unsustainable use of underground water and in many areas, irrigation is based on over-pumping of this water, *i.e.* pumping at a rate that exceeds the rate of aquifer recharge causing desertification of fertile agricultural land. On the other hand, in the canal irrigated areas, due to poor water management, many major problems have surfaced. Farmers located at the head-end of the canal system tend to appropriate more of irrigation supplies, to the disadvantage of their brother farmers at the end of the system. Besides reduction in maximum total output from canal irrigation, this unequal sharing of water leads to waterlogging and soil salinity problems, which are more visible in head-end reaches of a canal system.

Currently, a substantial proportion of water harnessed is lost (through evaporation, seepage and leakage) in conveying it to the points of end use, and there is much wasteful use of water at the users' end. However, salinity and water logging problems in the commands of major irrigation schemes can be minimized by recognizing them explicitly in the design stage and incorporating corrective measures into the scheme. Conjunctive use of surface and groundwater can go a long way in addressing this problem, and appropriate steps should be taken for its promotion. Appropriate pricing of water and electricity will also help in this direction. Further, proper drainage facilities, involving farmers' groups, need to be created where this problem has become acute.

Again, the change in the cropping pattern with the availability of irrigation water has brought an imbalance in the total agricultural output in the country, besides creating uncertainty in the long term growth of agriculture in the irrigated areas. It is imperative to adopt appropriate crop sequences in the irrigated areas so that agriculture in these areas becomes sustainable in the longer run without, however, compromising with the required growth rate in agriculture. Supplemental irrigation, deficit irrigation, rainwater harvesting, and some forms of precision irrigation need greater attention, particularly in the rainfed areas.

Lastly, but not the least, the efficiency of irrigation water use is an important, indeed more so, than the quantum of water delivered or area irrigated. The potential gains to effort and resources spent on improving would be both substantial and more assured than blind pursuit of irrigation water supply augmentation. There is much scope for improving the efficiency of water use in the existing systems which, if properly exploited, can ensure wider, more equitable and productive use of water.

6

Irrigation and Land Degradation

"So gladly do we go with you to the home for which you make us live. Waters! give us progeny"

—Rig Veda

Introduction

Land, which is the most precious heritage and the physical base of bio mass production of life supporting systems is finite. In this natural non-renewable endowment, the share of India is fixed at about 329 m.ha. It is not only inelastic but also heterogeneous in different parts and regions of the country with a definite set up, capabilities, suitabilities for different land resources. Conservation of land resources can promote sound land use to match with the land capabilities or suitabilities and to initiate correct land resources, development/ suitability in the country.

A close look at the present health of the soil and water resources reveals their wanton mis-use and degraded environment. About 173 m.ha. covering more than half of the area of the country is threatened by various types of degradation like salinity, alkalinity, waterlogging, ravenous and gullied lands, areas under ravages of shifting cultivation, desertification, etc. About 800 ha. of arable land are being lost annually due to ingress of ravines. There are specific problems of land degradation due to open-cast mining operations, using good productive land for brick kilns, coastal erosion and seawater ingress, excessive erosion and land slides in the crumbling hill areas. Our forests and grass lands have been over exploited. Frequent occurrence of floods and droughts in different parts of the country are evidence of improper land use in the catchments and inadequate conservation and use of rain water. The problem of land degradation has brought us face to face with the ever increasing depletion of the productivity and the basic land stock through nutrient deficiencies on the one hand and the ever growing demand for food, fodder, fibre, fuel, land based industrial raw materials and many non-farm land uses on the other.

IRRIGATION AS A CAUSE OF LAND DEGRADATION

Surface irrigation projects have direct and indirect effects (indirect being defined specifically as those operating through the concomitant exploitation of ground water) on cropping patterns, yields and cropping intensities. Cropping pattern changes affect the use of water relatively to land and yields, and cropping intensity in the absence of cropping pattern changes, constitutes a kind of scale effect which is land augmenting. There exists, a number of studies that examine this land augmenting nature of irrigation. Of late, a related aspect, that of land degradation has come to be recognized as significant. It is claimed that this degradation caused by over-irrigation is further accelerated by intensive use of fertilizers. In fact, land degradation is a major threat to our food and environmental security. Around 150 million hectares are degraded (loss in productivity) constituting nearly 45.6 per cent of total geographical area. The widespread erosion in the hilly catchments is resulting in excessive siltation of multipurpose reservoirs and other water bodies of the country at rates higher than the designed capacity (Table 6.1)

Table 6.1: Annual Siltation in some Reservoirs of India

Reservoir	*Catchment Area* (km^2)	*Silt deposition (tonne/ha.)m*	
		Assumed	*Observed*
Bhakra Nangal	56.90	6.40	9.20
Hirakud	82.70	3.80	5.40
Maithon	5.20	2.41	9.60
Panshet	9.80	3.70	15.00
Machkund	2.00	5.80	3.80
Tungabhadra	25.80	6.40	9.80
Mayurakshi	1.80	5.40	26.60
Gandhinagar	21.90	5.40	5.60
Ram Ganga	3.00	6.42	7.30
Ghod	3.60	5.40	22.90
Dantiwada	2.90	5.40	4.80

Source: V.V. Dhruvanarayana, and Ram Babu, 1983. Journal of Irrigation and Drainage Engineering, ASCE, 109,419

Author's Source: Indian Farming, November, 2002.

Estimates in the different parts of the world indicate that irrigated crop land has degraded significantly over the past decade through waterlogging and salinization. Waterlogging develops when the soil becomes saturated because of a high or perched water table. It often occurs in more humid regions and is generally caused by over irrigation or inadequate drainage. It leads to poor plant growing conditions. Salinization-or the accumulation of salts in the soil through water evaporation from the upper soil layers—can occur naturally but is generally a problem under irrigated conditions. This problem is more acute in arid areas because irrigation water evaporates more quickly and there is less natural leaching and drainage. Salinization generally leads to decreased production levels, although in some cases the problem may be severe enough to obstruct agriculture altogether.

Agriculture can contribute to water quality reduction in several ways, but it can also feel the effects of this pollution as reductions in water quality often lead to decreases in agricultural production. Water-induced soil erosion from cropland can be a major problem for water quality, leading to suspended solids in the water supply and siltation in the channel. Additional ground and surface water pollution from agriculture can be caused by fertilizer, pesticide, and animal manure inputs. Excess nutrients in the water supply can lead to entrophication, which affects the aquatic ecosystem by depleting oxygen through algae blooms. In addition, excess nitrates contributed by fertilizers and animal manure can have human health effects. These water quality problems can be intensified by increased water scarcity because there is less water to dilate the pollutions. However, in this chapter, we have discussed mainly about the problems of land degradation associated with irrigation.

We have mentioned that degradation of land that irrigation leads to takes the form of waterlogging and salinity. Large tracts of land go out of cultivation due to waterlogging. Alternatively, the productivity of cultivated areas decreases, with the consequences that an ecological imbalance is initiated that may result in a considerable dent being made in the benefits resulting from land augmentation. Salinity is another related form of land degradation. Its geo-chemical evaluation is a complex phenomenon. Soils become saline as a result of concentration of salts due to evaporation of ground water moving upwards. At times, this is the outcome of a rise of groundwater from a shallow water-table. In such cases, salinity appears together with waterlogging. Alternatively, it could also appear independently of any occurrences of waterlogging in part, accentuated by higher levels of fertilizer use. It is not, therefore, easy to determine when and whether salinity results from waterlogging. The crux of the problem appears to be a distortion of the natural equilibrium between groundwater discharge and recharge stemming from percolation losses in the irrigated areas, a phenomenon that hydrologists study with the help of water-balance equations.

These water-balance equation imply, in essence, that the rise or fall of the water-table depends on (a) specific yield, a property of the aquifer and the textural and structural characteristics of the soil, (b) precipitation or rainfall, (c) volume of irrigation water delivered to fields, (d) volume of water discharged at the land surface, (e) evaporation from land surface and in the zone of aeration respectively, and (f) increments in water storage at land surface and in the zone of aeration respectively. Each of theses components to be monitored in order to plan for an increase in the irrigation water to be provided to any region, and governments (who are almost always the managers of large surface water irrigation systems) do not try to enquire into the impact this increase may have on the hydrology of the region as illustrated by the water-balance equation. Consequently, if the soil and rainfall conditions also compound the impact of the increase in canal water flow, introduction of canal irrigation leads to widespread degradation of land.

Magnitude

No systematic survey was conducted in the past to map out the salt-affected and waterlogged areas in the country. Available estimates are largely based on the value judgment of various and concerned government and waterlogged areas varied from one source to another.

Several committees were constituted to provide precise estimates of area under various categories of land degradation. Since the constitution of the historic Indian Reh Committee in 1877, several committees, commissions, and working groups were formed to investigate causes and provide the extent of waterlogging and soil salinity in different parts of the country. Important committees/ commissions and working groups specifically constituted for salinity and waterlogging are listed in Table 6.2 below.

Table 6.2: Important Committees, Commissions, and Working Groups for Addressing Salinity and Waterlogging at the National Level

Year		*Committee/ Commission/ Working Group*
1877	–	Indian Reh Committee
1925	–	Waterlogging Enquiry Committee Irrigation Research Laboratory established
1927	–	Waterlogging Enquiry Committee expanded Chakamawali Reclamation Farm First salinity survey in waterlogged areas started
1928	–	Waterlogging committee abolished and Waterlogging Board was constituted
1937	–	Investigation on depth of water table started
1943	–	Salinity survey to entire irrigated area of Indus basin extended
1945	–	Land Reclamation Directorate was established
1969	–	Central Soil Salinity Research Institute was established.
1972	–	National Commission on Irrigation
1976	–	National Commission on Agriculture
1990	–	Working Group on salinity and waterlogging

Source: Joshi P.K. and Tyagi N.K. Salt-affected and waterlogged soils in India: A review, published in Strategic Change in Indian Irrigation, ed. Mark Svendsen and Ashok Gulati.

SOIL SALINITY

After independence, concerns were expressed with regard to the problem of soil salinity and water logging in India during the late 1960s. The Central Soil Salinity Research Institute, under the Indian Council of Agricultural Research, was established in 1969 to undertake research on salt-affected and waterlogged soils. There was no reliable statistics on the extent and the rate of spread. The first estimates of the National Commission on Irrigation (1972) were 4.84 million hectares, followed by the figure of the National Commission on Agriculture (1976) at 6.00 million hectares, and finally that of the Ministry of Agriculture (1984/85), at 8.53 million hectares. It is interesting to note how 6.00 million hectares was finally reported by the National Commission on Agriculture in 1976. Table 6.3 & 6.4 summarizes the information used by the National Commission on Agriculture (NCA) and their final estimates. Latest information available on Soil Degradation is as in Table 6.5. Looking at the statistics, it is observed that the NCA finally adopted the criteria of choosing the source that reported the highest area under waterlogging. No specific reason was given by NCA for adopting the highest estimates amongst different sources for none of the sources had undertaken any surveys to obtain the estimates of waterlogged area in different states. There is no mention about the type of methodology used by different authors. Interestingly, with the dramatic changes in all fields, including agriculture, irrigation, and infrastructure, the estimates of 1972 are still being reported in one form or the other.

Table 6.3: State-Wise Area Under Waterlogging

(1,000 hectares)

State	*Irrigation* Commission (1972)	*NCA (1976)*	*Ministry of Agriculture (1984-85)*	*Reconciled Estimates (1991)*
Andhra Pradesh	N.R.	339.00	339.00	266.40
Assam	N.R.	N.R.	450.00	N.R.
Bihar	N.R.	117.00	707.00	619.70
Gujarat	N.R.	484.00	484.00	172.59
Haryana	650.00	620.00	620.00	249.00
Karnataka	7.00	10.00	10.00	24.54
Karalla	N.R.	61.00	61.00	11.61
Madhya Pradesh	57.00	57.00	57.00	73.12
Maharashtra	28.00	111.00	111.00	15.35
Orissa	N.R.	60.00	60.00	196.26
Punjab	1,090.00	1,090.00	1,090.00	200.00
Rajasthan	348.00	348.00	348.00	179.50
Thamil Nadu	N.R.	N.R.	18.00	16.19
Uttar Pradesh	810.00	810.00	1,980.00	430.00
West Bengal	1,850.00	1,850.00	2,180.00	N.R.
Delhi	N.R.	1.00	1.00	N.R.
Total	4,840.00	5,986.00	8,526.00	2,455.96

Source: NCA= National Commission on Agriculture, N.R = Not Reported

Table 6.4: State-Wise Area Under Salt Affected Soils

(1,000 hectares)

State	*NCA*	*CWC*	*Working Group Group*	*Reconciled Estimates*
Andhra Pradesh	240.00	27.80	N.A	27.80
Assam	N.A.	N.A	N.A	N.A
Bihar	4.00	224.30	N.A	224.30
Gujarat	1,214.00	N.A	911.00	911.00
Haryana	526.00	N.A	N.A	197.00
Karnataka	404.00	51.30	51.40	51.35
Karalla	16.00	N.A	N.A	N.A
Madhya Pradesh	242.00	N.A	35.79	35.79
Maharashtra	534.00	N.A	5.35	5.35
Orissa	404.00	N.A	N.A	N.A
Punjab	688.00	N.A	490.00	490.00
Rajasthan	728.00	70.00	N.A	70.00
Thamil Nadu	4.00	47.60	140.30	140.00
Uttar Pradesh	1,295.00	N.A	1,150.80	1,150.00
West Bengal	855.00	N.A	N.A	N.A
Delhi	16.00	N.A.	N.A	N.A
Total	7,165.00	421.10	2,784.59	3,303.89

Source: NCA= National Commission on Agriculture, CWC—Central Water Commission; N.A = Not Available.

Similarly there were diverse estimates on the extent of salt-affected soils. The NCA (1976) reported an area of 7 million hectares as affected by salinity and alkalinity in the country. Later, the ministry of Agriculture put a figure of 9.08 million hectares under salt affected soils.

Table 6.5: Extent of Soil Degradation (Human Induced) under Different Degradation Types

Degradation type	*Degree of degradation*					*Area Affected*
	Slight	*Moderate*	*Strong*	*Extreme*	*Total*	*%*
Water erosion	27.3	111.6	5.4	4.6	148.9	45.3
(a) Loss of topsoil (wt)	27.3	99.8	5.4	–	132.5	40.3
(b) Terrain (Wd) deterioration	–	11.8	–	4.6	16.4	5.0
Water erosion	0.3	10.1	3.1	–	13.5	4.1
(a) Loss of topsoil (wt)	0.3	5.5	0.4	–	6.2	1.9
(b) Loss of topsoil/terrain deterioration (Et/Ed)	–	4.6	–	–	4.6	1.4
(c) Terrain deformation/ over blowing (Ed/Co)	–	–	2.7	–	2.7	0.8
Chemical deterioration	6.5	7.3	–	–	13.8	4.2
(a) Loss of Nutrient (Cn)	3.7	–	–	–	3.7	1.1
(b) Salinization (Cs)	2.8	7.3	–	–	10.1	3.1
Physical deterioration	–	–	–	–	116.6	3.5
Waterlogging (w)	6.4	5.2	–	–	11.6	3.5
Total (affected area)	36.8	137.9	8.5	4.6	187.8	57.1
Land not fit for agriculture Stable terrain					18.2	5.5
Under natural condition (Sn)					31.2	9.8
Total geographical area of India					328.7	100.0

Source: Data Book on Mechanization and Agro-Processing since Independence by Dr. G. Singh, Director, CIAE, Bhopal—462038, Dec., 1997.

Factors Causing Salinity and Waterlogging

A large number of factors are reported to be responsible for developing salinity and waterlogging conditions. Important among these are higher seepage losses; poor operation and maintenance of the canal network; introduction of canal irrigation in the marginal lands; mismanagement of canal irrigation, a high subsidy on canal irrigation; a higher acreage allocation in favour of high water requirements and high value crops, such as rice and sugarcane; and poor groundwater quality.

IRRIGATION SUBSIDY

The massive irrigation network in India has been developed through a huge investment and later supported by an enormous amount of subsidy, which has a social cost. The subsidy on irrigation is extended to both surface and groundwater. The water rates on canal irrigation are fixed at very low levels. These rates are arbitrarily fixed and do not cover even operational and maintenance cost. It was estimated that the all India average annual irrigation subsidy on major and medium projects during the years ending 1986-87 was as high as Rs. 65 billion, if capital cost was amortized at its replacement value.

The mounting subsidy on irrigation influenced the decision of farmers about crop choice and resource allocation. Such incentives, through subsidy, shifted the acreage under crops with high and moderate water requirements, such as rice, sugarcane, and cotton. These high water-requirement crops are relatively more profitable than low—water-requirement crops. In the rush to maximize profits, the private entrepreneurs, however, overlooked the consequences of their choices. Although the entrepreneurs were rational in maximizing short-run profits by equating marginal cost in terms of soil and water degradation in the long run. The historical trends have clearly shown that the crop substitution in favour of these crops resulted in overuse of canal water, causing a rise in the water table associated with the development of excess salts in the soil. With the development of these problems, the production function started shifting downwards and the same level of variable and fixed resources generated less and less returns over the years, and finally dropping to zero in many canal-irrigated areas.

To cite a specific example, sugarcane cultivation started in a big way in Maharashtra following introduction of surface irrigation. In Maharashtra, particularly in the western dry plateau region, the dominant crop under canal irrigation—not in terms of area, but in terms of share of total irrigation water used—was sugarcane. In the state as a whole, sugarcane claimed about 40 percent of the total irrigation water requirement. Although this crop covered only 11 per cent of the gross irrigated area, its high irrigation water requirement (181 centimeters per hectare) and high profit (Rs. 11,094 per hectare) conspicuously increased the total demand for water. Since sugarcane was the most profitable crop, farmers diverted resources in its favour and unknowingly threatened the sustainability of agriculture in this region. As the crop was gaining acreage in the state, reports were also pouring in about the rise in water table and soil salinity. The same thing happened in some canal commands in Gujarat where farmers shifted to sugarcane regardless of the consequences of this shift. As a result, a considerable area in the canal commands of Maharashtra and Gujarat has been waterlogged.

RISE IN WATER TABLE

The observation of water-table data over a period showed that in almost all the irrigation commands the water table has a rising trend after the introduction of canal irrigation (Table 6.6).

Table 6.6: Trends of Water-Table Rise in Selected Irrigation Commands of India

Irrigation Command	*State*	*Rise of water-table (m/year)*
Mahi Right Bank Canal Command (MRBC)	Gujarat	0.28
Rajasthan Canal Command (IGNP)	Rajasthan	0.29-0.88
Western Yamuna and Bhakra Canal Command	Haryana	0.30-1.00
Sirhind Canal Command	Punjab	0.10-1.00
Sharda Sahayak Canal Command	Uttar Pradesh	0.68
Malprabha Canal Command	Karnataka	0.60-1.20
Nagarjuna Sagar Irrigation Project	Andhra Pradesh	0.32
Sriram Sagar Irrigation Project	Andhra Pradesh	0.26

Source: O.P. Singh, 1998. Agricultural Salinity Management in India. Central Soil Salinity Research Institute, Karnal, pp. 61-77.

A heavy recharge in areas where quality of groundwater is poor and does not permit its recycling through irrigation causes a rapid rise in the water table and the associated soil salinity. Deep-percolation losses of applied irrigation water in crop need ranges from 30-40 percent. Obviously, these crops are a source of excessive recharge to groundwater. Since canal irrigation was largely introduced in arid and semi-arid regions, where quality of groundwater was poor, the examples of adverse consequences were also largely from such areas. A rise in the water table was also observed in parts of canal commands where good-quality groundwater was plentiful but farmers did not feel the necessity of using it because of availability of cheap canal water.

Strangely, irrigation planners and executives ignored the past experiences of canal irrigation in the country with references to the rise in the water table and secondary salinization. Thus, experiences of old irrigation projects did not a find place in new schemes, with the result that some of the best agricultural lands stand degraded today. The examples are parts of canal commands of such projects as the Nagarjunsagar Project, Ukai Karkrapar Project, Sri Ramsagar Project, Rajasthan Canal Project, Malprabha Irrigation Project, Tawa Irrigation Project.

Fall in Ground Water Table

Further, with the technological developments in pump manufacturing and deep-well drilling, excessive ground water extraction is becoming a common phenomenon. Presently, groundwater is used to irrigate almost 50 per cent of the total irrigated land in the country through about 5.5 million table-wells and more than 10 million dug wells. Unfortunately, these advantages have led to over exploitation of groundwater leading to fall in the groundwater-table in Haryana, Tamil Nadu, Rajasthan, and Delhi. Besides, intrusion of saline seawater into the inland aquifers due to over exploitation of groundwater is also affecting the agricultural productivity of coastal areas. Excessive lowering of water-table results in higher investment and energy consumption besides the possibility of arsenic contamination.

Impending Scenario

It has been mentioned elsewhere that India faces the herculean task of increasing its foodgrain production by over 50 per cent in the next two decades and water, as an input to agriculture, is vital for sustaining the food security. Farmers will find themselves unable to raise crop yields as quickly as in the past in the face of a decline in relative water supply. The projected relative crop yields for irrigated cereals show that scarce water is a significant cause of the slowdown in cereal yield growth in the country. Relative crop yield is the ratio of actual projected crop yield to the economically attainable yields at given crop and input prices under conditions of zero water stress. However, in the attempt to get enough water to grow their crops, farmers may extract increasing amounts of ground water for several years, driving down water tables. The cost of building new dams is likely to soar, discouraging new investments in many proposed dam sites. At other sites indigenous groups and non-governmental organization (NGOs) may mount opposition, often violent, over the environmental and human impacts of new dams. The protests and high costs may inhibit new investment in large and medium dams and storage reservoirs.

We have already discussed about the problems which have surfaced due to increased use of irrigation water. The withdrawal of ground water is much more than its recharge. We may reiterate that irrespective of the stage of development, agriculture claims a large proportion of useable land and even large proportion of useable water. The adverse effect of increasing use of water for irrigated crops by drawing more groundwater is glaring in Punjab. The central part of the state is fast emptying its sweet water reserves due to the water-intensive paddy-wheat rotation. The south-western zone is facing the problem of rising water-table which is brackish/alkaline/saline, neither fit for drinking nor for irrigation. This is one of prime reasons for land degradation. There is, thus, an urgent need to have an integrated approach for irrigation water use an watershed development.

The Strategies

Keeping in view the serious dimension of the problem of land degradation, the government have adopted the following strategies for conservation of soil and water:

- Adoption of project approach to the treatment of catchments.
- Increasing sectoral measures for comprehensive water development and maintenance.
- Consultation of treatment efforts through projectisation approach and proper choice of work areas.
- Emphasis on sustainability of treatment measures.
- Vegetative thrust for conservation measures.
- Involvement of the people dependent on a watershed in the planning and execution.
- Measures for sustaining benefits of treatment through institutional and operational mechanism.
- Promotion of skills of farmers and field functionaries through appropriate institutional linkages.
- Preparation of perspective plan for optimum utilization of land resources.
- Evolving a sound land use policy and oversee its implementation.

- Creation of necessary infrastructure both at the national and state level.
- Maintenance of environment hygiene.

Summing up

Potential irrigation demand is bound to grow along with increasing food requirement for the burgeoning population in the country. As a consequence, there may be further degradation of available land for growing foodgrains and other essential commercial crops. This calls for a systematic and scientific management of irrigation water resources. Stakes are high for those who collectively use, account and pay for it and claim their due share for system maintenance. There is an urgent need of certain policy directions before available options of controlling land degradation further narrow down. Development of institutional framework for an efficient use of existing and expanding supplies is central to enhance and sustain the economic and welfare contributions of scarce water and land resources in India.

Lastly, but not the least, we need suitable criteria for measuring the impact of irrigation and crop intensification on the water resources of an area, such as the rate of exhaustion of the aquifer, quality of groundwater, increase in salinity, etc. We are now face to face with the challenge of ensuring sustainable system of agriculture which attempts to balance the concern of both improved productivity and production from our basic assets, namely the land and so managing their resource that the soil health and water quality remain unimpaired.

7

Rural Water Problem

"Water, Water, Everywhere/Not any drop to drink."

—British Poet
Samuel Taylor Coleridge
(200 years ago)

Introduction

Water is the elixir of life. Literally, water is the source of life on earth. The human body is 70 per cent water. People begin to feel thirsty after a loss of only 1 per cent of bodily fluids and risk death if fluid loss nears 10 per cent. But freshwater is emerging as one of the most critical natural resource issues facing humanity. Its per capita availability in India (as well as in the world as a whole) is becoming less and less for the simple reason that the population is increasing at a fast rate whereas the total availability of water continues to remain constant. There is no more freshwater on earth now than there was 2000 years ago, when the population was less than 3 per cent of its current size. In a growing number of places people are withdrawing water from rivers, lakes, and underground sources faster than they can be recharged. Moreover, the supply of freshwater available to humanity is shrinking, in effect, because many freshwater resources have become increasingly polluted. In India, as in many other countries, lakes and rivers have become receptacles for a vile assortment of water including untreated or partially treated municipal sewage, toxic industrial effluents and harmful chemicals leached into surface and ground waters from agricultural activities.

Global Scenario—Problem Dimension

Each day before dawn millions of women in Africa and Asia begin the chore that will consume many of their waking hours-fetching water for their families. They may walk for miles to collect water from a river, spring, or well. Often, the water they carry home is bringing their families not life and good health, but sickness and even death. These women are among the 1.1 billion people around the globe who lack access to any improved water source that is, a pipe, well, or protected spring. The developing world, especially Africa and Asia, remains in the midst of a water crisis that is exacting millions of casualties. According to the World Health Organization, the parasites, bacteria, and chemicals in unsafe water are killing 3.4 million people, mostly children, each year. Many experts agree that each person requires a minimum of about 50 litres a day for drinking, cooking, bathing, and sanitation. Only about 5 to 10 litres of that water must be clean enough to drink. Yet many poor people get much less than 50 litres a day, of which virtually none is technically safe for drinking. Increasingly, a minimum amount of safe drinking water is recognized as part of a fundamental human right to food.

Although water scarcity is worsening worldwide, this is not necessarily the main reason so many people lack access to safe water. After all, compared with other water uses, the amount of water required for domestic use is small.[1] In 1995 about 8 percent of the water used worldwide went for household needs, compared with about 86 per cent for irrigation. On a national basis, access to safe drinking water depends less on a country's available water resources than on its level of wealth and its political commitment to safe water. World Bank figures show, for instance, that 70 percent of people in South Africa have access to an improved water source, compared with only 32 percent in its poorer neighbor Mozambique.[2]

Even though household water use is small the devil is in the details of ensuring that water is safe and getting it to the people who need it in an affordable manner. And people's quality of life depends on their access to water. In 2000 the member states of the United Nations agreed to work toward cutting in half the number of people who do not have access to safe water and sanitation by 2015, as one of the Millennium Development Goals. At the September 2002 World Summit on Sustainable Development in Johannesburg, these countries reaffirmed this goal. Achieving even this limited goal, under which millions of people will still be without safe water, will require heavy commitments of money, time, and political will. Of course, failing to achieve the goal carries a heavy price as well.

Millennium Development Goal

This highly participatory approach to household water projects may appear to be at odds with the Millennium Development Goals of halving the number of people without access to clean water and sanitation by 2015. Deciding on and implementing sustainable local solutions, community by community, takes time. But providing clean water and sanitation to another half billion to a billion people in the next 13 years will require speedy action. One key to greatly expanding coverage of clean water and sanitation is to replicate successful solutions on a larger scale. This scaling up presents its own risks, however. A water system that works well in one community may be inappropriate for another. A sanitation system that succeeds on a small scale may present environmental risks when replicated thousands of times. Yet development professionals have little choice but to forge ahead, balancing the need for sustainable solutions with the need for greatly expanded water supplies and sanitation services.

On a positive note the majority of countries' supports are not starting from scratch. They have many years of experience. The risks to sustainable coverage and facility usage are well recognized by most. In the end, the hope of many is that the Millennium Development Goals can help stimulate action towards solving a problem that has lasted far too long. The goals are admirable, though they will require major public and private investments and creative institutions. In any culture the poisoned well is a symbol of crime at its worst. In reality many of the poor in rural and urban areas are daily victims of the crime of inaction on the drinking water front. The needed investments will come about only through strong economic incentives combined with strong public advocacy.

IFPRI'S ASSESSMENT OF FUTURE SCENARIO

International Food Policy Research Institute, Washingto D.C. in their study entiled 'World Water and Food to 2025—Dealing with Scarcity' have indicated three possible futures for global water and food markets. These are:

1. The business-as-usual scenario (BAU) projects the likely water and food out-comes for a future trajectory based on the recent past, whereby current trends for water investments, water prices, and management are broadly maintained.

2. The water crisis scenario (CRI)) projects deterioration of current trends and policies in the water sector.
3. The sustainable water use scenario (SUS) projects improvement in a wide range of water sector policies and trends.

In this context, IFPRI points out that a scenario is a coherent, internally consistent, and plausible description of a possible future state of the world. A scenario is however, not a forecast but rather a snapshot of how the future could unfold.

Domestic Sector

IFPRI has assessed the total domestic water demand in 1995 and 2025 under the three scenarios, which is shown in Table 7.1. Total domestic demand under CRI is 160 cubic kilometers in developing countries including India where it will be 28 cubic kilometers (28 per cent and 32 per cent less than under BAU respectively); 63 cubic kilometers in developed countries (8.5 per cent less than under BAU); and 223 cubic kilometers globally (28 per cent less than under BAU). Under SUS in 2025, total domestic demand is 10 per cent less than under BAU for developing countries, 4 per cent less for developed countries, and 9 per cent less globally. On the contrary, demand in India under SUS will be marginally higher by 3 per cent than under BAU

Table 7.1: Domestic Water Consumption under Business-as-Usual, Water Crisis, and Sustainable Water Use Scenarios, 1995 and 2025

Region/Country	*Domestic Water Demand (km^3)*			
	1995 Baseline Estimates	*2025 Projections*		
		BAU	*CRI*	*SUS*
Asia	79.1	156.7	113.0	143.9
China	30.0	59.4	42.3	54.3
India	21.0	40.9	27.7	42.0
Southeast Asia	13.9	30.4	23.6	23.8
South Asia excluding India	7.0	16.2	11.1	15.3
Latin America (LA)	18.2	30.7	24.7	22.8
Sub-Saharan Africa (SSA)	9.5	23.9	15.2	23.8
West Asia/North Africa (WANA)	7.1	13.1	9.9	11.2
Developed countries	58.7	68.6	62.8	65.8
Developing countries	110.6	221.0	159.7	198.7
World	169.3	289.6	222.5	264.5

Source: IMPACT-WATER assessments and projections, 2002.

Notes: BAU Indicates business-as-usual scenario; CRI, water crisis scenario; and SUS, sustainable water use scenario.

Author's Source: World Water and Food to 2025—dealing with scarcity; Mark W. Rosegrant, Ximing Cai and sarah A. Cline; IFPRI, Washington D.C.

Table 7.2: Per Capita Potential Domestic Water Demand for Connected/Unconnected Households in Rural and Urban Areas under Business-as-usual, Water Crisis, and Sustainable Water Use Scenarios, 1995 and 2025

Region/Country	Connected (m^3/yr/person)				Unconnected (m^3/yr/person)			
	1995 Baseline	*2025 projections*			*1995 Baseline*	*2025 projections*		
	Estimates	*BAU*	*CRI*	*SUS*	*Estimates*	*BAU*	*CRI*	*SUS*
Rural Areas								
Asia	27.1	29.7	27.5	33.5	17.6	18.9	17.1	n.a.
China	25.7	27.6	26.7	36.9	17.1	18.4	17.8	n.a.
India	26.8	27.8	24.2	31.4	17.9	18.4	16.4	n.a.
Southeast Asia	29.6	38.2	33.1	34.7	19.2	23.6	20.0	n.a.
South Asia excluding India	25.7	28.2	22.1	27.1	17.3	18.9	15.3	n.a.
Latin America (LA)	27.2	29.7	27.5	33.4	17.6	18.9	17.1	n.a.
Sub-Saharan Africa (SSA)	18.8	19.5	16.5	21.5	12.8	13.1	12.0	n.a.
West Asia/North Africa (WANA)	18.3	17.0	16.2	19.9	10.6	10.0	8.0	n.a.
Developed countries	47.0	48.8	47.2	51.2	22.3	33.9	24.1	n.a.
Developing countries	25.2	27.0	24.7	30.2	16.9	17.6	15.8	n.a.
World	31.0	29.3	29.2	33.7	17.0	17.7	15.8	n.a.
Urban Area								
Asia	41.8	52.4	47.0	33.5	25.5	27.8	24.4	n.a.
China	44.3	64.4	57.6	36.9	26.9	28.7	28.0	n.a.
India	38.7	43.0	31.0	31.4	23.4	24.1	21.1	n.a.
Southeast Asia	43.3	56.8	55.3	34.7	26.1	32.0	28.8	n.a.
South Asia excluding India	36.4	36.0	34.0	27.1	23.2	27.5	17.0	n.a.
Latin America (LA)	41.7	52.4	46.9	33.4	25.5	27.8	24.3	n.a.
Sub-Saharan Africa (SSA)	29.2	29.1	19.8	21.5	17.8	18.0	12.5	n.a.
West Asia/North Africa (WANA)	25.7	27.0	26.2	19.9	17.8	15.8	12.3	n.a.
Developed countries	49.3	55.7	51.2	51.2	28.6	34.8	23.0	n.a.
Developing countries	39.1	45.5	41.3	30.2	24.7	24.9	21.1	n.a.
World	43.4	48.1	45.3	33.7	24.8	25.1	21.4	n.a.

The reasons for these shortfalls in domestic consumption globally (except in India under SUS), are, however, very different for SUS than for CRI as seen in Table 7.2 which compares the per capita domestic water demand for connected and unconnected households in rural and urban areas in 1995 and 2025 under each of the three scenarios. As per the IFPRI assessment, domestic water supply for disadvantaged households improves under SUS through the universal extension of household water connections, while the initially connected household reduce consumption in response to high prices and improved water savings technology. In the rural areas, this leads to an actual increase in the overall per capita domestic consumption compared with BAU. In urban areas overall per capita consumption declines because of the greater weight of initially connected households in urban areas.

Under CRI, however, domestic water supply conditions continually worsen because of the proportion of population in households connected to the water supply decines sharply compared with BAU. Per capita demand under CRI for both connected and unconnected households is significantly lower than under BAU in both rural and urban areas of most regions including the developed and developing worlds as a whole. In the developing world, per capita demand declines by 9 per cent for connected rural households, by 9 per cent for connected urban households, by 10 per cent for unconnected rural households, and by 15 per cent for unconnected urban households.

THE HEALTH COST OF POOR WATER

The health problems associated with unclean water are enormous. Of the 3.4 million people killed each year by water-related diseases, 2.1 million, mostly children, die from diarrheal diseases stemming from lack of access to clean water, inadequate sanitation, and poor hygiene.[3] In some areas, childhood diarrhea is perceived as normal. If everyone has it, people do not perceive it as a problem. IFPRI research has shown that diarrheal diseases can greatly exacerbate malnutrition because diarrhea prevents the body from using whatever nutrients are consumed. The bottom line is that water is food.

Therefore clean water is essential for food security and food safety. Efforts to improve people's nutrition can be futile without attention to their water consumption. Even where the quantity of water supplied has been improved, unexpected quality problems have arisen, at great human cost. In the 1980s and 1990s, several million tube-wells were installed in Bangladesh, offering what was thought to be a safe water supply to millions of people who had previously been at risk from microorganisms in their water sources. Then in 1993, it was discovered that many of these tube-wells were producing water contaminated with arsenic, a mineral found underground in many areas in Bangladesh that causes debilitating skin lesions and cancer when ingested. Recent estimates say that between 35 and 77 million people in Bangladesh are at risk of drinking this contaminated water.

According to an article in 2000 in the Bulletin of the World Health. Organization, "Bangladesh is grapping with the largest mass poisoning of a population in history". As serious as the arsenic problem is excessive fluoride in water supplies which may have even more severe consequences. Excessive fluoride is a serious problem in China, India, and the Rift Valley of Africa. We know a little bit of fluoride is good because it protects teeth, but if we consume too much, it produces gross skeletal deformities by the time people reach young adulthood. So just when people should be most productive, it takes them out of the productive work force quite effectively. It is crucial to reduce the severity and duration of people's exposure to these chemicals but removing them from dispersed water sources is not easy or straightforward. In Bangladesh, studies show that one pump may be contaminated with arsenic, while another just meters away may be safe. The governmental has painted many of the tubewells-red for contaminated and green for clean but the wells must be regularly retested, as their status can change quickly.

The Link to Sanitation and Hygiene

The issue of clean drinking water is inextricably linked to sanitation and hygiene. if human waste is not treated or disposed of properly, it can contaminate a water source that is used for drinking. if a woman places clean water from a well into a dirty container, it is no longer clean. If people do not practice hand washing, they may lose whatever benefits they might have derived from clean drinking water and sanitation facilities. Research has shown that providing clean water can reduce childhood diarrhea by 15 to 20 percent, but getting people to wash their hands after using the toilet and to handle food safely can lower it by 35 percent. It is increasingly standard practice for governments and NGOs to integrate water supply, sanitation, and hygiene projects. Improving hygiene is the most difficult thing we do.

It is necessary to break oral-fecal transmission. If we can get people to wash their hands after using the toilet, we get major health benefits. The trick is to make it an automatic behaviour.

Hygiene improvements are more effective when accompanied by sanitation facilities, which 2.4 billion people globally still lack. The problem is often most obvious in cities, where raw sewage runs through slum neighborhoods. In rural areas, where people may go out into the bush, the problem is less noticeable but no less serious. Human waste can seep into surface water and groundwater. In West Bengal, a popular marketing slogan is no marriage without a toilet. The toilet is being promoted as an essential facility in a newlywed's house.

Getting Clean Water to the Poor

Many countries have responded to people's sense of entitlement to water by subsidizing it. One might expect this approach to benefit poor people, but all too often it does not In much of India, water is basically a free good. But those who have the best access to it are the wealthy farmers and urban dwellers, because they are the ones who are connected to the water system. Poor people in rural areas or urban slums, who are not connected to a piped water supply, in fact pay a higher price for their water, in terms of time spent collecting it or money spent buying it from small-scale private water vendors. So the government pays to supply water to the well off, whereas the poor pay the price for their water themselves. One school of thought says that if people don't pay for water, they won't conserve it. But others say that even very low prices can be burdensome for very poor households.

The actual cost of supplying more clean water to a growing world population will certainly rise over the coming decades because of the cost of developing new water sources and increasing competition for water between farmers, cities, and industries. Water suppliers are going to have to raise prices even for their existing supplies. What makes the most sense is for water suppliers to raise prices for all users and then subsidize water for the poor. Some areas are trying this approach, or a variation. The city of Chennai, Tamil Nadu, secured its water supply by clamping down on unaccounted for water lost to leaks or unauthorized users and then made a commitment to supply a minimum amount of water to all users. According to Paul Appasamy, director of the Madras School of Economics in Chennai, who has studied the system, "the quantity provided through the public system is extremely limited and has to be augmented in other ways. The poor are given water from public tankers that traverse the city. Better-off people get water from their won bore wells or from private vendors". The city is also promoting the harvesting of rainwater for people's household needs.

Clean Water is a Boon for Women

Girls and women are the biggest beneficiaries when their families get access to clean water and sanitation services. In developing countries, obtaining water for domestic use is usually the responsibility of women and girls. The hours that women and girls spend fetching water is time that is not available for child care, productive activity, and schooling. Moreover, when someone in the household becomes sick, it is the women who are responsible for care giving.

Partly because of women's inherent interest in safe water and sanitation, NGOs are making women important players in developing and maintaining these services. The Indian NGO, Gramalaya, for example, has helped build communal latrines and water points in urban slums in India. These community toilets are managed by SHE (Sanitation and Hygiene Education) teams-made up of members of women's self-help groups who collect fees from users and ensure that the toilets remain in working order. Water Aid's Turner explains, "Women are used as caretakers because they are at the center of their household. Men are more likely to have paid employment and may be away from home, while women do not have the same mobility and so remain caretakers for longer. As women are the main collectors of water, their motivation to maintain water points is stronger".[4] This job can also serve as a valuable source of income for poor women.

PRIVATIZING WATER

Some people argue that only the private sector can meet the vast infrastructure requirements of supplying safe water to billions of people. Others see privatization of water supplies as a violation of people's basic right to water and argue that because private water companies are ultimately seeking profits, they will not serve the water needs to the poor. In itself privatisation is neither a terrible thing nor a solution to all that ails us. Either a public or a private system can be managed efficiently, but it is necessary to recover costs in order to maintain efficient services and expand services to meet the needs of rapidly growing populations, and ensure that the system is meeting the needs of the poor.

Buenos Aires, for example, privatized its water system in 1993. The move was highly controversial at first, but later it appeared to be generally successful. The water company significantly expanded water service to the edges of the city and to poorer neighbourhoods, and the pricing structure was such that the poor could afford the water. In other cases, privatisation has been less successful. What's often missing is that they have to supply social goods that is, they have a responsibility to provide clean water even to the poor. It is the job of government to ensure that private water companies fulfill this responsibility to the poor, even while they seek profits. The critical issue is the government's contract with the private water supplier-how it's written and whether it's enforced. The debate over privatization of water supply is probably moot in most of Sub-Saharan Africa. For over 90 percent of the people, the public sector is going to be the water provider. There isn't a private sector company that will take this on, because it just isn't profitable enough. So the real question is: How do we strengthen the public sector?

INDIA'S PITIABLE SITUATION

Despite more than five decades of planned development, safe drinking water remains a problem area for the country. So much so, that nearly 45 per cent of India's rural population does not have access to potable drinking water in adequate quantity. Worse still, about 10 per cent villages do not even have a single safe public drinking water source in their vicinity (1994 figures.) Naturally, the government is concerned. It has already envisaged a water authority, called Rajiv Gandhi National Drinking Water Mission Authority, to pace up implementation of rural drinking water programmes. Besides, it stepped up the outlay for rural drinking water supply in 2002-03 to Rs. 2,110 crore.

But that was just a drop in the bucket. The magnitude of the problem questions the suitability of the existing approach: whether or not it is the optimum one to tackle India's rural drinking water problem? There are two main issues in this regard. The first is the dependence only on the limited resources of the states for funding drinking water systems, and its consequent uncertainties. The second is the sustainability of drinking water systems created, using the present approach. Both the issues are interlinked, since the first affects the second.

As far as fund availability is concerned, given the government's resource constraints, it is time to seriously consider the question of implementing cost recovery-partial for the capital costs and full for operational and maintenance from the users. Implementation of a cost recovery policy would automatically address the second issue, of sustainability. Sustainability problems arise because the present policy is almost entirely 'supply driven.' The user community is rarely involved in the planning, implementation and operation and maintenance of the drinking water system created for them. Hence, demand-based policies-*i.e.* in terms of the type of system desired by the community or the level of service demanded are rarely adopted.

The present approach is primarily a topdown one, the emphasis being on engineering interventions, largely dependent on the availability of funds. Typically, for piped water schemes, state Public Health Engineering Departments (PHEDs) or water utilities design large-scale systems (not necessarily the least cost solution) which are equally fairly expensive to build and even more costly to operate and maintain. With limited funds available with the state for operation and maintenance, coupled with poor recovery of water tariffs from rural consumers, many rural piped water systems are working much below the capacity or lying unused. As for the deep bore India Mark 11 hand pumps, though their maintenance is much easier than that of a piped water scheme, a large number of them are perpetually out-of-order. Because, the PHED mechanic does not go too often to the far-flung areas, and the user community is unable or unwilling to do simple repairs.

Experience worldwide has shown that the sustainability of rural drinking water systems is closely linked to the use of a 'demand-driven' approach in devising the system. True, the Eighth Five-Year Plan document stated that water should be treated as a commodity based on the principle of effective demand. But the reality is, most government-funded schemes are based on a 'supply-driven' deal, where the user community has little role in its planning or management.

One of the fallacies underlying the existing approach is that it does not rate drinking water as an economic good, but as a commodity to be distributed free of charge (except for a nominal water fee, which is rarely collected in rural area) without any economic value. This virtual 'free meal' syndrome needs to be cured. The fact is, easy availability of safe drinking water in rural areas leads to better health, and saving of time, from carrying water over long distances, especially for women. The opportunity cost involved is high, since the time saved could be used for income generating activities. So, easy access to safe drinking water has some direct and indirect economic spin-offs. But there is a catch. Rating water as an economic good means that it should carry a price tag. If a demand-based approach (*i.e.* if it is ensured that a rural community gets a drinking water scheme if it is interested in) is adopted, then the user group should be willing to pay a price for the economic good being provided.

A demand-based approach and cost recovery for drinking water go hand in hand, since the best proxy for demand is willingness to share the costs of the water supply system. While sharing costs, it is essential that the community is informed about the technological options available and the likely costs (including operation and maintenance) of the different options. Thereafter, it is upto the user group to decide what type of drinking water system they would prefer. Presumably, since they share the cost, they will go in for the least cost solution, leading to both sustainability and efficiency. An issue which often crops up whenever cost-sharing is raised is that of the 'need' versus 'demand' debate. In a cost-sharing setup, will a needy village get left out because of its inability to share the costs?. The answer is no suite, exceptional cases obviously deserve special treatment, such as some relaxation in the cost sharing policy.

As a rule, however, the appropriate strategy to deal with village communities which have the highest need for safe drinking water is to prioritize by adopting a system of 'eliminators' and 'selectors'. Such an approach would first shortlist the water scarce villages and prepare a list of such habitations. Thereafter, from amongst this universe of shortlisted villages, self-selection could take place on the basis of a demand-led system, where villages agreeing to the cost-sharing policy would get priority in obtaining drinking water schemes. Within the selected village, the question of how much should each family contribute would be settled by consensus within the community. Cross-subsidization would take place, since the really poor families may not be able to contribute as much as some of the others.

While the government has initiated some policies towards cost recovery, they mainly pertain to maintenance of hand pumps by the community. In any case, even this still remains, for the most part, more a policy than practice. No decision has yet been taken to universally introduce even a token percentage of capital cost recovery from the user community or make them totally responsible for operation and maintenance. A strong political will is required to enforce sensitive cost-sharing policies, However, the policy need not be implemented across the board in the beginning. In any case, if the management of such schemes is to be decentralised, then communities have to be properly trained. This takes time. Initially, therefore, pilot projects using different cost recovery models, could be started on a small scale. Successful models could thereafter be upscaled and replicated around the country.

Magnitude of the problem in a few selected States/Regions

(*a*) *Kutch region of Gujarat*

The north-western region of Gujarat is well known as water scarce and drought prone. Owing to its semi-arid character, scarcity of water is however not a recent phenomenon in Kutch. What is recent is the problem of drinking water despite having one of the largest piped water supply networks in Gujarat, covering 92 per cent villages of Kutch. But despite huge investments in piped water supply, this region of Gujarat reels under a severe drinking water crisis throughout the year. More than 20 per cent villages remain under tanker water supply. The annual expenditure on tanker water supply exceeds Rs. 20 million besides large capital and operational expenses of the public water supply schemes. In this region, there is failure of modern and traditional technologies and wells and bores are drying up rapidly. Even tube-wells are fast turning saline and drinking water quality is deteriorating. This is happening not in one village or in a taluka, but all over Kutch.

With the limited surface water potential and its poor efficiency in Kutch, groundwater became the main source of water for all purposes. New technologies were introduced to tap groundwater. Diesel pumps could draw water much faster than the 'KOS' (traditional water lifting device) and were introduced as early as in 1950. As the water table went down with time, bores were drilled in the dried wells and electric motors replaced the diesel pumps. Finally, deep tube-wells with submersible pumps were introduced to tap water from deeper aquifers. As groundwater withdrawal became possible, well, bore-well and tube-well-based irrigation came into existence. The process of drilling bores and tube-wells for agriculture has never stopped since then. Official numbers of dug wells and dug-cum-bore-wells increased from 18,000 in 1960-61 to more than 32,000 by 1993-94. More importantly, the technology for withdrawing water catalyzed a much faster depletion of groundwater. The over-extraction of groundwater has led to a rapid depletion of the water table and sharp deterioration in the quality of water across the district (Table 7.3).

Table 7.3: Changing Groundwater Status in Kutch

Taluka	*Groundwater Status*	
	1984	*1991*
Mandvi	White	Dark
Nakhatrana	White	Grey
Bhuj	White	White
Bhachau	White	OE
Lakhpat	White	White
Anjar	White	OE
Rapar	White	Grey
Mudra	White	Grey
Abdasa	White	Grey

Notes:

1. These categories are based on the groundwater withdrawal as per cent of recharge.
2. White: Withdrawal less than 65 per cent of recharge.
3. Grey: Withdrawal between 65-85 per cent of recharge.
4. Dark: withdrawal between 85-100 per cent of recharge.
5. Over Exploited (OE): Withdrawal more than recharge.

Source: Central Groundwater Board, Ministry of Water Resources, 1999 and Report of Committee on Estimation of Groundwater, Potential, 1991.

The above Table 7.3 indicates that, seven out of nine talukas moved to higher and unsustainable extraction categories. Two talukas, Anjar and Bhachau, came under the over-exploited category; Mandvi advanced into the dark zone whereas Nakhatrana, Rapar, Mundra and Abdasa came under the grey category. Over-extraction of groundwater had led to multifarious impacts both on the quantity and quality of groundwater resources at the regional level. Over-extraction along coastal areas has resulted in seawater ingression into the inland freshwater aquifers making them saline. Irrigation with saline water has led to large-scale salination of agricultural lands. In the process, the soils began losing productivity and water extraction became more expensive making agriculture unviable. The livestock-based pastoral economy that flourished at one time is in a shamble now as the fodder and water resources are shrinking.

(b) Punjab—the Granery of India

The emerging water crisis pose a grave threat of Punjab State. The estimates of scientists reveal that good quality water is available to the extent of 3.12 million hectare meters (25.34 million acre feet, MAF), whereas the normal requirement amounts to 4.90 million hectare meters (39.75 MAF). The net deficit comes out to 1.78 million hectare meters (14.41 MAF). The possible addition to surplus of water after treating the marginally fit water available in different parts of the state is estimated to be 0.24 million hectare meters (1.94 MAF). The amount of deficit is so substantial that unless a very serious view is taken of the matter and immediate corrective policy measures adopted, the state is heading towards a critical water famine situation. The fast-changing cropping patterns have precipitated the water crisis since ushering of the green revolution in the mid-60s. Consequently, the water table has been going down for the past four decades. A recent study by Dr. Joginder Singh, Head, Department of Economics & Sociology, Punjab Agricultural University, Ludhiana indicates that the situation is pretty grim in the central or sweet water region of the state and also in the sub-mountaneous or the kandi zone, whereas in the south-western zone the rising water table has posed a different set of problems like salinity and water-logging. It is apprehended that a situation may arise when the water table may go down to such an extent that lifting water would require heavy capital investment in the form of high power electric motors on submersible pumps and also increased energy consumption. At a certain point it may even become technically and economically an unfeasible proposition. Yet, another equally worrisome possibility may be that when a sizable quantity of presently good quality water gets mined out from the upper ground water strata, the subsequent lower level suppliers may turn out to be unfit both for drinking and irrigation.

The trend analysis of data collected by hydrologists reveals that the water table which was 14 meters in 1981-2001 has not changed much in these two decades. The water table, however, as the above study reveals, is going down by 1 cm per year in Gurdaspur, 8 cm per year in Ropar and is up by 2 cm to 6 cm per year in Hoshiarpur. The central or sweet zone, which is highly productive with a well-knit irrigation system, mainly dependent upon tube-wells, presents a serious challenge because of extensive and intensive rice cultivation. In this zone, according to Dr. Joginder Singh, the water table is falling at an average rate of 0.23 meter per year, during the past 15 years. The water table which was 5 to 6 meters in 1981, has now shown a fall of 24 cm to 25 cm per annum.

In the south-western zone, which comprises almost one-fourth of the state's cultivated area, the groundwater is brackish and unfit for crops and humans. The water table in this zone which was 11 to 12 meters deep in 1981 is rising continuously at the rate of 9 cm per year in Mansa, 21 cm in Bhathinda and 17 cm in Ferozpur. This has compelled the farmers to give up cotton cultivation and now sow paddy.

It may be reiterated that Punjab's water problem is very serious. It deserves serious action. The solutions are not cheap. To tackle the impending crisis, a multi-pronged approach is required. To arrest the decline in the underground watertable, it is necessary to undertake water harvesting measures to conserve rainwater. The cropping pattern has to be changed. The existing water sources has to be protected. At the city, town, village and individual levels, everyone has to make efforts to save water and avoid its wastage.

However out of a total of 12,402 villages in Punjab, 7,700 had already been provided water supply, while projects were being completed to ensure drinking water supply in 2118 villages during the current (2003-04) financial year and Rs. 102 crore have been earmarked for the purpose. To improve living conditions in villages, the state government has submitted a project report of "Water Reserves Management Project" to the World Bank authorities for sanctioning loan to carry out development works in the rural areas.

(*c*) ***Rajasthan—A desert State***

The potable water crisis in the state is so serious that the Rajasthan High Court has recently (January, 2003) ruled that the right to receive potable water is a basic and fundamental right of the people guaranteed by Article 21 of the constitution and the court has directed the state government to expedite completion of drinking water schemes. As in some districts, the presence of fluoride, chlorine and TDS content in potable water was much higher than the tolerance limit, the court observed that "people deserve a life worth living, free from diseases which can be prevented by the state by pressing into service the state-of-the-art technology.

(*d*) *Ramtek's (a tehsil in Maharashtra) water woes*

In tribal dominated Ramtek tehsil, agriculture is the main source of livelihood. Despite claiming to have spent crores of rupees on agricultural and welfare schemes, precious little has been done for the uplift of the tribals of this tehsil. Due to scant rainfall, absence of dams, wells and bore-wells and neglect of proper crop planning, the level of groundwater is going down in this region.

Hence, digging bore-wells and wells may not solve the water problem. Ramtek's tanks were built using traditional knowledge and techniques. The expertise required in building such tanks and distribution systems was available locally and any kind of emergencies was addressed immediately to minimize the losses. Moreover these structures were need-based and it was within the capacities of the local people to address the maintenance and repair requirements of related structures. The technologies adopted were simple and could be utilized by local inhabitants. Thus the local community was not dependent on external agencies in tank-building, repairs and maintenance. Traditional practices ought to find wider promotion, acceptance and application.

Traditional technologies, ideal for the time in which they were devised, can be augmented by modern science. For example, run-off from catchments can be increased. Recharging efforts can be made more efficient. In fact, traditional technologies can also be used to deal with the present-day water quality problems like flouride and arsenic in groundwater. Groundwater recharge, through the construction of traditional water harvesting systems like small ponds and tanks in catchment areas, has revived rivers. The emphasis should be on greater decentralization of the maintenance and revival of tanks and ponds around which village life is centered.

(*e*) *Delhi State*

Inadequate and inferior quality water supply position in Delhi state has become a matter of serious concern. It has been accepted by the Mayor of Delhi that the civil authorities have been able to provide potable water to hardly 40 percent of Delhiites. Apart from the 1,600 unauthorized colonies and 1,100 slums, a large chunk of the city's rural belt which fall in the water deficient zone is facing acute water scarcity. Delhi's peak summer demand is estimated to be about 850 million gallons daily (mgd.), but the Delhi Jal Board manages to provide just over 650 mgd. The river Yamuna which passes through Delhi has become a dirty drain. The faecal coliform in the river water has risen to an alarming level of 11.8 crore per 100 ml. of water as against the permissible level of 5,000 per 100 ml. Coliform cause serious diseases affecting the digestive system. Further, on most occasion, the dissolved oxygen level is found to be zero or below the level of 4 mg. per litre in many places through which the river flows. Delhi's total requirement of drinking water is 1,480 cusecs, but the availability from all sources is only 1,221 cusecs.

Tata Energy Research Institute (TERI) in their recent study claims that the water table in most parts of the metropolis is falling rapidly and in many places in South Delhi is almost bone-dry. In Mehrauli zone, the groundwater may get exhausted in the next 10 years. A survey of the trend in water levels between 1960 to 2002 show a constant decline. During the 60's, ground water levels were approximately four to five metres and in several parts waterlogged conditions existed. Between 1960-2001, water level have declined by two to six meters in most parts of the alluvial areas. A further decline of 8-20 metres has been recorded in south-west district and in the southern district, the decline has ranged between 8-30 metres. The prime instrument in this case have been the tube-wells. While there are court restrictions on the boring of new tube-wells, those that are already operating are wreaking havoc on the precious little groundwater reserve.

Faced with such alarming situation, today the only solution seems to be rainwater harvesting in large scale and thereby groundwater recharging. Groundwater recharge can be done through various methods like recharge pits, trenches, abandoned dugwells, hand pumps, recharge wells, recharge shafts, lateral shafts with bore-well and so on. The government is taking many steps in popularizing these methods. Even the Delhi Development Authority (DDA) felt the need for saving water through water harvesting and undertook a major rain water harvesting project for recharging 1 million cubic metres of rain water annually by covering an area of 190 hectares.

Finally, it may be stated that the average daily water consumption of Delhite is highest in the country at 240 litres. Residents of European cities like Copenhagen consume only 111 litres a day. Residents of Delhi end up wasting more water than other cities. If people start using water judiciously, the gap can be bridged to a great extent. The city needs 150 gallons of water more than it gets to meet its daily demand. But the crisis Delhi faces is far more acute than what such a shortfall should create. That is because 50 per cent of the supply is lost in distribution. Estimated water loss in distribution is nearly double of the shortfall.

Baolis can Help Solve Water Problem

The dusty pages of history might hold the answer to the Capital's water shortage. While the High Court recently directed the civic authorities to maintain lakes, ponds and reservoir as water bodies, traditional sources of water have been left out of its purview once again. The historic Hauz-i-Shamsi in Mehrauli is only one such example. According to local folklore, the Prophet came to Illtutmish in his dream riding a white horse and indicated where he should build his tank. Existing since 1,229, this old water body today is choked with plastic bags.

Providing the much-needed breathing space in the heart of congested Mehrauli, this tank is functional even today. But instead of being an oasis, it is doused with all the sewage from Mehrauli. The land around it is relatively empty and has catchment area in the West and the South, so it could be made thrice as big and be used to supply water to villagers around the area. There are many lesser known tanks dotting the city which can be put to similar use. But unfortunately they are being filled in an built over. Besides tanks, wells and baolis were good supplies of water for Delhiites for centuries and can be used as an alternative source, experts feel. Due to depletion of the water-table and construction of high-rise buildings around baolis, many of them have dried-up or their capacity is reduced. But they can still be put to original use as they were designed to be recharged. As people only use 1 per cent of water to drink, these baolis have tremendous potential for non-potable purposes.

But with the short-sighted attitude of the civic agencies, these traditional sources are often not even considered. These baolis have been looked at in isolation. The Archeological Survey of India only desilts them for maintenance. It should look at the feeder canals and see how these functioned, clues can be found to recharge them. The problem is of course encroachments over the years. These baolis are trapped in a different world where the land mafia is at work and agencies are nervous to touch them. About 400 baolis dotted the cityscape at the turn of the century but now only a handful remain. Filled up with "malba" or earth, they have been lost forever.

With the world turning to traditional wisdom to solve problems, neglecting the past in an urban jungle like Delhi might make Shekhar Kapoor's "reel version" on water wars a reality.

(f) Soibug-a village in Budgram, in Central Kashmir

Jammu and Kashmir, one would imagine, has no shortage of water. For, the Indus, the Jhelum and the Chenab flow through the state. Yet, the village Soibug faces a critical drinking water problem. With a population of over 6,000, the village is now facing an acute water shortage problem. The villages are being forced to drink water having a very high iron content. It is alleged, since independence, no water supply scheme has been implemented in the village, Attempts were however made to dig tube wells in the village, but unfortunately bore no water. The villagers have no alternative but to drink the polluted water from small streams.

PRESENT STATUS AND GOVERNMENT INITIATIVES

Access to Safe Drinking Water in Households in India

As per Census of India, if a household has access to drinking water supplied from a tap, hand-pump/tube well within or outside the premises, it is considered as having access to safe drinking water. Census 1991 reported the coverage of households in India having access to safe drinking water at 62 percent, comprising over 81 per cent of urban households and around 56 per cent of rural households (Table 7.4). The 54th Round of National Sample Survey (July 1999) on drinking water, sanitation and hygiene estimated that 50 per cent of rural households were served by a tube well/hand pump, 26 per cent by a well, and 19 per cent by taps.

Essentiality of Safe Drinking Water

It may be reiterated and the national government recognized that safe drinking water and improved sanitation play a major role in the overall well being of the people, with a significant bearing on infant mortality rate, death rate, longevity and productivity. The poor, both in rural and urban areas, bear a disproportionately higher burden of the non-availability of water, as well as its poor quality. Seasonal disruption

Table 7.4: Access to Safe Drinking Water in Households in India

Sl No.	*States/Union Territories*	*1981*			*1991*			*Average Annual Change*		
		Rural	*Urban*	*Combined*	*Rural*	*Urban*	*Combined*	*Rural*	*Urban*	*Combined*
1.	Andhra Pradesh	15.12	63.27	25.89	48.98	73.82	55.08	22.39	1.67	11.27
2.	Arunachal pradesh	40.16	87.93	43.89	66.87	88.20	70.02	6.65	0.03	5.95
3.	Assam	–	–	–	43.28	64.07	45.86	–	–	–
4.	Bihar	33.77	65.36	37.64	56.55	73.39	58.76	6.75	1.23	5.61
5.	Delhi	62.26	94.91	92.97	91.01	96.24	95.78	4.62	0.14	0.30
6.	Goa	8.57	52.31	22.50	30.54	61.71	43.41	25.64	1.80	9.29
7.	Gujarat	36.16	86.78	52.41	60.04	87.23	69.78	6.60	0.05	3.31
8.	Haryana	42.94	90.72	55.11	67.14	93.18	74.32	5.64	0.27	3.49
9.	Himachal Pradesh	39.56	89.56	44.50	75.51	91.93	77.34	9.09	0.26	7.38
10.	Jammu & Kashmir	27.95	86.67	40.28	–	–	–	–	–	–
11.	Karnataka	17.63	74.40	33.87	67.31	81.38	71.68	28.18	0.94	11.16
12.	Kerala	6.26	39.72	12.20	12.22	38.68	18.89	9.52	-0.26	5.48
13.	Madhya Pradesh	8.09	66.65	20.17	45.56	79.45	53.41	46.32	1.92	16.48
14.	Maharashtra	18.34	85.56	42.29	54.02	90.50	68.49	19.45	0.58	6.20
15.	Manipur	12.91	38.71	19.54	33.72	52.10	38.72	16.12	3.46	9.82
16.	Meghalaya	14.26	74.40	25.11	26.82	75.42	36.16	8.81	0.14	4.40
17.	Mizoram	3.57	8.79	4.88	12.89	19.88	16.21	26.11	12.62	23.22
18.	Nagaland	43.43	57.18	45.63	55.60	45.47	53.37	2.80	-2.05	1.70
19.	Orissa	9.47	51.33	14.58	35.32	62.83	39.07	27.30	2.24	16.80
20.	Punjab	81.80	91.13	84.56	92.09	94.24	92.74	1.26	0.34	0.97
21.	Rajasthan	13.00	78.65	27.14	50.62	86.51	58.96	28.94	1.00	11.72

(Contd.)

Table 7.4 (*Contd.*)

Sl No.	*States/Union Territories*	*1981*			*1991*			*Average Annual Change*		
		Rural	*Urban*	*Combined*	*Rural*	*Urban*	*Combined*	*Rural*	*Urban*	*Combined*
22.	Sikkim	21.70	71.93	30.33	70.98	92.95	73.19	22.71	2.92	14.13
23.	Tamil Nadu	30.97	69.44	43.07	64.28	74.17	67.42	10.76	0.68	5.65
24.	Tripura	22.17	67.92	27.33	30.60	71.12	37.18	3.80	0.47	3.60
25.	Uttar Pradesh	25.31	73.23	33.77	56.62	85.78	62.24	12.37	1.71	8.43
26.	West Bengal	65.78	79.78	69.65	80.26	86.23	81.98	2.20	0.81	1.77
27.	Andaman & Nicobar Islands	36.35	91.95	51.64	59.43	90.91	67.87	6.35	-0.11	3.14
28.	Chandigarh	94.39	99.39	99.09	98.11	97.68	97.73	0.39	0.17	-0.14
29.	Dadra & Nagar Haveli	16.85	54.35	19.35	41.17	90.97	45.57	14.43	6.74	13.55
30.	Daman & Diu	46.42	67.04	54.48	55.87	86.76	71.42	2.04	2.94	3.11
31.	Lakshadweep	0.97	3.65	2.19	3.41	18.79	11.90	25.15	41.48	44.34
32.	Pondicherry	76.88	84.18	80.59	92.86	86.05	88.75	2.08	0.22	1.01
	All India	26.50	75.06	38.19	55.54	81.38	62.30	10.96	0.84	6.31

Note: All India figure excludes Assam in 1981 and Jammu & Kashmir in 1991.

Source: Housing and Amenities, Paper 2 of 1993; Census of India, 1991

Author's Source: Economic Survey 2002-03, Government of India.

of water supply is common, especially during summer months. Fetching of water for domestic use, sometimes from far-flung sources, is time-consuming physical burden borne by women, particularly in the rural areas.

Apart from repercussions on health, this also affects their overall well-being. Moreover, 70-80 per cent of illnesses are related to water contamination and poor sanitation. Women and children are particularly vulnerable to the effects of water contamination. It is also a matter of concern that despite the progress achieved in provision of water supply, the level of water-related sickness continues to be high.

RURAL WATER SUPPLY

The Department of Drinking Water Supply under Ministry of Rural Development has been mandated to provide safe drinking water in all rural habitations, by April, 2004. To achieve this objective the programmes like the Accelerated Rural Water Supply Programme (ARWSP) and the Pradhan Mantri Gramodaya Yojana—Rural Drinking Water (PMGY-RDW) are being implemented. Prior to that the Government had launched the National Drinking Water Mission in 1986, renamed as Rajiv Gandhi Drinking Water Mission in 1991. Considerable success has been achieved in meeting the drinking water needs of the rural population. With an investment of over Rs. 34,000 crore, 91.06 percent of rural habitations have been fully covered (FC) with drinking water facilities and 7.93 per cent are partially covered (PC).

Accelerated Rural Water Supply Programme (ARWSP)

The Accelerated Rural Water Supply Programme (ARWSP), currently implemented through the Rajiv Gandhi National Drinking Water Mission in the Department of Drinking Water Supply, has been in operation since 1972-73 to assist the States and UTs to accelerate the pace of coverage of safe and adequate drinking watersupply facilities to the rural population. The programme focuses on the coverage of all rural habitations specially the un-reached ones, to ensure sustainability of the systems and sources, to tackle the problem of water quality and institutionalize water quality monitoring and surveillance through a Catchment Area Approach. Necessary reforms have been introduced in 1999 so as to gradually replace the government oriented, centralized, and supply driven programme by a people-oriented decentralized, demand driven and community based one. The reform initiative now under implementation in 67 selected pilot districts is being scaled up through 'Swajaldhara' where Block/Panchayat/Village/Beneficiary Group can also avail the benefit of the reforms with 10 per cent contribution by the community in cash upfront and full responsibility for operation and maintenance.

The provision of potable drinking water to all villages has been identified as one of the priority tasks of the Government. During the Ninth Plan, 72,859 Not Covered (NC) and 3,52,423 Partially Covered (PC) habitations have been covered with water supply facilities. As on January 2,82,003 there were 12,95,504 Fully Covered (FC) habitations, 1,12,804 Partially Covered (PC) with a balance of 14,356 habitations as Non Covered (NC) ones. The State wise coverage of habitations under Rural Water Supply is at Table 7.5.

An area of concern is the re-emergence of uncovered villages in view of non-involvement of people in design/operation/maintenance of assets and fast depletion of ground water levels leading to problems of arsenic and fluoride content. This necessitates addressing of the issues of control on ground water withdrawal, water harvesting, soil and water conservation and community water as an economic asset.

Swajaldhara

An ambitious programme for providing drinking water to all the villages in the shortest possible time at a minimum cost, named as Swajaldhara was launched by the Prime Minister on 25th December, 2002. The Swajaldhara scheme envisages people's participation in the progrmme in the form of contribution to the extent of 10 per cent of the project cost and ownership and maintenance of the project itself. However, Panchayats/Vilages where SC and ST population is at least 50 per cent, need to share only five per cent of the cost. Under the scheme, the Ministry of Rural Development is increasing its share of the funds for financing the water supply schemes to 90 per cent. The Swajaldhara programme is an extension of reform initiative in rural drinking water supply where the community will plan, implement, operate, maintin and manage the asset. To begin with, 882 projects worth Rs. 87 crore have been sanctioned. Initially the project is being launched in eight states. They are Andhra Pradesh, Orissa, Himachal Pradesh, Haryana, Maharashtra, Madhya Pradesh, West Bengal and Uttar Pradesh.

Table 7.5: Status of Coverage of Habitations Under Rural Water Supply

Sl No.	*States/Union Territories*	*Status of Habitations*				*Percentage Distribution*			
		NC	*PC*	*FC*	*Total*	*NC*	*PC*	*FC*	*Total*
1.	Andhra Pradesh	0	13,844	55,888	69,732	0.00	19.85	80.15	100.00
2.	Arunachal pradesh	325	880	3,093	4,298	7.56	20.47	71.96	100.00
3.	Assam*	441	17,585	52,643	70,669	0.62	24.88	74.49	100.00
4.	Bihar	0	0	1,05,340	1,05,340	0.00	0.00	100.00	100.00
5.	Chhattisgarh	0	0	50,379	50,379	0.00	0.00	100.00	100.00
6.	Goa	8	40	348	396	2.02	10.10	87.88	100.00
7.	Gujarat	50	1,428	28,791	30,269	0.17	4.72	95.12	100.00
8.	Haryana	0	48	6,697	6,745	0.00	0.71	99.29	100.00
9.	Himachal Pradesh	820	9,277	35,270	45,367	1.81	20.45	77.74	100.00
10.	Jammu & Kashmir	1,709	3,576	5,899	11,184	15.28	31.97	52.74	100.00
11.	Jharkhand	235	42	99,819	1,00,096	0.23	0.04	99.72	100.00
12.	Karnataka	1	17,832	38,849	56,682	0.00	31.46	68.54	100.00
13.	Kerala	782	6,879	2,102	9,763	8.01	70.46	21.53	100.00
14.	Madhya Pradesh	0	0	1,09,489	1,09,489	0.00	0.00	100.00	100.00
15.	Maharashtra	1,999	24,019	59,912	85,930	2.33	27.95	69.72	100.00
16.	Manipur	4	183	2,604	2,791	0.14	6.56	93.30	100.00
17.	Meghalaya	327	788	7,524	8,639	3.79	9.12	87.09	100.00
18.	Mizoram	0	454	457	911	0.00	49.84	50.16	100.00
19.	Nagaland	349	569	607	1,525	22.89	37.31	39.80	100.00
20.	Orissa	0	0	1,14,099	1,14,099	0.00	0.00	100.00	100.00
21.	Punjab	1,393	2,114	9,942	13,449	10.36	15.72	73.92	100.00
22.	Rajasthan	5,631	2,521	85,794	93,946	5.99	2.68	91.32	100.00
23.	Sikkim	0	290	1,389	1,679	0.00	17.27	82.73	100.00

(Contd.)

Table 7.5 (*Contd.*)

Sl No.	*States/Union Territories*	*Status of Habitations*				*Percentage Distribution*			
		NC	*PC*	*FC*	*Total*	*NC*	*PC*	*FC*	*Total*
24.	Tamil Nadu	0	0	66,631	66,631	0.00	0.00	100.00	100.00
25.	Tripura	93	20	7,299	7,412	1.25	0.27	98.48	100.00
26.	Uttar Pradesh**	0	0	2,43,633	2,43,633	0.00	0.00	100.00	100.00
27.	Uttaranchal	119	913	29,976	31,008	0.38	2.94	96.67	100.00
28.	West Bengal	0	9,060	69,976	79,036	0.00	11.46	88.54	100.00
29.	Andaman & Nicobar Islands	0	121	383	504	0.00	24.01	75.99	100.00
30.	Dadra & Nagar Haveli	30	241	245	516	5.81	46.71	47.48	100.00
31.	Daman & Diu	0	0	32	32	0.00	0.00	100.00	100.00
32.	Delhi	0	0	219	219	0.00	0.00	100.00	100.00
33.	Lakshadweep	0	10	0	10	0.00	100.00	0.00	100.00
34.	Pondicherry	40	70	157	267	14.98	26.22	58.80	100.00
35.	Chandigarh	0	0	18	18	0.00	0.00	100.00	100.00
	All India	14,356	1,12,804	12,95,504	14,22,664	1.01	7.93	91.06	100.00

NC: Not Covered, PC: Partialy Covered, FC: Fully Covered

Note: As per reports received upto January 28, 2003 by Department of Drinking Water Supply.

* In case of Assm F.C. habitations include 114 habitations which are not in existence now.

** In case of U.P, habitations include 125 habitations merged in urban areas.

Author's Source: Economic Survey 2002-03, Government of India.

Besides meeting rural drinking water requirement, the scheme is considered as an important step towards decentralization of the power in favour of the local bodies. Under the scheme, the Gram Panchayats/ Villages/Communities can plan, implement, and operate maintain and manage the drinking water scheme. The project proposals involve mini pipe water supply, bore-well, water harvesting and regeneration of water bodies. An amount of Rs. 500 crore is expected to be spent on the scheme this year (2002-03). The Ministry has been mandated to make provision of safe drinking water to all habitations by 2004. Under Accelerated Rural Water Supply Programme (ARWSP), the budgetary support has been increased from Rs. 1,960 crore in 2000-01 to Rs. 2,235 crore in 2002-03, which is more than 30 per cent increase over 1998-99. Out of the total number of 14.22 lakh rural habitations in the country, almost 12,84,555 rural habitations are fully covered with drinking water facility, about 1,23,014 habitations are partially covered and only about 15,095 habitations are not covered.

A Review of the Rural Water Supply Progrmmes

It has been acknowledged that the task of providing safe drinking water to all is a difficult one. We have already mentioned about the plethora of policies and progrmmes, which have sought to accelerate the pace of coverage of drinking water supply in India's villages. A significant investment has been made by the central as well as state governments, which has led to the creation of about 37 lakh hand pumps and 1.45 lakh piped water supply schemes. This has enabled to provide at least 40 litres of water per capita per day in 12.80 lakh rural habitations. There is to be a water source for every 250 persons; one within 1.6 km. of a settlement in the plains and within 100 mm elevation in the hills. Yet, available reports based on micro studies indicate persistent paucity of pure drinking water in many villages in different parts of the country.

A recent project undertaken by Tata Energy Research Institute (TERI) with United Nations Population Fund (UNFPA) appears to indicate that while these programmes have made a dent, they have not truly delivered. Even where water schemes have been put in place, various factors deter any significant improvement in the quality of lives of the people (TERI 2002-Impact of Population Growth on Water and Quality of Life). The study entailed micro studies in 20 villages in four districts—Solan (Himachal Prdesh), Thiruvananthapuram

(Kerala), Raichur (Karnataka) and Bikaner (Rajasthan). The study identified various factors for inter-region and inter village variations in the nature and extent of water stress. These include:

Natural Factors: For instance, arid and semi-arid conditions in Bikaner and Raichur; hilly terrain in solan which makes the chore of water collection more arduous; proximity to the coast in Thiruvananthpuram which causes salt water incursion in wells-all exacerbate water-related concerns.

Demographic Force: Of the demographic forces, population density of settlements was an important factor affecting not only per capita water availability, but also the quality of water at source because drinking water sources were being used for washing of livestock as well in some cases.

Socio-Economic Factors: Poverty is evidently an aggravator of water related stress, given the inability of poor households to invest in water collection, storage or purification. Further, caste continues to play a role in determining access to water in some villages, as the study found in Bikaner.

Problems with Water Supply Scheme: The study observed that the existing water supply had various constraints: (i) few sources relative to the population served; (ii) water availability restricted by inadequate and erratic power supply in the villages; and (iii) poor maintenance—broken taps, cracks in tanks, and unclean storage tanks commonly found.

Expectations from Swajaldhara

The question is raised whether the well-intentioned and full of promises, will Swajaldhara be able to deliver what the earlier schemes could not? Swajaldhara is stated to have the following key elements: (i) demand-driven and community participation approach; (ii) panchayats/communities to plan, implement, operate, maintain and manage all drinking water schemes; (iii) partial capital cost sharing by the communities upfront in cash; (iv) full ownership of drinking water assets with gram panchayats; and (v) full operation and maintenance (O & M) by the users/panchayats. Undoubtedly, the involvement of the local communities in all aspects of the water schemes including cost-sharing is an important step towards the impending water crisis in India. Nevertheless, the scale and complexity of the problem calls for tripartite ventures involving the government, NGOs and the local villagers. Further, given the limited resource, it may be worthwhile for each state also to identify priority areas for the provision of water for various uses—drinking, other household activities and for irrigation. Several important issues related to water supply must be addressed for ensuring that the Swajaldhara initiative improves the quality of life for those most affected by water stress, viz. poor households, women and children. This requires either an enhancement of the scope of Swajaldhara or simultaneous implementation of other programmes as complementary activities.

Summing up, it has been universally observed that when there is an improvement in the access to water there is a decline in the perceived value of water, leading to wastage and inefficient use. Therefore, awareness creation about the value of the resource and the need for conservation should lie at the core of water initiatives like Swajaldhara. It needs to be borne in mind that the supply of water is by and large limited by natural availability; technologies for harnessing, recycling and re-using can enhance supplies, but only to a limited extent. Therefore, the crux of water management lies in managing demand and it is, therefore, essential to integrate Swajaldhara as a part of the rural development planning process and customize the initiative to local needs and priorities.

Other Initiatives—Examples of a Few States

Scheme to Provide Potable Water in Kitchens in Rohtak District (Haryana)

The government of Haryana has launched an ambitious project also called "swajal dhara" to provide potable piped water in the kitchens of rural population. This canal-based water supply project will cost approximately Rs. 25 lakh, 10 per cent of which would be contributed by the villagers. The scheme has been launched in 82 out of 147 villages in Rohtak district. Thus over 54 per cent of the rural population in the district would soon get potable water in their kitchens. The project is being executed and maintained by the village panchayats, which would also take care of its launching. The villages have been advised to install boosters so that the houses at the tail end of the scheme also get piped water. Priority has been given to those villages that have been facing water scarcity for the past many decades, nay centuries. The underground water in most parts of the district is brackish and unfit for human consumption. The womenfolk in these villages have to fetch water from far off places to meet the daily needs of water. After the implementation of the scheme, these women will benefit the most.

The NCR Development Board has approved the project submitted to it for augmenting water supply in Rohtak town. The project costing Rs. 1,123.48 lakh is expected to be completed in four years. Following the completion of this scheme, water supply in the town would increase by 30 lakh gallons per day. Water supply in the town is canal based. At present, the residents are being supplied 110 litres per head per day water as against the norms of 150 litres per capita per day. A campaign has also been launched to disconnect unauthorized water supply connections. The public health authorities, who manage the water supply in the town and the rural areas have been directed to intensify the campaign against illegal water users.

A People-Oriented Rural Water Project in Kerala

A people-oriented, decentralized, demand driven scheme for water supply where the onus of operation and maintenance rests with the users, the Kerala Rural Water Supply and Sanitation Agency (KRWSSA) has designed its own Jalnidhi project. The World Bank has chipped in with Rs. 300 crore. About 30-40 schemes are being implemented with capital expenses shared by the government of Kerala, the Gram Panchayat (GP), and the beneficiary group. The ratio being 75:10:15 for water supply schemes and 70:0:30 for ground water recharge programmes. Support organizations or NGOs are helping to monitor and implement. The schemes include installation of combined rainwater harvesting and pumping systems, 1,000—litre ferro cement tanks and rainwater collecting equipment for each household and private connections at homes.

The first phase (pilot, started in 1999) of the four district project covering Palakkad, Trissur, Malappuram and Kozhikode has been completed in Erimayur, Pookkottur, Kunnumel and Nediyiruppu. While these GPs were selected by the KRWSSA based on scarcity of potable water, latrine coverage, presence of marginalized capacity levels, applications are being invited for the next phase. By 2006, 100 GPs will be covered. Kerala has over 900 gram panchayats. Each beneficiary houshold including SCs has contributed 15 per cent of the cost, the figure ranges between Rs. 1,000—Rs. 2,500 depending on the features of the scheme. Those who could not afford 15 per cent in cash, contributed the balance through labour.

Important is the absence of contractors in the project. Support for this community-driven project is from the women. They have done the plumbing, constructed rainwater tanks, handled money as treasures and guided schemes to completion as secretaries. It is much more dignified to take on these responsibilities than to walk long distance for a pot of water. Some works that they could not handle was done through labour contracts.

Rural Water Supply: Simple Technology, Big benefits for the poor—A World Bank Project in Tamil Nadu

"We like handpumps because they give us safe, clean water. They also free us of the worry of power cuts and contamination that affect motorized pumpsets and open wells". So reports a member of punchayat, or local governing body of Thiruvalanakadu, one of the villages in the northern part of the Indian state of Tamil Nadu benefiting from the rural water supply component of the World Bank-assisted Tamil Nadu Agricultural Development project.

Located in southern India, the state of Tamil Nadu has a population of about 58 million, or just over 6 percent of India's total. Although Tamil Nadu is one of India's most industrialized states, its economy depends largely on agriculture. About 60 per cent of its population is rural. Its gross domestic product (GDP) per capita lies just below the all-India average of US$333, and about 37 per cent of the state's population remains below the poverty line. Tamil Nadu has achieved dramatic improvements in social indicators in recent years, and its government is committed to achieving even greater improvements in the health of the state's population by increasing the effectiveness of health services. In this regard, it is attempting to provide universal year-round access to safe drinking water because it recognizes that limited or no access to safe water increases the risks of serious waterborne diseases such as cholera, typhoid fever, dysentery, and jaundice, as well as exposure to vector-borne diseases such as malaria and elephantiasis.

A 1995 study suggests that less than 50 per cent of the state's rural inhabitants has access to adequate safe drinking water throughout the year. Facing severe water scarcity problems and rising demands for water from agricultural, industrial, and domestic users, Tamil Nadu finds that supplying safe, protected drinking water is becoming ever more difficult. However, the state puts a high priority on doing so, particularly in rural areas. The Tamil Nadu Agricultural Development Project is helping in this effort by supporting a part of the state's broader program in rural water supply. In line with the state's commitment to fighting rural poverty, the project is emphasizing the provision of safe water to the poorest segments of rural society-the scheduled castes and tribes, which make up about 20 per cent of Tamil Nadu's population and represent the largest part of its agricultural work force. Concentrated in rural areas, Tamil Nadu's poor often have to rely on contaminated rivers or open wells for meager water supplies. In cases of extreme scarcity, the state is currently supplying water by tank trucks to poorer villages at very high cost.

Safe Water—and More

Started in 1991, the Tamil Nadu Agricultural Development Project aims at strengthening state-level institutions to provide more cost-effective rural-based services, thus raising agricultural production and rural incomes and enhancing the quality of life for the rural poor. The project supports selected investments in agricultural and rural development within the state's current programs, but in the process it is introducing a number of important innovations in technology, public administration, and policy. Besides improving the rural water supply, the project is encouraging more effective farmer extension services, seed production, integrated watershed management, livestock development, forestry, rural road infrastructure, and state-level agricultural planning. The US$ 133 million project is being supported with an International Development Association (IDA) credit equivalent to US$92.8 million and a World Bank loan of US$20 million to supplement financing from the government of Tamil Nadu and beneficiaries. The rural water supply component represents about 17 per cent of total costs. The project is a broadly based initiative-the first of its kind in India-and aligns well with the Bank's goals of assisting India to accelerate agricultural growth, reduce rural poverty, and improve the health of its rural population.

The project is making possible the installation of 12,500 simple borewells operated by handpumps in rural areas of seven relatively arid districts. In the past, state policy was to install one handpump per cluster of 500 persons in rural areas, but this policy often proved arbitrary and missed many of the poorest persons in need. In the project, the clustering criterion has been relaxed, and need-based on current access to a reliable supply of safe water-is given more importance. As a result, scheduled castes and tribes have benefited particularly because they typically live in separate hamlets within village communities.

How the Rural Water Supply Component Works

Under the project, a local panchayat is responsible for identifying those areas in its village that lack adequate year-round access to safe water. The district administration then consolidates this information from all panchayats under its jurisdiction and advises the state's Department of Rural Development. The department dispatches a survey team to a village to determine the likely soil depth at which water will be struck, the soil profile, and the likely water output. This team surveys common land around a household cluster to select the best location for the borewell and to ensure access for all intended beneficiaries.

The department packages a number of borewell projects together and requests bids for their installation on a turnkey basis. In the past, the government supplied all the necessary equipment to contractors. In the project, in contrast, contractors supply all needed equipment according to the department's specifications. Each contract includes installation of the borewell pipe and handpump and construction of a simple concrete platform and soak pit. Newly introduced into Tamil Nadu's rural water supply program, soak pits prevent stagnation of spilled water-an extra health safeguard. The borewell and handpump represent simple, easily maintained, internationally recognized technology available in India. Units can take from one day to three weeks to install, depending on whether they are located in relatively easy-to-work hard rock or more difficult mud, sand, or loose stone. Complete costs range from about US$1,000 for units in hard-rock areas to US$2,500 for those in sedimentary areas. Borewells-averaging about 100 meters deep-ensure year-round supply.

After the unit is installed, the department tests its water for contaminants. So far, less than 3 per cent of the project's borewells have yielded contaminated water. Once certified, the handpump is turned over to the local panchayat. Beneficiaries select a women volunteer from their group to protect the pipe, clean the surrounding areas, and do minor repairs and maintenance. For this she receives training from the Department of Rural Development. For major repairs, a mechanic is dispatched from a regional panchayat union office. The woman volunteer also organizes public health campaigns for pump beneficiaries. Health care workers from the panchayat union office come to these meetings and provide information to local residents on the safe use of water, the need to avoid water stagnation around the pump, and the importance of hygiene in its vicinity.

Outcomes and Reasons for Success

At the project's completion in 1998, 3.5 million persons are expected to be benefiting from project-financed handpumps, 30 per cent of whom will be from scheduled castes and tribes. The project is helping beneficiaries to get safe, dependable water close to their doorsteps, thereby reducing the drudgery of hauling water from open wells and rivers. Village women report that it takes them half the time formerly needed for getting daily water supplies for their households, providing more time for more productive activities such as day labor or childcare. Beneficiaries also tell of dramatic reduction in water-related diseases once the pumps go into operation. In addition, simple technology means that the pumps can be easily operated and maintained by local villagers.

These are the key factors in the project's success:

- Strong state government support
- Good interdepartmental coordination by the project's monitoring and coordination cell
- A keen sense of ownership and commitment from each department for its part of the project. The project has effectively involved all the departments within the state's Ministry of Agriculture
- Effective funding arrangements.
- Proactive supervision by IDA using persons familiar with the sector and the state's government.

The rural water supply component represents an effective marriage of initiatives from government departments and beneficiaries: locally elected governing bodies identify groups of villagers who need improved water supplies, and these groups take ownership of and maintain the completed handpumps. The government departments provide key technical expertise to locate the best sites, supervise construction, certify the water as satisfactory for drinking purposes, and provide spare parts.

Tenth Plan Strategy

The Tenth Five Year Plan envisages the provision of potable drinking water to every settlement in the country on a sustainable basis and the pursuit of all possible measures for the rapid expansion and improvement of sanitation facilities in rural and urban areas.

Conclusion

The water problem, particularly in respect of potable water in the rural areas has remained unsolved despite more than five decades of sustained efforts. Though the Ministry of Rural Development has been mandated to provide safe drinking water in all rural households by 2004 and to achieve this objective, many programmes like Accelerated Rural Water Supply Programme (ARWSP) and Prime Minister's Gramodaya Yojana—Rural Drinking Water (PMGY-RDW) are being implemented, from the progress so far, it appears that resolving safe drinking water crisis in all rural habitations may continue to remain a far cry. The guaranteed sustainable supply of water in the rural areas, the quality of which is also good enough to meet requirements of people, is an enormous challenge to the government. The rural water problems include among other things, the rising costs of tapping new water sources, the wasteful use of those already made available, soil degradation in irrigated areas, the overuse or exhaustion of groundwater resources, the release of harmful substances (pesticides, fertilizers, human and animal faeces etc.) into the water with negative effects on the health of water users and massive subsidies for providing water, which militates against the sustainable use of water resources.

There are in essence two strategies which need to be adopted to meet these challenges; supply management and demand management. While supply management is no doubt a difficult task in the water scarce areas, with economic growth, increasing competition for water resources and the increase in the value of water, demand management is certainly becoming more important than supply management. The recently introduced Swajaldhara programme is an attempt in the direction of both supply and demand managements. The community participation approach in implementation, operation, maintenance and management of all drinking water schemes may go a long way in achieving the target of providing drinking water to rural inhabitants, if not by 2004, within a foreseeable future.

REFERENCES

1. Global Water Outlook to 2025: Averting an Impending Crisis, a new food policy report from IFPRI.
2. World Bank
3. Population Report, Series M, Number 14, February 1999,
4. Stephen Turner, Deputy Director of Water Aid, a London-based nongovernmental organization (NGO)
5. Krishnan, Rekha: Swajaldhara: More Empty Promises? *Economic & Political Weekly*, March 8, 2003.

8

Urban Water Problem

"The lack of fresh water and sanitation is the greatest obstacle to sustainable development and the most visible symbol of the growing gap between the rich and the poor"

—UN Report

Introduction

A long-standing belief in development studies holds that, on the whole, living conditions in developing countries are superior for residents of large cities than for persons living in smaller cities, towns and villages. The concept of big cities as islands of privilege is fundamental to otherwise discrepant theories of modernization, dependency, world systems of cities, and the global division of labour, each of which posits long-lasting disadvantages for populations outside the major urban centers.

Even as late as 1990, nearly two-thirds of the population of the developing world were living in villages, and just about one-third in towns and cities. In other words, the degree of urbanization—the percentage of population living in urban areas—is quite low in the developing world compared to that in the developed world. However, it is said that nearly all population growth worldwide will take place in urban areas of developing countries. By 2015, the United Nations (UN) projects, there will be 21 "megacities" of at least 10 million people each—all but 4 in developing countries[1].

In India, however, the proportion of population in urban areas is still lower than the overall position of all developing countries. By 1991, India had just about one-fourth of its population as urban. While the level of urbanization in the developing countries is low, the urban patterns in India are marked by high and increasing concentration of population in big metropolitan cities. In India, nearly one-fourth of its urban population lived in its metropolitan cities in 1990, and this percentage has recorded a steep increase over the years. This process of increasing concentration of urban population in big towns is largely due to very rapid growth of all urban areas in India, a pattern that holds good for other developing countries too. From 1970 to 1990, the urban growth rate in the country was three times as high as that in developed regions of the world.

Worldwide, about three-fourths of all current population growth is urban. Cities are gaining an estimated 55 million people per year. Rapid urban growth in the developing world reflects migration of people to cities (over 1 million of new residents every week) as well as natural population increase among urban residents. Rural areas have virtually stopped gaining population. There are at least two measures of urban growth—the urban/rural growth differential (URGD) and the rate of urbanization—that clearly indicate this. URGD is the difference between the growth rate of urban population and rural population while rate of urbanization is the rate of growth of degree of urbanization, that is, the rate at which the proportion of population living in urban areas increases. Both these measures for India, show an increase in 1970-90 compared to 1950-70, suggesting that migration from rural to urban areas continued to play a significant role in the country's urban process[2].

While there are substantial differences in the reasons behind and characteristics of urban growth, overall in developing countries rapid urban population growth reflects three basic factors: (i) migration from rural areas and from other urban areas, (ii) natural population increase (births minus deaths) among urban residents; and (iii) reclassification of previously rural areas as urban as they become built up and change character. During the initial phases of urbanization in accounting, migration from rural to urban areas tends to play a greater role than natural population increase in urban areas. As a greater share of the total population lives in cities, however, natural population increase within them surpasses migration in importance. As natural population increase slows, migration can once again play dominant role in urban population growth—for example, if economic opportunities in urban areas expand rapidly while those in rural areas do not.

The rapid growth of cities in the developing world including India puts them in the forefront of the struggle for improved living standards and protection of the environment. Urbanization dramatically increases per capita freshwater use, as millions of households gain access to piped water, as industry increases, and as large-scale irrigated agriculture replaces subsistence farming. Caught between growing demand for freshwater on one hand and limited and increasingly polluted supplies on the other, many countries face difficult choices. Finding solutions requires responses at local, national, and international levels-a 'Blue Revolution' that focuses on integrated management of watersheds and shared water basins[3].

The reason for India's growing thirst is the rate at which its population is increasing, more so in the cities and towns. While it may be one of the reasons, water crisis in the urban areas of the country is also the culmination of myopic planning, muddled policies and misguided perceptions. As cities and towns sprouted, no thought was paid to emerging mismatch between demand and supply. In the quest for food security, groundwater was pushed as a solution-it was cheaper and quicker—while storage and distribution projects were neglected. The race for industrialization ignored checks on wasteful technology and pollution of water resources. So the industry is not obliged to re-use water and continues to be the biggest polluter alongwith pesticides/fertilizer-ridden discharge from fields. Further, as India developed and urbanized at a rapid pace, traditional systems of managing water resources were dismantled and thrown by the wayside.

Water is the biggest crisis facing India in terms of spread and severely affecting one in three people. Even in Chennai, Bangalore and Delhi water is being rationed and India's food security is under threat. A greater need to mitigate urban water scarcity is to rope in consumers to conservation beyond saving to augmentation. As dependence on groundwater increases in virtually every town and city, there is an urgent need to institute changes on use of groundwater and recharge groundwater via roof/rainwater harvesting at individual building/house level. While cities like Delhi, Chennai, Bangalore and Hyderabad have scripted this requirement for new building and houses, it needs to be made mandatory all over the country, if necessary, by stringent fiat.

Defining "Urban"

What is urban? What is a city? The terms "city" and "Urban" are often used interchangeably, and there is no international agreement on their definition. Almost all national governments agree that settlement of 20,000 or more people are urban, but some consider smaller settlements to be urban as well, with various cut-off points. Few, however, would consider a small urban center of 1,000 to 2,000 inhabitants to be a city[4]. Thus, while the term "urban" can refer to settlements of all population sizes, most people reserve the term "city" for urban centers with large populations.

Urbanization statistics depend to a certain extent on how countries define urban settlements, especially countries with large populations, such as China and India. For example, most of India's rural population lives in villages with 500 to 5,000 inhabitants. If the government of India classified settlements of this size as urban, as some countries do, instead of using the cutoff of 5,000 or more inhabitants, India would have a predominantly urban population[5].

Most governments define urban settlements based on one or a combination of criteria, including population size, population density, and social and economic factors, such as the proportion of the labor force engaged in nonagricultural activities; the administrative or political status of a locality, such as national, provincial, or district capitals; or census designations. In the UN World Urbanization Prospects 1996 Revision, for example, 46 per cent of the represented countries defined "urban" based on administrative criteria; 22 per cent used population size and sometimes population density; 17 per cent used other criteria; 10 per cent had no definition; and 4 per cent defined their country as either entirely urban or entirely rural[6].

In the Indian context, there are three basic features that distinguish an urban settlement (namely, a town) from a rural settlement (namely, a village): the size of the settlement in terms of its population; the density of population in the settlement; and the nature of workforce in the settlement. A town, compared to a village, in general, is relatively larger in size, is more crowded, and has a greater percentage of its workers in non-agricultural activities. The Indian Census recognizes these important dimensions and defines an urban place as:

(i) any place with a municipality, corporation, or cantonment, or notified town area; or

(ii) any other place which satisfies the following criteria;

 (a) a minimum population of 5,000

(b) a population density of at least 400 per sq. km
(c) at least 75 per cent of the male working population in the non-agricultural sector.

Nature of Urbanization in India

Before dwelling on the issue of urban water problem, it is necessary to analyze the nature of urbanization and the pattern of urban population growth. In India, nearly one-fourth of its urban population lived in its metropolitan cities in 1990, and this percentage has recorded a sharp increase over the years. This process of increasing concentration of urban population in big towns is largely due to very rapid growth of all urban areas in India, a pattern that holds also for other developing nations. In fact, the main features of urban growth pattern in the country, as in the case of other developing countries, relate to the low level of degree of urbanization combined with a relatively high rate of urban growth and a high and increasing concentration of urban population in large cities. This pattern of urbanization has important implications for adequate fresh water supply in urban areas and therefore, it has been felt useful to briefly discuss this link.

However, while above indicates only the overall pattern of urbanization in India, it is essential to consider the variations in the pattern of urbanization across the length and breadth of the country. Given that the process of development is uneven across space, the patterns of urbanization are also not uniform. The uneven development of the process of urbanization is clearly evident in Table 8.1. While the overall level of urbanization in the country was about 28 per cent in 2001, there is wide variation across states.

On the basis of the variation with regard to the major aspects of urbanization, the three distinctly different patterns of urbanization, across the states and Union Territories of India have been identified and are shown in Table 8.2. States that are grouped in Pattern—1 exhibit relatively high levels of urbanization, better spread of towns, and a higher extent of concentration of urban population in large cities. The rate of growth of urban population is also quite high and the urban/rural growth differential is much higher than the average rate for the country as a whole. This indicates that rural-urban migration plays a significant role in the urban process here. Pattern 2 exhibits low levels

Table 8.1: Salient Features of Urbanization Across the States and Union Territories, 1991 and 2001

Sl. No.	*State/ Union Territory*	*Degree of Urbanization*		*Town Density*		*Percentage of Population in Large Cities*		*Rate of Growth of Urban Population*	*URGD*
		1991	*2001*	*1991*	*2001*	*1991*	*2001*		
1	Andhra Pradesh	26.89	27.08	0.96	0.76	40.96	50.21	1.37	0.09
2.	Arunachal Pradesh	12.80	20.41	0.12	0.20	0.00	0.00	7.25	5.82
3.	Assam	11.10	12.72	1.19	1.59	23.49	24.67	3.14	1.59
4.	Bihar	13.14	13.36	1.56	1.45	31.48	33.93	2.59	0.19
5.	Goa	41.01	49.77	8.37	11.89	0.00	0.00	3.38	3.60
6.	Gujarat	34.49	37.35	1.35	1.23	48.37	53.54	2.87	1.27
7.	Haryana	24.63	29.00	2.13	2.40	20.56	37.92	4.19	2.30
8.	Himachal pradesh	8.69	9.79	1.04	1.02	0.00	0.00	2.85	1.34
9.	Jammu & Kashmir	23.83	24.88	N.A.	0.74	N.A.	57.50	3.14	0.59
10.	Karnataka	30.92	33.98	1.60	1.41	43.51	52.18	2.57	1.42
11.	Kerala	26.39	25.97	5.07	4.09	28.49	33.08	8.84	3.86
12.	Madhya Pradesh	23.18	24.98	1.05	1.10	33.04	40.84	2.83	1.01
13.	Maharashtra	38.69	42.40	1.09	1.23	69.81	74.03	2.99	1.57
14.	Manipur	27.52	23.88	1.39	1.48	0.00	38.56	1.21	–1.95
15.	Meghalaya	18.60	19.63	0.54	0.71	0.00	0.00	3.21	0.69
16.	Mizoram	46.10	49.50	1.04	1.04	0.00	52.08	3.33	1.40
17.	Nagaland	17.21	17.74	0.54	0.54	0.00	0.00	5.42	0.38
18.	Orissa	13.38	14.97	0.80	0.89	29.72	35.32	2.64	1.34
19.	Punjab	29.55	33.95	2.38	3.12	41.96	44.32	3.24	2.08
20.	Rajasthan	22.88	23.38	0.65	0.65	39.73	47.33	2.75	0.29

(Contd.)

Table 8.1 (*Contd.*)

Sl. No.	*State/ Union Territory*	*Degree of Urbanization*		*Town Density*		*Percentage of Population in Large Cities*		*Rate of Growth of Urban Population*	*URGD*
		1991	*2001*	*1991*	*2001*	*1991*	*2001*		
21.	Sikkim	9.10	11.1	01.13	1.27	0.00	0.00	4.95	2.29
22.	Tamil Nadu	34.15	43.86	3.61	6.40	36.72	35.40	3.63	4.16
23.	Tripura	15.30	17.02	1.72	2.19	0.00	0.00	2.56	1.30
24.	Uttar Pradesh	19.84	21.02	2.56	2.68	41.74	46.45	2.88	0.74
25.	West Bengal	27.48	28.03	4.30	4.23	41.78	52.33	1.86	0.28
1.	Andaman & Nicobar	26.71	32.67	0.12	0.36	0.00	0.00	4.50	2.95
2.	Chandigarh	89.69	89.78	43.86	8.77	88.67	100.00	3.46	0.09
3.	Dadra & Nagar Haveli	8.47	22.89	2.04	4.07	0.00	0.00	15.71	12.73
4.	Daman & Diu	46.80	36.26	17.86	17.86	0.00	0.00	1.89	-4.54
5.	Delhi	89.93	93.01	21.58	41.81	88.63	78.88	4.23	4.08
6.	Lakshadweep	56.31	44.47	125.00	93.75	0.00	0.00	-0.77	-4.83
7.	Pondicherry	64.00	66.57	22.36	12.20	39.28	74.51	2.29	1.15
	INDIA	25.71	27.78	1.51	1.63	44.57	50.13	2.75	1.08

Notes:

1. Figures for Jammu & Kashmir for the year 1991 are on the basis of estimated population
2. Figures for the new states of Jharkhand, Chhattisgarh, and Uttaranchal for 2001 have been added to their original states, to facilitate comparison over time.
3. Town density is number of towns per 1,000 sq. km.
4. Large cities are cities with a population size of 2,00,000 and above
5. N.A = Not available.

Source: Census of India 1991 (a); Census of India 1991 (b); Census of India 2001 (a); Census of India 2001 (b)

Table 8.2: Patterns of Urbanization Across the States and Union Territories, 1991 and 2001

Urban Pattern	*State/ Union Territory*	*Degree of urbanization*		*Town density*		*Percentage of population in large cities*		*Rate of growth-urban population*	*URGD*
		1991	*2001*	*1991*	*2001*	*1991*	*2001*		
Pattern 1	Maharashtra	38.69	42.40	1.09	1.23	69.81	74.03	2.99	1.57
	Gujarat	34.49	37.35	1.35	1.23	48.37	53.54	2.87	1.27
	Tamil Nadu	34.15	43.86	3.61	6.40	36.72	35.40	3.63	4.16
	Karnataka	30.92	33.98	1.60	1.41	43.51	52.18	2.57	1.42
	Punjab	29.55	33.95	2.38	3.12	41.96	44.32	3.24	2.08
	West Bengal	27.48	28.03	4.30	4.23	41.78	52.33	1.86	0.28
	Andhra Pradesh	26.89	27.08	0.96	0.76	40.96	50.21	1.37	0.09
	Kerala	26.39	25.97	5.07	4.09	28.49	33.08	8.84	3.86
	Haryana	24.63	29.00	2.13	2.40	20.56	37.92	4.19	2.30
	Delhi	89.93	93.01	21.58	41.81	88.63	78.88	4.23	4.08
	Chandigarh	89.69	89.78	43.86	8.77	88.67	100.00	3.46	0.09
	Pondicherry	64.00	66.57	22.36	12.20	39.28	74.51	2.29	1.15
	Lakshadweep	56.31	44.47	125.00	93.75	0.00	0.00	-0.77	-4.83
	Daman & Diu	46.80	36.26	17.86	17.86	0.00	0.00	1.89	-4.54
	Goa	41.01	49.77	8.37	11.89	0.00	0.00	3.38	3.60
	Andaman & Nicobar	26.71	32.67	0.12	0.36	0.00	0.00	4.50	2.95
	Dadra & Nagar Haveli	8.47	22.89	2.04	4.07	0.00	0.00	15.71	12.73
	Pattern 1	**32.78**	**36.23**	**1.88**	**2.13**	**49.30**	**54.83**	**2.71**	**1.56**
Pattern 2	Madhya Pradesh	23.18	24.98	1.05	1.10	33.04	40.84	2.83	1.01
	Rajasthan	22.88	23.38	0.65	0.65	39.73	47.33	2.75	0.29
	Uttar Pradesh	19.84	21.02	2.56	2.68	41.74	46.45	2.88	0.74
	Orissa	13.38	14.97	0.80	0.89	29.72	35.32	2.64	1.34

(Contd.)

Table 8.2 (*Contd.*)

Urban Pattern	*State/ Union Territory*	*Degree of urbanization*		*Town density*		*Percentage of population in large cities*		*Rate of growth-urban population*	*URGD*
		1991	*2001*	*1991*	*2001*	*1991*	*2001*		
	Bihar	13.14	13.36	1.56	1.45	31.48	33.93	2.59	0.19
	Himachal Pradesh	8.69	9.79	1.04	1.02	0.00	0.00	2.85	1.34
	Jammu & Kashmir	23.83	24.88	N.A.	0.74	N.A.	57.50	3.14	0.59
	Pattern 2	**18.64**	**19.68**	**1.29**	**1.31**	**36.81**	**42.71**	**2.80**	**0.68**
Pattern 3	Mizoram	46.10	49.50	1.04	1.04	0.00	52.08	3.33	1.40
	Manipur	27.52	23.88	1.39	1.48	0.00	38.56	1.21	–1.95
	Meghalaya	18.60	19.63	0.54	0.71	0.00	0.00	3.21	0.69
	Nagaland	17.21	17.74	0.54	0.54	0.00	0.00	5.42	0.38
	Tripura	15.30	17.02	1.72	2.19	0.00	0.00	2.56	1.30
	Arunachal Pradesh	12.80	20.41	0.12	0.20	0.00	0.00	7.25	5.82
	Assam	11.10	12.72	1.19	1.59	23.49	24.67	3.14	1.59
	Sikkim	9.10	11.10	1.13	1.27	0.00	0.00	4.95	2.29
	Pattern 3	**13.83**	**15.45**	**0.77**	**0.97**	**13.22**	**21.23**	**3.16**	**1.33**
INDIA		25.71	27.78	1.51	1.63	44.57	50.13	2.75	1.08

Notes:

1. Figures for Jammu & Kashmir for the year 1991 are on the basis of estimated population
2. Figures for the new states of Jharkhand, Chhattisgarh, and Uttaranchal for 2001 have been added to their original states, to facilitate comparison over time.
3. Town density is number of towns per 1,000 sq. km.
4. Large cities are cities with a population size of 2,00,000 and above
5. N.A = Not available.

Source: Same as Table 8.1

of urbanization combined with a relatively lower spread of urban settlements. Concentration of urban population was at moderate levels in 1991 but increased rapidly over the decade and was quite high by 2001. In Pattern 2, while urban areas have grown rapidly, the difference between urban and rural areas in terms of population growth is quite low. This suggests that here urban growth is essentially related to the natural growth of urban population and rural-urban migration does not seem to be an important contributory factor to urban growth. Pattern 3, essentially the northeastern States along with Sikkim, shows very low levels of urbanization, extremely low levels of urban concentration, and a low level of urban spread. While Pattern 3 is similar to Pattern 2 with regard to levels or urbanization, the crucial differences between the two is with regard to the absence of metropolitan cities as well as a much lower town density in the latter.

We would like to note that the patterns we have identified are to be taken as broad patterns, reflecting the broad contours of urbanization. Within each broad pattern we can identify some States that deviate from the rest in one way or the other. For instance, it is well known that Kerala's urban pattern, which is an urban rural continuum, is quite different from that exhibited by any other State in the country. Even though Kerala does not have a high level of urbanization or very high concentration of urban population, we think it fits in better with the States of Pattern 1 than with the other patterns. Similarly, the Union Territories in Pattern 1, with the exception of Dadra and Nagar Havelli, have a high level of urbanization but with regard to other aspects of urbanization, such as urban concentration and urban-rural growth differential, they do not quite exhibit the same pattern. As the Union Territories are geographically contiguous to the major States that exhibit Pattern 1 and as the urban population of all Union Territories account for only about 5 per cent of the country's urban population and since they have high levels of urbanization, these can be grouped in Pattern 1. Similarly within Pattern 2 all the States do not exhibit a homogenous urban pattern. Uttar Pradesh has a better urban spread compared to the other States in this group but it is similar to the other States with regard to the level of urbanization and URGD. Himachal Pradesh in Pattern 2 and Mizoram and Manipur in Pattern 3 may be considered as outliers but has been given weightage to geographical contiguity.

Dimension of Urban Water Problem

Urbanization and rapid growth in urban population can dramatically increase per capita use of freshwater. Fast population growth with accelerated urbanization, combined with scarce water supplies means that the governments all over the world often cannot supply enough water to meet demand. The number of urban residents without access to improved water sources rose from 113 million in 1990 (5 per cent of the total urban population to 173 million in 2000 (6% of the total urban population), according to a study by WHO and UNICEF (Table 8.3)[7]

Table 8.3: Access to improved* Water in Urban Areas by Region, 1990 and 2000

	Population Without Access (in million)		*Population with Access (in million)*		*% of Population with Access +*	
	1990	*2000*	*1990*	*2000*	*1990*	*2000*
World	113	173	2,149	2,672	95	94
Africa	31	44	166	253	84	85
Asia	57	98	972	1,254	94	93
Latin America & Caribbean	26	29	287	362	92	93
Ocenia	0	0	18	21	100	98
Europe	0	3	522	542	100	100
North America	0	0	213	239	100	100

Notes:

* The WHO/UNICEF Joint Monitoring Programme for Water Supply and Sanitation defines "Improved" sources as those that are better than previous sources, although not necessarily safe for household use.

\+ Due to rounding, figures may not total 100 per cent even if the population without access is shown as O.

Source: WHO/UNICEF Joint Monitoring Programme for Water Supply and Sanitation, 2001,

Author's Source: Population Reports, Series M, Number 16 May-June 2003.

Water is clearly the single biggest crisis facing urban India and the stress is very much revealing. In fact, from being a necessity, water has now become a luxury in the cities and metropolis. With pipes running empty, our cities are turning into villages serviced by tankers.

We have mentioned earlier that as India developed and urbanized at rapid pace, traditional systems of managing water resources were dismantled and abandoned. With more tankers muscling their way into urban India, the situation is set to deteriorate. The Ministry of Water Resources estimates that the demand for water in the top 35 cities is expected to double to 12,906 million cubic metres as population of these cities shoots up from 107 million 2001 to 202 million in 2021, while availability remains stagnant at 2001 level.

In this context, it may be pointed out that in 1970s and 1980's, the government had focused only on demand—getting pipes into homes—and though city councils are setting up new treatment plants and laying pipelines, the crux of the problem, supply has not been properly addressed. Most cities are not based on river banks, and the rapid pace of urbanization has led to the drying up of traditional water sources like tanks and lakes. In Bangalore (Karnataka), for instance, 181 of the 262 tanks have dried up and the biggest tank has been converted into the Kanteerava Stadium. Similarly, the Hauz Khas tank in South Delhi built by Alauddin Khilji and resurrected by the third Tughlak, Firoz Shah Tughlak has practically disappeared. Today, while there is little water, the lake's catchment area is still very evident. The catchment region has been replaced by the kikar and other trees and sizable part of the area has been paved thus making it impossible for the rainwater to seep through strata to refill the ground water.

The first sign of population boom and water stress were visible in the 1980s, but most municipalities and corporations focused on the immediate tapping of groundwater resources in and around the city. For instance, as of now, Aiyer[8] of India Today points out that the Delhi Jal Board depends on its 1,250 tubewells, while Chennai outsources water from wells of 250 farmers on the outskirts of the city and plans to lease water from 250 more. The people too have joined the rush and bore wells for water. However, except in Chandigarh, there is no levy on drawing water and little control over boring wells. Expectedly, the pressure on ground water has shown up. Tubewells are now routinely dug at a depth of over 200 m in South Delhi, while in and around Chennai, water is being mined and pumps are being sunk three to six metres deeper every year.

Aiyer further mentions that desperate, many cities are now being forced to get water from over 100 km. away. Delhi's irony is unmistakable. Despite being on the banks of the Yamuna, it depends on the Ganga water from Haryana and Uttaranchal. The water supplied to Bangalore—across 100km. from the Cauvery and lifted over 1,000m gradient—is perhaps the most expensive. But replication of the long distance model is constrained by two factors. Not many cities are located at a feasible distance and the cash—strapped municipalities can ill-afford the cost of such an endeavour. Worse, conservation has not figured in their scheme of things—neither directly through steps like water harvesting, nor indirectly through levying.

Urban Water Supply Status—An Appraisal

According to the 54th round of National Sample Survey (NSS) an estimated 70 per cent of urban households reported being served by tap and 21 per cent by tubewell or handpump. Sixtysix per cent of urban households reported having their principal source of drinking water within their premises, while 32 percent had it within 0.2 km. Fortyone per cent had sole access to their principal source of drinking water, which means that 59 per cent were sharing a public source. Fifteen per cent of households did not get sufficient drinking water from their principal source, between April and June, May being the worst month. In the aggregate, 91 per cent of urban households reported the quality of drinking water served by their principal sources to be satisfactory, 18 per cent reported using some supplementary source of drinking water and 96 per cent reported storing their drinking water[9].

Since independence there has been sustained effort at providing, on a priority basis potable water to all habitations, and this has been successful to a large extent as seen from the NSS data. Almost all urban centers have been provided with some public water supply scheme which is the primary source of potable water for most urban inhabitants. State level data show that on an average 88 per cent of the urban population were provided with water supply through a public water supply facility, with many States reporting 90 per cent and more coverage. However, official reports tend to give greater weight to physical and financial progress rather than the quality, reliability and sustainability of services. The picture will be incomplete and misleading if the many and significant lacunae in the existing scenario are not highlighted[10]:

What the 'Coverage' Statistics do not Reveal

Statistics of 'coverage' of around 90 per cent, and bare figures of quantity of water supplied in the cities as claimed by authorities, tend to hide several realities regarding both the operations of the system, and the experience of consumers.

- The 'coverage' figure may relate to installed capacity. What is relevant for the consumer is the actual operating capacity of the water supply system, or the average actual supply over a sustained period.
- 'Coverage' by itself does not reveal the adequacy of the system, especially of the adequacy of storage, treatment and distribution arrangements, both in terms of capacity and operational features. Thus, there can be wide variations within a city in the quantity and quality of water supplied.
- Coverage figures do not indicate the actual functioning of the system and allowance may need to be made for breakdowns which may deprive the consumers of water for several days;
- Coverage figures also do not reveal the regularity or duration of supply, or even whether the supply is made on a daily basis or less frequently;
- Coverage does not reveal the financial sustainability of the system which may be low due to low tariff and poor recovery. In many towns, poor finances result in poor maintenance, leading to more frequent breakdowns and degeneration of the system leading to major expenditure on renovation, which could have been avoided by timely maintenance. Due to inadequate revenue raising, defaults in payment of electricity bills are quite common, making non-payment of water supply dues a major source of loss to the State Electricity Boards.
- Coverage does not reveal the year-round performance, like water availability in summer. In a number of schemes dependent on ground-water as well as surface sources, availability reduces in summer, causing serious disruption of normal life;
- Figures on treated and non-treated water supply are not available. Many urban centers lack treatment facilities, and where they do exist, they are often not used or used without quality control and testing;
- Coverage figures do not reveal information regarding the number of hours of supply in the case of household

connections, and for public stand-posts, the distance, time taken to collect water, number of users of each stand-post, etc. New connections are delayed or sometimes denied, while new areas, especially unauthorized developments, remain unserviced;

- Typically, the urban residents need to supplement public supplies with water obtained from private sources, and this is usually much more expensive. The extent to which this is required to be done—the proportion of water procured privately against the quantity determined as norm—is not brought out when only 'coverage' figures are provided;
- There are no accurate figures for unaccounted for water and the quantity actually reaching consumers. The supply figures often include unaccounted for water. Unaccounted for water also increases costs to consumers as the cost relating to water that is lost has to be made up in the tariff charged to legal connections;
- Most importantly, the coverage figures say nothing about the equity of distribution, while it is well-known that the poorer areas are provided with less water whereas the influential rich will get a more satisfactory service.
- Coverage figures say nothing about unauthorized connections, and contamination of supply due to poor maintenance and mixing with drainage and sewerage waters, nor about inequity in tariff due to flat rates applied equally to those who obtain superior service and those who get poor service.

Accelerated Urban Water Supply Programme (AUWSP)

Considering the water scarcity and the narrow base of the small towns having a population of less than 20,000 as per 1991 census, the Government of India initiated a scheme known as the Accelerated Urban Water Supply Programme (AUWSP) during March 1994 with the objective of solving the drinking water problems in these small towns. The scheme is funded by the Central and the State Governments in the ratio of 50:50. Till mid-December, 2002, 726 Water Supply Schemes for identical number of towns were approved at a total estimated cost of Rs. 937.61 crores. Out of these, 281 schemes were reported to have been completed and commissioned (some partially)[11].

The scheme, no doubt is a right step in mitigating the problems of water supply in the smaller towns. However, solution to water supply problem ought not to be seen from the angle of capacity addition alone, rather than operating the existing capacity more efficiently. Efficient operation of existing water supply schemes is the critical first step in any move to make the schemes operate in a viable fashion, by increasing availability, improving reliability and customer service, and reducing cost. While additional scheme will augment availability, it will not improve either the viability of the schemes, or lead to greater consumer satisfaction as the quality of service may remain the same. Since the low quality of service is the single biggest obstacle to the levy of reasonable user charges, efficient operation will also help improve acceptability of higher user charges.

Before initiating augmentation project, it will be prudent to undertake a detailed study of the needs of the consumers, and the possibility of managing the available capacity more efficiently. It is also necessary to study the extent of utilization of the capacity already created, and to assess the losses due to leakages and misuses. Such exercise should be carried out for every town and city.

Norms of Water Supply

The Manual on Water Supply and Treatment, published in May, 1999 by the Central Public Health and Environmental Engineering Organization (CPHEEO), Ministry of Urban Development and Poverty Alleviation, specifies norms and standards of municipal water supply for domestic and non-domestic needs, such as institutional, commercial, fire-fighting and industrial requirements. The recommended minimum per capita water supply levels for designing schemes are:

1. Towns with piped water supply but without sewerage system: 70 litres per capita per day (lpcd)
2. Cities with piped water supply and existing or planned sewerage system: 135 lpcd.
3. Metropolitan and megacities with piped water supply and sewerage: 150 lpcd.
4. Public stand posts: 40 lpcd.

The existing general norms are by no means very high though they have hardly been met in most cities. The norms are not a ceiling but only a minimum. They should not therefore come in the way of any municipal body or water supply authority taking up projects for further augmentation of supplies over and above the norms. However, the first and foremost priority should be to plug leakages, update existing water works, and conserve water through proper planning of supply and maintenance of the system.

Safe Drinking Water

Water, which is essential for life, growth and health can also be a source of diseases and cause of ill-health, if contaminated or improperly handled and stored. Safe drinking water and improved sanitation play a major role in the overall well-being of the people, with significant bearing on the infant mortality rate, death rate, longevity and productivity. The urban poor, like that of their brethren in the rural areas, bear a disproportionate burden of non-availability of water, as well as of poor quality. They often supplement public source of water with supplies obtained at high prices. And poor women and children are particularly vulnerable to the effects of water contamination. In fact, 70-80 per cent of illnesses are related to water contamination and poor sanitation. Water supply and sanitation programme can hardly be successful if water-related illnesses are not reduced. It is indeed a matter of concern that despite the progress made with urban water supply, the level of water-related sicknesses continues to be high.

Urban water supplies often are contaminated from a variety of sources, including discharge of untreated industrial wastes, leading from waste dumps in to sewage, and poor solid management. Few cities in developing countries have adequate sewerage systems, and they often are limited to more advantaged areas. Purification and recycling of wastewater in sewage treatment plants is rare. In Asia, for example, treatment plants process only an estimated 35 per cent of wastewater, and in Latin America and the Caribbean, about 14 per cent[12]. Worldwide, two-thirds of the sewage from urban areas is pumped untreated into lakes, rivers and coastal waters[13].

Safe drinking water is one of the most important indicators of food absorption. Many water-borne infections spread due to the use of unsafe drinking water. Worldwide, about 2.3 billion people suffer from diseases that are linked to water problems[14]. Water related diseases kill millions of people each year, prevent millions more from leading healthy lives, and undermine development efforts. Nearly half of urban residents in Africa, Asia and Latin America suffer from one or more of the main diseases associated with the inadequate provision of water and sanitation[15]. Thus, it needs to be remembered that the augmentation of water supply alone does not ensure good health; proper handling of water and prevention of contaminations are equally important.

Data provided by census of India, 1991 indicate that in 1991, slightly more than 81 per cent of the Urban households at the all India level had access to safe drinking water which was an improvement from the position prevailing in 1981 when it was only 75 per cent (Table 8.4). In some of the states, the access to safe drinking water was extremely satisfactory, such as Delhi (96.24%), Haryana (93.18%), Himachal Pradesh (92%), Maharashtra (90.50%), Punjab (94.24%), Sikkim (92.95%), Andaman & Nicobar Islands (91%), Chandigarh (97.68%), and Dadra & Nagar Haveli (91%). On the other hand, the accessibility was deplorably poor in the states of Mizoram (20%), Nagaland 45%), Manipur (52%), Assam (64%), Goa (62%), and in the Union Territories of Lakshadweep (19%) and Daman & Diu (37%). Further, over the decade (1981-91), there was no significant improvement in the accessibility to drinking water in the states and Union Territories except in Mizoram (12.6%), Dadra & Nagar Haveli (6.774%) and Lakshadweep (41.5%). In Kerala, Nagaland, Andaman & Nicobar Islands and Chandigarh, there was a marginal decline in the number of households having accessibility to drinking water.

Status of Water Supply Service

The urban population faces an increasing demand-supply gap, wholly inadequate and unreliable supplies particularly in settlements of the urban poor, and deteriorating financial and technical performance of systems. To take only the metropolitan cities, according to a World Bank study of 27 Asian cities with populations of over one million, Chennai and Delhi share the same rank as the worst performing cities in terms of hours of water availability per day, while Mumbai is ranked as the second worst performer and Kolkata as the fourth worst.

The problems of the sector are manifold. Transmission and distribution networks are old and poorly maintained. Consequently, physical losses are high, ranging from 25 to 50 per cent. Low pressures and intermittent supplies allow back-siphoning, which results in contamination of water in the distribution network. The utilities are overestaffed. Intermittent supplies for only two to eight hours place a burden on the women who have to fetch water from public taps. The urban poor, who constitute one-fourth of the urban population, and the slumdwellers, who may range from one-third to half of the population in certain cities, are the worst affected. Ensuring equity in distribution of available supplies is one of the key challenges of the urban water supply sector.

Table 8.4: Access to Safe Drinking Water in Per Cent of Households in Urban Areas

Sl. No.	*States/Union Territories*	*1981*	*1991*	*Average Annual Change*
1.	Andhra Pradeh	63.27	73.82	1.67
2.	Arunachal Pradesh	87.93	88.20	0.03
3.	Assam		64.07	
4.	Bihar	65.36	73.39	1.23
5.	Delhi	94.91	96.24	0.14
6.	Goa	52.31	61.71	1.80
7.	Gujarat	86.78	87.23	0.05
8.	Haryana	90.72	93.18	0.27
9.	Himachal pradesh	89.56	91.93	0.26
10.	Jammu & Kashmir	86.67		
11.	Karnataka	74.40	81.38	0.94
12.	Karalla	39.72	38.68	-0.26
13.	Madhya Pradesh	66.65	79.45	1.92
14.	Maharashtra	85.56	90.50	0.58
15.	Manipur	38.771	52.10	3.46
16.	Meghalaya	74.40	75.42	0.14
17.	Mizoram	8.79	19.88	12.62
18.	Nagaland	57.18	45.47	-2.05
19.	Orissa	51.33	62.83	2.24
20.	Punjab	91.13	94.24	0.34
21.	Rajasthan	78.65	86.51	1.00
22.	Sikkim	71.93	92.95	2.92
23.	Tamil Nadu	69.44	74.17	0.68
24.	Tripura	67.92	71.12	0.47
25.	Uttar Pradesh	73.23	85.78	1.71
26.	West Bengal	79.78	86.23	0.81
27.	Andaman & Nicobar Island	91.95	90.91	-0.11
28.	Chandigarh	99.39	97.68	-0.17
29.	Dadra & Nagar Haveli	54.35	90.97	6.774
30.	Daman & Diu	67.04	36.76	2.94
31.	Lakshadweep	3.65	18.79	41.48
32.	Pondicherry	84.18	86.05	0.22
	All India	**75.06**	**81.33**	**0.84**

According to various studies, most cities are unable to operate and maintain the existing systems to the full capacity. The capacity utilization has been reported to be less than 50 per cent in 40 per cent of the towns, and less than 75 per cent in a further 20 per cent of towns. There is grave danger that cities may, in many instances slip back to lesser levels of water supply, due to poor maintenance and depletion of sources even as the population continues to grow. This may lead to a situation where the per capita availability of water by 2020 may actually decrease, unless corrective action is taken expeditiously.

There are wide variations among towns in terms of quantity of water supplied, per capita supply, achievement of normative supply, duration of supply, etc. In the case of slums, which are generally provided public stand-posts, there are wide variations in the availability and average number of households per stand. Wherever ground water is the main source, the over-exploitation of aquifers, depletion of water resources, and pollution by urban human wastes, are causing serious problems. Water supply and sanitation are, above everything else, issues in environmental health. There is a close link between water and sanitation requirements because availability of adequate water is essential for proper sanitation practices.

Demand Management

Demand for water is rising and is estimated to have risen six to seven times from 1900 to 2001 in India, more than the rate of population growth. It is a rise, which seems likely to accelerate in the future, because the population in India is expected to reach 1.3 billion by the year 2025 AD and 1.6 billion by 2025 A.D. A tentative study indicates that total annual requirement of fresh water for various sectors in the country will be about 1,093 b.cu.m by 2025 A.D. This fresh water requirements by 2025 AD will be almost at par with exploitable water resources including both surface and ground water. Thereafter additional supply will be necessary or else scarcity conditions would prevail in majority of our river basins.

Access to clean water for drinking, cooking bathing, and other household needs is fundamental, but people still lack safe domestic water supplies and adequate sanitation. Compared to other uses, the volume of water required for basic domestic needs is not great;

municipal and industrial demand are growing much faster. By 2025, urban populations in India are expected to be around 50 per cent of the total population posing additional challenges for municipal water supply. While water availability is not a major constraint for domestic use, chemical and biological contamination threatens the quality of water for human consumption.

According to some estimates, Indian cities need nearly 4,000 million litres more water per day. This huge requirement makes water supply a real messy problem.

- In most Indian cities-says the World Bank—the availability is less than three hours everyday. Not because they are drying up or overcrowded, but due to waste and inefficiency.
- Political control, mismanagement and corruption have ensured that civic bodies which provide water are bankrupt. They are monoliths, with excess employees with no urge or incentives to deliver. Most efficient systems have 2-3 people per 1,000 connections. India has 10 plus; Delhi has 21.4 people per connection and cost: Rs. 355/person. Water cess recovery on the other hand is poor being only about 60 per cent.It needs strong political will to reduce staff, cut operations costs and increase efficiency. Unless the subsidized middle class is made to pay the true costs of water and sewage, the man-made water scarcity will continue to haunt the poor.
- Water tariffs are absurdly skewed. According to the Urban Development Ministry, the cost of delivering water to a metro home is Rs. 15 per 1,000 litre, but the customer pays only Rs. 1.50.
- There is criminal waste-there is upto 55 per cent unaccounted for water in Indian cities, says the World Bank. Such low tariffs can only encourage wasteful utilization of water. No wonder, some 30 per cent of potable water gets flushed through the toilets. This is equivalent to 195 million gallons a day-or the amount required by the Delhi Jal Board to reach the un-serviced colonies and slums.
- Mismanagement, rather than scarcity, rules. "Delhi is a leaking bucket," says Junaid K Ahmad of the World Bank, which can be plugged with institutional reforms.

- The national cost of fetching water is 150 million woman days, resulting in a national loss of Rs. 10 million every year, according to eco-scientist Vandana Shiva.

While inept handling of the situation leads to irregular supplies, the mineral water industry gets the most of the situation. If current trend is any indication, beleaguered municipalities will soon find replacement in the private operators. The resultant tariff hike will push the poor to the margins again.

Studies indicate that the poor pay anywhere between 8-20 times what the rich pay to get water from unreliable sources. In contrast, what the rich pay for piped water is a fraction (less than 10 per cent) of the actual cost of producing potable water. It is clear that the poor face the brunt of this inequality. On one extreme are a sizeable number of urban poor who don't get assured supply and on the other are the urban rich whose consumption is subsidized by the state. The poor have remained at the end of the 'pipe dream' of getting water. In the successive five-year plans, the benefit of subsidized water delivery system has failed to reach the poor. Since a majority of the poor, for whom the subsidy has primarily been directed, have not been linked to the supply lines, the benefits have flown to the privilege class. The process of water distribution has continued to encourage inequality. Not only have the rich been served at the cost of the poor, they have literally been getting it free. For instance, at the cost of 1,600 unauthorized colonies and 1,100 slums in Delhi, who have yet to get piped supply, the privileged class pays Rs. 1.6 per cubic metre of water-the lowest in the country. The rich in India can support a Rs. 1,800 crore bottled water industry and boost purifier sales, but cannot pay commercial rates for water to enable cities and towns to distributes clean water. For the poor, the subsidy acts like a double-edged sword. First, they don't seem to benefit from it and second, the sources of water they depend on get polluted by untreated sewage that is flushed out by the rich. Can the water utilities ensure equity in water distribution by reaching out to the unreached? Can the poor be extended subsidy by readjusting water tariff upward of the prevailing rates for the rich? Unless intention to change get reflected, electoral politics will seize every conceivable opportunity to promise water, through mammoth projects like the 'interlinking of rivers'.

Ever since water has been moved from the socio-cultural domain of community control to the techno-economic sphere of bureaucratic management, it seems to have transformed from being seasonally scarce in the past to chronically unavailable now. Part of the problem may be rooted in rising demand due to increasing population but mismanagement remains the core issue. For the ruling elite, water scarcity is quite simply a political tool. With water under full jurisdiction and control of the government under existing laws, communities are dependent on the state for providing better access during lean periods. However, successive governments have learnt to evade the crucial governance issues. No one risks opening the issue of rationalization of water tariff at the cost of losing electoral base. Consequently, inefficient municipalities have already been stretched to their limits. With mounting debts these civic bodies are on the verge of collapse everywhere.

Is privatization the way out of this messy maze? "Our focus in India is inefficiencies in operations". Veolia, Suez and a handful of Indian companies along with agencies like the World Bank, say privatization is the answer. Politicians, bureaucrats, water utility employees and activists say no.

No matter which side of the debate one is on, reform is the only way out. The World Bank-backed formula for Indian cities emphasize private-public partnerships: limited government control, entry of private companies which can work at arm's length with it and the presence of a strong regulatory mechanism (like in telecom) as the broad matrix. The challenge is to redesign water institutions in a way that assigns the responsibility for policy formulation, delivery and regulation to separate entities in order to create transparent checks and balance in the system.

If activists are wary of MNCs for profiteering, the MNCs are tired about the lack of political will. Yet, they look at India as a very serious market. While the world water market is estimated at one trillion dollars, after the rumbling of tech stocks, Fortune magazine has rated water as the most profitable business for investors. Other water giants working in Indian cities include Veolia Water (formerly Vivendi), Bechtel, involved in the Tirupur project in Tamil Nadu along with Indian companies like IL & FS, L & T and Thermax. Thames Water, Biwater and Anglian Water of UK are pitching hard. According to Vandena Shiva, in 2000, the business of safe water provision was estimated to have reached $300 mn in India and Mexico.

Privatization experts have suggested different models for local governments:

- *Management contracts:* The service provider awards a contract to a private company for a fixed period to manage the water utility. The investment is public, the commercial risks either public or shared.
- *Lease contracts:* The contract is offered on lease for a finite period, the risks are shared.
- *Concession contracts:* Using private investment and taking greater risks than management contract.
- *Divestiture:* The sale of utilities where everything from ownership to risk belongs to private companies.

The Need for Better Policies

Freshwater is the liquid that lubricates development. It is as much an economic input as are infrastructure, energy, and human know-how. Without better management of scarce water supplies, sustainable development will be a far cry. Four principal policy weaknesses, in general, underscore the world's inability to manage freshwater supplies for sustainable development, according to Ismail Serageldin, Vice President for Environmentally Sustainable Development at the World Bank[16]. These are:

1. *Water management is fragmented among sectors and institutions*. There are too many agencies with their fingers on the water tap and virtually no coordination of policies between sectors of the economy. Issues of water quality and health often go entirely unaddressed because they do not fit within the mandate of any single government agency.
2. *Governments depend too much on centralized administration to develop, operate, and maintain water system*. Agencies charged with managing water supplies often are already overextended and lack technical competence. At the same time, there is little stakeholder involvement and community participation in setting water policies and regulating use, so projects often do not meet people's needs.

3. *Most countries undervalue freshwater as a resource and do not price it at its economic value.* Many heavy users of water, such as farmers, receive government subsidies and in effect are encouraged to waste water that they otherwise probably would not waste. Most governments have found it politically more palatable to develop new water supplies than to charge heavy users the full costs of water.
4. *Water Management policies do not link the quality of water to human and environmental health.* In most cases water is viewed as a resource to be pumped around and used as often and as much as needed for any purpose. Without adequate consideration of water's key role in human and environmental health, it is little wonder that water resources are degraded nearly everywhere.

Freshwater is the universal necessity—providing all forms of life with sustenance, nurturing natural ecosystems, and transporting and diluting wastes. Without assured supplies of freshwater, living standards decline, people suffer, and development becomes more difficult. Insuring sustainable development increasingly will require wise policies and effective strategies that not only conserve and protect freshwater sources but also manage them equitably to meet the needs of consumers, industry, and agriculture.

For conservation, augmentation and recycling of urban water the Planning Commission, Government of India has recommended following action[17]:

1. Water tariff should be set at levels that discourage excessive use. Water-efficient systems for flushing should be made mandatory. Conservation of water should be a recurring theme for both users as well as those managing the water supply system.
2. Leakages and unaccounted for water is another constraint in cities. Due to old and rusted pipes or poor maintenance of the system, these losses sometimes go up to 50 per cent. These must be controlled and brought to the minimum level. Severe penalties should be levied on those found responsible for leakage and wastage of water.

3. Reuse of treated sewage must be given priority in view of the fact that water is going to become more scarce in the near future. With tertiary treatment, water from treated sewage can be used even for air-conditioning, industrial cooling and other non-potable uses. This should be made a thrust area.
4. Use of potable water for purposes like washing of vehicles, maintenance of gardens, etc. should be prohibited. In urban concentrations, dual supply of potable and non-potable water should be undertaken. For non-potable domestic uses, tubewells should be permitted to be sunk, subject to the construction of a percolation structure in the premises. This will conserve potable water without affecting ground water availability.
5. Rainwater harvesting should be implemented widely. This has been taken up as a thrust area in Chennai and Delhi and must be given priority in all towns in the country. The Delhi Jal Board has taken up more than 80 works to harvest rain water and intends to cover about 200 buildings. The Delhi Government has approached the Ministry to Urban Development and Poverty Alleviation to amend building bye-laws to make rain water harvesting mandatory in the Capital. During the Tenth Plan, it shall be made obligatory for all urban areas to adopt rainwater harvesting as a part of the building bye-laws.
6. Central Ground Water Board (CGWB) is engaged in developing techniques for artificial re-charge of ground water, which should be implemented where conditions are appropriate. Similarly the exploitation of ground water in urban areas must be constantly monitored in order to prevent the drying of this important source.

GROUND WATER IN URBAN ENVIRONMENT

Ground Water plays a critical, but complex and often unrecognized, role in the urban environment. It has a major importance in municipal, industrial and domestic water supply. In any urban center ground water conditions are of primary significance in construction and maintenance of the sub-surface engineering constructions (such as tunnels, sewage, underground storage etc., and more generally in urban drainage.

Urbanization profoundly effects ground water flow and quality creating serious impact upon urban infrastructures and may lead to a spiral of socio-economic and environmental decline. It is also intimately linked to economic development and may present a challenge for environmental management in near future as urban dwellers will continuously increase in the future. In rapidly developed economics, the urban explosion and indiscriminate solid and liquid waste disposal result in increasing scarcity and deteriorating quality of ground water. This also results in escalation of water supply costs and seriously threatens human health.

It is essential to know urban ground water process, geo-technical and construction problems, overall ground water based supply system, ground water pollution hazards and ground water management in the urban scenario. Ground water management should be based on sound scientific knowledge. The role of scientist in this paradigm is to address the problems and develop solutions. The resource managers and scientists need to work together to achieve more technology transfer and better ground water management.

An Asset as well as a Liability

Urban ground water is both an asset and a problem. It is an asset because it meets the requirements of drinking water as well as industrial use. Low capital intensive ground water extraction makes it equally valuable as water supply for high densities of small domestic wells which are very common in the upcoming cities round the world. On the other hand, urban groundwater poses problem from pollution. Over-extraction of groundwater causes decline in water tables, necessitating continual changes in the water supply infrastructure as the wells must be deepened or replaced. It causes a more acute problem when shallow tube-wells are replaced by deep tubewells, increasing cost of pumping and lowering pore pressures in deep, fine-grained sediments. Falling water table may also cause land subsidence, which has been a major problem in many cities around the world coming up on unconsolidated sediments.

Urban areas affect the behaviours and characteristics of groundwater below and around them. In many urban and peri-urban areas, there are cases in which local aquifers are not able to meet the quantity and quality of water needs of the burgeoning population. Urban development is generally believed to reduce recharge due to high proportion of impermeable surfaces. However, increase storm runoff may well be at the expense of evapo-transpiration following reduced plant cover in cities.

The increased abstraction of ground water to meet the growing urban requirements not only decreases well yield of water and increases the groundwater abstraction cost, it may also introduce changes in the surface water-groundwater relationship and may induce infiltration from rivers and losses to sea. When large areas progressively dry up, it may affect the subways, sewers and underground spaces provided for car parking, cellars and service chambers. Depleted groundwater may also cause saline water intrusion.

Again, if groundwater abstraction is reduced due to replacement of surface water based distribution system over a large area, it may result into groundwater level recovery. Leakage from water distribution network may also cause increase in the recharge, especially in the old town area. Irrigation of public and private gardens and green belt areas may be another source of recharge. Further, since many urban areas are near or on alluvial plains, irrigation on the farmlands in peri-urban areas, well drained lands may contribute significantly to aquifer recharge. Rising water table may damage sub-surface structures as a result of hydrostatic uplift or reduced bearing capacity, inundation of surface facilities, excessive ingress of ground water to sewers and chemical changes on concrete foundation of buildings and underground structure. On the contrary, the progressive urbanization and use of the flat areas for peripheral services like roads, motorways parking and storage lots, factories, etc., reduces recharge to aquifers.

Changes Occurring in Urban Groundwater Quality

Urbanization has a series of effects on groundwater quality, many of which are poorly known. A common result is in increase of dissolved nitrogen compounds, and also some increase in total dissolved solids as a consequence of additional salts in the infiltrating used water or due to concentration in irrigated fields and gardens.

One of the principal urban services, besides provision of water supply, is the 'waste water elimination and solid waste disposal'. This service may have limitation where soil infiltration capacity is low due to shallow water table condition or to relatively impermeable superficial strata. The major adverse impact of waste water discharge and solid waste disposal is on ground water quality.

Ground water pollution may assume serious conditions in certain areas due to washing by rain water of soluble salts, heavy metals and organic (mostly hydrocarbons) deposited on buildings and streets as also from point sources such as leakages from service areas, workshops, storage tanks and leaching of different stored substances and from disposal into ground of contaminated water and solid wastes. Complex situation may occur when rural areas are taken by industrial settlements which are later converted into residential areas.

Quality changes in the ground water are of different types. Oxidation of organic matter in the water and soil produces carbon dioxide and increase water hardness. Nitrogen compounds are oxidized to nitrate. If, on the other hand, reducing conditions are strong, some sulphate reduction is possible as well as iron and manganese solubilization. Although the aquifers may maintain its quality from a lithological point of view, mixing of other biologically polluted water in shallow aquifer is frequent due to poor well design, construction, operation and maintenance.

Urban Ground Water Management

Water management decision may influence aquifers in urban areas. Urban groundwater management needs are to be assessed from the rather different perspective of water supply provision, waste water/ solid waste disposal and engineering infrastructure development and maintenance. An effective management of urban aquifers has to reconcile the demands of water supply, waste disposal and urban sub-surface engineering, and an integrated approach to avoid long term problems interrelating all the above three has to be developed. The major ground water issues should address the two underlying causes—inadequately controlled ground water abstraction and excessive sub-surface contaminated load.

The goal of urban groundwater management should be to strike a reasonable balance between maintaining water supply availability and quality, preserving the urban infrastructure and ensuring the safe disposal of wastes. In the case of already over-exploited aquifer, abstraction control may be needed. Recharging of the aquifer through roof-top and rain-water harvesting may be of much help in mitigating this problem, and 'use planning' in urban complexes is inescapable for effective preventive control.

Any urban management practice must stem from a coherent understanding of the local ground water/aquifer system. Each city is unique in its particular hydrogeological setting and effective management of ground water will depend upon hydrogeological awareness. Hydrogeological data are required to prevent water quality deterioration and the critical overdraft of the aquifer with potential deleterious secondary effects. These secondary effects include subsidence, salt-water intrusion, degradation of ground water quality as well as aquifer depletion.

New and continually evolving technologies create more opportunities for the use of ground water in urban areas which include aquifer storage and recovery, ground water heat pumps, integration of ground water supply into dual distribution system, artificial recharge and using aquifer to reclaim treated sewage water. Detailed hydrogeological mapping and analysis must be done before key decisions are made in this regard.

Institutional and Legal Framework

A strong institutional framework will be required to achieve the above objectives. The ideal framework would include legislation to render ground water resources the property of the country with the granting of licenses for their legal exploitation in an approved manner. Other necessary activities for the regulatory agency would be to conduct appropriate levels of resource monitoring, surveillance of potable water supply quality and effluent discharges and inspections of potentially polluting activities.

In the case of already over exploited aquifers, abstraction control will often need to include measures to prohibit the construction of need wells and to reduce abstraction in existing wells. However, the control of ground water abstraction may not be always achieved by means of regulation. Carefully considered economic instruments can often play an important constraining role. In fact, a combination of financial incentive and users-group endorsed regulation is likely to be the most effective means of achieving sustainable abstraction for public and private users alike. Effective urban ground water management has to be active, not passive and to be effective requires a strong institutional framework to bring together ground water stake holders, uniform water-use/interest group and enact regulatory and economic instruments.

Ground Water Status and Development Prospects in a Few Important Cities

The Central Ground Water Board, has recently (December 2000) made a maiden attempt to understand the hydrological conditions prevailing in the major cities in India, its development potential and quality hazards which are endangering human health in sprawling urban centers. The highlights of the findings[18] of a few important cities are indicated below:

Kolkata

Kolkata is one of the biggest cities of the world. The area covered by Kolkata corporation is 1,877.33 sq. km. with 141 wards. The western part of the city is bordered by mighty Hooghly river, southern and eastern part is bordered by south 24 Parganas district. The eastern and northern sides are bordered by North 24 Parganas district. The area lies between North latitude 22° 28'00" and 22° 37' 30" and East Longitude 88° 17'30" and 88° 25'. The demographic density during 1981 was 22,260 persons per sq. km. and during 1991 this rose to 23,670 persons in the city area. It is reported that the pressure of population on land has further increased significantly since then.

Water Supply and Demand

Both surface water and ground water sources are being harnessed for meeting the water supply to the Kolkata Metropolitan area. The river Hooghly is the only source of surface water supply. The total quantity of surface water now available in Kolkata area is estimated to be about 9,90,000 KLD (320MGD). The rated capacities of the water works located at Palta and Garden Reach are proposed to be increased significantly and the augmented water supply is expected to meet the increased water demand. But the main problem in the water supplies in the city is the loss due to leakage in the distribution system and wastage. It is estimated that the total loss amounts to nearly 30 per cent of the rated supply.

Ground Water Resource

Kolkata metropolitan area is located on the lower deltaic plain of the composite Ganga-Bhagirathi delta and is covered by the sediments deposited by the river system flowing through the area during quaternary era. The sub-surface geology of the area is completely obscured by blanket of recent sediments. The upper 300 m of sediments are of quaternary era and consists of alternations of clay, silt, sand and occasional pebble beds.

Ground water occurs in the city area both under unconfined and confined conditions. Ground water in the shallow zone occurs under unconfined conditions. The direction of ground water flow indicates that the recharge area of Kolkata region lies in the north and western part of the contiguous Greater Kolkata region. The major aquifers of the northern part of Greater Kolkata are inter-connected with the near surface ground water sand beds at Kalyani—Ranaghat—Shantipur area in Nadia district.

In the west, the Cis-Damodar belt similarly has extensive near surface sand beds of considerable thickness forming favourable zones of intense recharge. The average rainfall of 140-152 cm. in the Cis-Damodar belt ensures considerable percolation to these near surface and surface sand beds for recharge of aquifers in the east at a great depth. Potential fresh water aquifers also occur in the depth range of 40-100 m in the northern part, within 60-160 m in the central part and within 180-300 m in the south—western part of the Kolkata Municipal Corporation area.

Ground Water Withdrawal and its Impact

Due to large scale withdrawal of ground water from the confined aquifers, a depression of the piezometric surface in Central and South Central Kolkata has developed. The magnitude of depression is 6-8 metres over a period of about 40 years (1958-98) in the core sector covering Narkeldanga—Park Circus-Bhowanipur. As a result, the general southerly flow in the confined aquifers has become radial in a much larger area surrounding the cone of depression. The Central part of the Kolkata Municipal Corporation area is drawing water from all directions resulting in its radial flow. The total quantity entering from the immediate vicinity of KMC area into the central depression zone comes to 204 mld (45.3 mgd). In the absence of precise census data of the drawal structures it is difficult to evaluate precise ground water draft in KMC area.

However, according to the available information from KMC authorities, there are 325 heavy duty tubewells (203 mm dia) and 4,834 private tubewells (101 mm dia) which function on an average 7 hrs. Besides, there are over 12,000 tubewells fitted with hand-pumps (40mm dia) distributed throughout the KMC area including Garden Reach, Jadavpur, Behala area.

Withdrawal from big dia tubewells has been estimated to be 34 mgd (153 mld), considering 325 big dia tubewells functioning at the rate of average discharge of (15,000 GPH) $68m^3$/hr. for 7 hours of daily pumpage. The yield from 4,834 private tubewells (101 mm dia.) functioning at the rate of average discharge of (6,000 GPH) $27m^3$/hr for 7 hours of daily pumpage is (203 mgd) 914 mld.

The yield from 12,000 nos. shallow hand pump fitted tubewells comes out to be 12.6 (56.7 mld) at the discharge of (150 gph) 0.68 m^3/hr for 7 hours of daily run. Thus the total ground water withdrawal from Government heavy duty tubewells, private tubewells and hand pumps came to (249.60 mgd) 1,123.70 mld during 1998. Since daily withdrawal is far exceeding the daily flow *i.e.* 204 mld (45.3 mld) gradual lowering of the piezometric surface is taking place resulting in expansion of the area under depression.

Groundwater Quality

The nature of chemical characteristics of the ground water in the depth range of 60-250mbgl as analysed by the CGWB reveals very high level of chloride concentration varying between 250 and 500 mg/l in the different parts of the Kolkata Municipal Corporation area. The quality of ground water is alleged to be problematic. However, ground water in south-western part is fresh and suitable for various purposes.

In the eastern part of the city, industrial effluent from the tanneries cause serious environmental hazards due to pollution of water bodies, wet lands and agricultural lands. Toxic trace elements like Cr, Co, etc. are found (above 0.01 mg/l) in the shallow aquifers (within 20 m.bgl) in the area making water unfit for drinking purposes. However, analysis of water samples from deeper aquifer (80 to 200 m depth) shows that concentration of Cr, Co and other heavy metals is below permissible limit.

Sum-up

The computations have established that the withdrawal of ground water from KMC area substantially exceeds total ground water flow through the area. This has manifested in the lowering of piezometric surface in central part of the KMC area in the range of 5-8 m. Reversal of ground water flow in the southern part attributed to the lowering of piezometric surface has imparted critical environmental hazard to the area. Thus, it is necessary to have effective control and exercise appropriate regulatory measures for the withdrawal of ground water.

In Kolkata area further exploitation of ground water needs to be avoided. Withdrawal from existing tubewells should be regulated by phasing out tubewell operation time in the area where recession of piezometric levels has taken place conspicuously. Where the recession of piezometric level is not so conspicuous the present status of ground water development may be maintained. It may be prudent to examine if full utilization of balance ground water available in the northern part of KMD area can supplement/augment surface water based supply system in the area. Well fields to the north of Srirampore-Barrackpore-Barasat may be developed for large scale transfer of ground water to the deficient areas. Close monitoring system should be envisaged and developed to record the decline in the piezometric levels at regular intervals of time vis a vis withdrawal of ground water from the deep tubewells in the entire Kolkata area.

It is also emphasized that in order to ensure supply of surface water to Kolkata, it is essential to control rate of siltation and salinity of the river Hooghly. For this purpose, release of adequate quantity of fresh water from Farakka is an inescapable necessity. The river running through the Kolkata area and in close proximity to the affected areas can be considered as the source of surface water for recharging the ground water through properly designed injection wells. This would however involve adequate treatment of water before recharging to make it suitable for drinking purpose as per standard norms.

Bhopal

Consequent to the reorganizations of states in 1956, the new state of Madhya Pradesh, which included Bhopal State also along with other regions came into being on 1st November, 1956. Bhopal was made a separate district and constituent unit of Bhopal division. It attained very rapid urbanization and industrialization and at present Bhopal city encompasses an area of 296 Sq. Km. The rapid rate of urbanization has increased generation of waste waters, majority of which is being discharged into upper and lower lakes through drains/nalas. These means of disposal of waste waters and other human activities are leading to environmental deterioration of the surface water bodies. The waste waters and domestic wastes are also being drained in existing nalas which pass through municipal area and cause pollution of Ground Water. Besides, industrial effluents are also not properly treated and allowed to flow through the unlined drains. This is also polluting the ground water.

Bhopal Municipal Corporation includes 56 municipal wards. The population of Bhopal city, in 1951 was 1,02,333. The population has increased tremendously after Bhopal became capital of Madhya Pradesh. It has increased from 1.855,374 in 1961 to 10,63,622 in 1991 indicating and increase of 474 per cent over a period of 30 years. The population of Bhopal city as on 1998 has been estimated to be 14,35,000. About one fifth of this population lives within 15 municipal wards which are draining into either Upper Lake or Lower Lake. It is reported that 27 per cent at the city population lives in scattered settlements. The rate of industrialization has substantially increased in the city.

Rainfall

The Bhopal city is endowed with good rainfall. The normal annual rainfall of Bhopal city is about 1,260.2mm. The southern part of the city receives more rainfall than northern part of the city. The rainfall decreases as we move from south to north and towards north-west. About 92 per cent of the annual rainfall takes place during the south-west monsoon, *i.e.* from June to September. Only 6 per cent and 2 per cent of the annual rainfall takes place during winter (November to February) and summer season (March to May) respectively. Hence only about 8 per cent if the annual rainfall takes place from November to May. Therefore the surplus water is available only during the south-west monsoon period. From November to May, in general, no surplus water is available for ground water recharge.

Water Supply and Demand

A quantity of 231 million liters per day (mld) water is being released from surface water sources and 23 mld through tube wells and large diameter dug wells to the customers for domestic purposes at present. After accounting for distribution and generation losses the net water supply from surface water sources is about 210 mld. The, total water supply available to the consumers from both surface water and ground water sources is 233 mld against the present demand of 281 mld. Thus the water supply presently available falls short of 48 mld. There is proposal to check the leakages in the distribution system in order to increase the water supply.

The water supply to Bhopal city is available from Upper Lake and Kolar reservoir through a number of filtration plants, about 350 tube wells and a few large diameter dug wells.

The demand for water supply to Bhopal city is increasing fast. There is a proposal for releasing 45.42 mld (10 MGD) additional water from Kolar reservoir to meet the increasing demand.

The Bhopal Municipal Corporation has estimated that the water demand in the year 2015 will rise to 500-525 mld to cater the domestic needs of projected population of about 35 lakh. To meet out this projected water demand there is a public demand to divert water of Narmada river through pipe line to Kolar reservoir.

Groundwater Resource

In order to study the behaviour of ground water levels the CGWB recorded water levels from dug wells tapping different water bearing formations in the city. The dug wells generally tap the phreatic aquifer. In the southern part of Bhopal city the depth to water level during the pre-monsoon period varies from 4 to 6 mbgl while in the northern part it rests between 6-8,8-10 and more than 10 mbgl. Depth to water level in the major part lying in the southern part of Bhopal city rests between 2-4 mbgl while in the area lying around Govindpura, TT Nagar, Anna Nagar and Chunabhatti it lies less than 2mbgl during the post monsoon period. The depth to water level in the same period rests between 4-6 mbgl in the extreme southern part. The water level fluctuation in the major area lies less than 4 m. The higher fluctuation in water levels is recorded in Vindhyan sandstones due to low permeability except Nabibagh area which is underlain by Deccan trap.

Ground Water Quality

The chemical quality of surface water bodies, mainly Upper and Lower lakes play a vital role in assessing impact of urbanization on them and also on quality of ground water in adjacent areas. Regular inflow of integrated sewage in the lakes through unlined and open drains has produced high magnitude of nutrients. As a result, excessive growth of aquatic plants has taken place in the lake waters. This has resulted to eutrophication to hyper-eutrophication of Upper and Lower lakes, which has made them polluted. Other undesirable human activities like washing of clothes, vehicles, disposal of domestic wastes and agricultural residues and religious activities like immersion of idols and Tazias (only in Upper lake) are also making the lake waters polluted. The chemical analyses substantiate the above phenomenon. The means of disposal of untreated sewage into the lakes is also leading to pollution of ground water.

In major part of the city the quality of ground water is suitable for various needs, leaving patches where nitrate concentration is high. The main pollution hazards in the central and northern parts are salinity with respect to higher concentration of chloride in ground water. The nitrate pollution in ground water occurs almost in every part of the city in patches and is a matter of concern. The high nitrate concentration is due to disposal of untreated sewage through open and unlined drains/ nalas and indiscriminately dumping of solid wastes without considering hydrogeological situations. It has polluted the ground water in such areas. There is immediate need to adopt suitable measures for proper disposal and treatment of sewage at identified sites. For wells constructed for drinking water, it is much more important to ensure that the intake portion is not directly in connection with pollution from ground surface.

The main pollution hazards are salinity with respect to higher concentration of chloride in the northern and central parts of the city and nitrate concentration in general. Nitrate pollution is mainly due to disposal of untreated sewage through open and unlined drains and wrong siting of land fill sites on permeable formations thereby polluting ground water gradually. The higher concentration of nitrate is mainly found in the tubewells and hand pumps which have been installed near the drains carrying waste waters and in the vicinity of land fill sites without providing salinity protection. There was spurt in the recent past in drilling of boreholes and installation of hand pumps to augment domestic water supply.

Chandigarh

Chandigarh, the capital of Union Territory of Chandigarh is located at about 250 km north of Delhi at the foot hills of Himalayas. The city has the distinction of being the joint capital of Punjab and Haryana States also.

Chandigarh UT is highly urbanized. The Union territory comprises Chandigarh city, Notified area committee Mani Majra and 22 villages. The area extent of the U.T. is confined within north latitudes 30° 40' and 30° 48' and east longitudes 76° 42' and 76° 51' falling in Survey of India toposheet No. 53B.

The U.T. is bounded by Siwalik hills in the north and north east, by the Ropar district of Punjab in west and north west, by Panchkula district of Haryana in south east and by Patiala district of Punjab in south.

Total population of the Chandigarh U.T. was 6,42,015 as per 1991 census. The U.T. is urban in character with the urban rural population ratio being as high as 43:7 against all India ratio of 6:19. In the decade 1981-91, Chandigarh has shown population growth of 42.16 per cent. The literacy rate in Chandigarh U.T. is among the highest in the country with total literates being 77.81 per cent. The density of population is 5,631 person per sq. km. The statistical data pertaining to land use indicate a gradual decline in net area sown and total cropped area due to large scale urbanization of the territory.

There are two major streams which originate from Siwalik hill ranges and form the natural drainage of U.T. Of the two streams, the Sukhna choe drains the eastern part of the U.T. and joins the river Ghaggar. Ther other important stream is Patiali Rao, which drains the northern and western part of U.T. Both these streams are ephemeral in nature and carry huge flows during monsoon.

There is no large natural surface water body in Chandigarh though small ponds do exist in the area. The Sukhna choe has been dammed in north west side of U.T. which has given rise to a artificial lake covering an area of about 1.62 sq. km. The lake is perennial with water holding capacity of 5 MCM.

Water Supply and Demand

The U.T. of Chandigarh has limited area for agriculture. The main source of irrigation is ground water through 39 deep tubewells installed by Chandigarh Administration and 90 private shallow tubewells. Around 23 per cent of the area of U.T. is covered under forest.

Ground Water Resource

The principal aquifers in the area are pebbles, cobbles, gravels and sand of various grades of Quaternary age. Rainfall, seepage from surface water bodies and return flow from applied irrigation are the main source of recharge to ground water in the area.

The ground water occurs under both the unconfined and confined conditions depending upon disposition of subsurface formations. In 'Kandi' belt ground water occurs under unconfined condition and thickness of aquifer is more compared to alluvial formation.

The depth of water level during pre monsoon period in the U.T. of Chandigarh varies between less than 5 m in southern part and more than 20 m in the northern 'Kandi' zone. However, in general it varies between 10 and 25 m bgl.

Ground Water Quality

The chemical quality of ground water in the Union Territory is fresh and within the drinking standards prescribed by Bureau of Indian Standards (1991).

Sum-up

Initially only 55 sq. km. are out of the total area of 114 sq. km. of the territory was planned to be developed as urban area. Now the urbanization has increased beyond this limit due to development of industrial area and Mani Majra area. Therefore, now about 80 sq. km. has been considered to be under urban area. Thus a large part of the area is covered by impermeable surface like buildings and roads. The covered area is about 60 per cent in urban area and 5 per cent in the rural parts. On the whole 40 per cent of the total area can be considered paved and accordingly rainfall infiltration has reduced considerably.

Based on the above assumption the rainfall recharge is estimated to be 14.0 MCM. The subsurface inflow to the U.T. is 15.66 MCM. Therefore, the total annual recharge to the U.T. is 29.66 MCM. The annual ground water draft from various sources is 21.26 MCM. The subsurface out flow is 3.28 MCM per year the total discharge from the U.T. is thus estimated to be 24.54 MCM, leaving a balance of 5.12 MCM per year for future development. The stage of ground water development is 82.7 per cent.

However, there exists possibilities of further augmenting the ground water resource in urban areas of the territory. The monsoon runoff which goes as waste due to 40 per cent of the paved area of the territory, can be utilized by roof top harvesting. Chandigarh has a large number of private and public buildings with considerable roof top areas. The water falling on roof top of these buildings which is nearly silt free can be diverted to recharge well/shaft for artificial recharge to the desired aquifer system. Such utilization of rainfall runoff will also help in checking the flow of storm water and flowing of areas which will lead to increase the life of man made structures like roads, drains etc. and will also check silting problem caused by heavy flow of rain water during monsoon.

Delhi

The water requirement of National Capital Territory of Delhi (NCT of Delhi) is mainly to fulfill the drinking water needs of its growing population. The NCT of Delhi occupying an area of 1,485 Sq. Km. has six administrative blocks.

Out of the total 1,485 Sq. Km., the urban area including the new settlements in rural habitants account for about 496 Sq. Km. As per 2001 census, the population of Delhi was 1.38 billion. Besides, a floating population of 3 to 4 lakh is a common feature of the territory. According to a recent survey, more than 50 per cent of Delhi's population lives in about 1,304 unauthorized colonies, 1080 J.J. Clusters, 44 resettlement colonies and 209 rural villages. Based on the present water availability, Delhi has less than 15 per cent of its own water resources to meet the needs—the bulk of the water supplies come from neighbouring states. The rapid growth of population exerts increasing stress on the water supply infrastructure of the territory.

The river Yamuna is the only perennial river flowing in the territory. Eastern and Western Yamuna canals taking off from Tajewala and the Agra canal taking off from Okhla are the three major canal system originating from the river Yamuna.

The ground water resources of NCT of Delhi are being developed to a greater extent but are not adequate to meet the major water requirement of Delhi. However there is further scope to augment the ground water resources with suitable management techniques. This would provide additionality to the total water availability.

Rainfall

The rainfall increases from the south west to north east. About 81 per cent of the annual rainfall is received during three monsoon months of July, August and September. The rest of the annual rainfall is received as winter rains and as thunderstorm rain in the pre and the post monsoon months.

The Normal Monthly Rainfall for NCT Delhi

	Jan.	*Feb.*	*March*	*Apr.*	*May*	*June*	*July*	*Aug.*	*Sept.*	*Oct.*	*Nov.*	*Dec.*
A	14.5	13.2	9.9	5.5	9.2	38.8	191.6	197.4	105.3	19.3	2.8	4.3
B	1.2	1.0	0.8	0.5	0.8	2.1	7.4	7.9	4.0	0.8	0.1	0.4

Total annual normal rainfall 611.8 mm
A—normal rainfall in mm;
B—average number of rainy days with rainfall 2.5mm or more.

The precipitation over NCT of Delhi generates surface water run off through streams, drains and as sheet flow. Delhi being highly urbanized, the run off is high due to extensive paved area. Considering a run off coefficient of 30 per cent in urban areas and 12 per cent in other areas, the total surface run off works out to be 162 mcm. The major part of this run off generally contributes to Yamuna flow in the mid and down stream part of the river.

Water Supply and Demand

The Master Plan 2001 for Delhi suggests the norm of 363 litres per capita per day (80 gallons per capita per day (gpcd) with the following break up.

i)	Domestic	225 lpcd
ii)	Industrial, Commercial and community requirement based on 45,000 lts per hectare per day	47 lpcd
iii)	Fire protection based on 1% of total demand	4 lpcd
iv)	Garden based on 67 litres per hectare per day	35 lpcd
v)	For floating population and special uses like Embassies, Hotel etc.	52 lpcd
	TOTAL	363 lpcd

However, taking into account transmission losses for bringing raw water from distant sources, the requirement of raw water in Delhi would be about 400 lpcd (88 gpcd). The total requirement of water in 2001 considering the population of Delhi as 122 lakh would be 4.88 million cubic metres per day (mcm/day) (10.74 million gallon per day (mgd).

Present Raw Water Availability

The water availability from surface water sources viz. Yamuna, Ganga and Bhakra system is of the order of 1,150.2 mcm. The Yamuna river contributes substantial part of the total 724 MCM water available (NCT Delhi's share). The flood water is of the order of 580 mcm. About 50 per cent of this amount is not being used and flows out of Delhi.

The total ground water available in the territory is of the order of 291.54 mcm/yr. The present use is for irrigation, drinking and industrial purposes. The area irrigated by ground water is 47,500 ha. As regards the withdrawal for other uses, it is difficult to assess the quantity withdrawn from ground water reservoir. Besides, officially estimated withdrawal of ground water, substantial quantity of ground water is being withdrawn by private houses, hotels, hospitals, etc. Thus the contribution from ground water reservoir is quite substantial. A rough estimate indicates that about 142 mcm withdrawal is for drinking and industrial uses and 118 mcm for irrigation purposes.

Ground Water Situation

The ground water availability in the territory is controlled by the hydrogeological situation characterized by occurrence of alluvial formation and quartzitic hard rocks. The ground water occurs in weathered and fractured/jointed parts of these rocks. The shallow aquifers mainly consist of weathered residum, whereas the joints and fractures constitute deeper aquifers.

The alluvial tract occupying large parts constitute the potential ground water reservoir in the territory. The characteristics and potential of ground water reservoir, ground water movement and occurrence show a distinct variation even in the alluvial aquifers due to their manner of deposition. The aquifers in western alluvial plains of the territory are distinct from the aquifers of Chattarpur basin and those occurring in Yamuna flood plains. The hydrological parameters and the potential of these aquifers varies widely. The variation of quality of ground water in space as well as depth adds another dimension to the complex ground water situation in the territory. The presence of saline water aquifers at shallow depths varying from 20 to 50 m below ground level in many parts presents a typical ground water scenario. The ground water occurs under phreatic condition in the shallow aquifer zone whereas semi confined to confined conditions of ground water are quite common in deep aquifers.

In the Yamuna flood plains and the Chattarpur basin the shallow fresh water aquifers within depth range of 40-50 m behave as a single unconfined aquifer system. In western part of territory in Kanjhawala and Najafgarh blocks semi confined and confined aquifers occur at depth.

A generalized ground water flow map depicts the altitude of water table in the range of 200 to 250 m. above mean sea level. The north northeast-south southwest trending quartzite ridge acts as a divide to the ground water flow. The water table slopes westwards towards Najafgarh in the western side of ridge. In the eastern side ground water flows towards river Yamuna. Local ground water depressions have also been created due to excessive ground water pumpage in Chattarpur basin, Cantonment area, South Delhi etc. In Chattarpur basin the ground water flow is towards central part of the basin from all other sides mainly due to depression created from large scale pumping. East of ridge the ground water flow is towards the river Yamuna.

Ground Water Quality

The quality of ground water is variable in space and depth in alluvial formation whereas it is generally fresh in hard rock formations. In alluvial formations, the quality of ground water deteriorates with depth which is variable in different areas. The ground water is fresh at all depths in the areas around the ridge in the city and Mehrauli blocks and also in the Chattarpur básin. In the areas west of the ridge, in general, the thickness of fresh water aquifer decreases towards north-west, west and south-west the thickness of fresh water zones being limited in major parts. In the flood plains of Yamuna, in general, fresh water aquifers exist down to 30 m to 40 m.

Sum-up

The growing population and continuously increasing demand for water has put the total water supply system of NCT of Delhi under great strain. The demands have outstripped the present supply. The present installed capacity for drinking water supplies is 22.38 mcm/day (525 mgd) which is not adequate to meet the requirement of a population of 100 lakh of the territory. Considering the population as 122 lakh by 2001 the requirement would go up to 4.88 mcm/ day (1,074 mgd) leaving a wide gap in water demand and supply.

There is thus a need to augment the water availability to release the pressure on the strained water supply system of Delhi. The ground water resource though being developed extensively, hold further scope for augmentation which can bring additionality to the water supplies.

The ridge and adjoining areas can be taken up for rain water harvesting through construction of check dams and percolation tanks. The scheme taken up on micro watershed basins may not bring in substantial additionality but would provide sustainability to existing ground water structures which are already under strain due to declining water level.

Roof top rain water harvesting offers a reasonable solution to conserve the rainfall runoff in the ground water aquifer in urban area. The roof top harvesting in Delhi urban area can bring in additionality of 67.6 mcm/year to ground water reservoir. Considering even 10 per cent of total roof top rain water harvesting in institutional area the augmented ground water reservoir can provide about 0.0186 mcm/ day (4 mgd) to Delhi Water Supply.

Hyderabad

Hyderabad, the capital of Andhra Pradesh state is the fifth largest city in India. Rapid urbanization in the city has created stress on the environment. Migration of population and high concentration around urban centers has posed problem in providing water supplies in adequate quantity of assured quality standards. The urbanization and industrialization has greatly influenced the ground water quality of the area. Hyderabad district surrounded by nine municipalities falling in Ranga Reddy district forms the Hyderabad Urban Agglomeration. This area is of 1,000 Sq. Km. in arial extent and spreads on the North and South bank of the river Musi, a tributary of Krishna. The estimated population as on date is 57 lakh while it was 42.7 lakh as per 1991 census. It is expected to increase to 80 lakh by 2011 and 100 lakh by 2021 as per the growth rate. A large number of large and medium sized industries are located in the area which has greatly influenced the ground water quality. Indiscriminate drilling, exploration and innumerable borewells tapping ground water resources resulted fall in ground water levels. Hyderabad Metro Water Supply and Sewerage Board is presently supplying 660 MLD from surface water source which is found to be inadequate and the deficit is met from ground water development worked out to 89 per cent and is classified as dark category. Ground water draft is increasing steadily whereas recharge has remained almost constant. So it is essential to initiate ground water management practices within and outside the urban area in order to establish a balance between draft and recharge.

Water Supply and Demand

With increasing population growth and economic developments demand for water also increases, therefore it becomes imperative to plan optimum resources development in Urban areas. Both surface and ground water sources are fully utilized to meet the need of Hyderabad and environs. The surface water sources are Himayat Sagar, Osman Sagar, Manjira barrage and Singur dam.

The total storage capacity of these surface water source is 38.367 TMC. The demand of water for various establishment/purposes is as follows:

Domestic Demand (5 lakh × 135 LPCD)	=	770 MLD
Industrial Demand	=	220 MLD
Non-Domestic Demand	=	70 MLD
		1,040 MLD

There exists a deficit of water supply to the tune of about 380 MLD, part of this deficit is met through ground water source. Hyderabad Metro Water Supply and Sewerage Board (HMWSSB) is supplementing surface water supply through bore wells by drawing ground water. There are 5,399 number of 113 dia bore wells spread over the entire twin cities. In addition to this majority of the residents in the city have their own bore wells to meet their complete requirement.

The ground water resources evaluation indicate the ground water recharge in the area is of the order of 73.3 mcm and ground water draft is estimated at 65.7 mcm. The ground water balance stands at 7.6 mcm. In view of the fact that the ground water utility for various purposes is fast increasing, it is most essential to initiate projects on artificial ground water recharge.

Ground Water Availability

Ground water occurs under phreatic conditions in shallow weathered zone and under semi-confined to confined conditions in the fractured zones at deeper levels. The thickness of weathered zone varies from 5 to 25 m. Ground water development in the area is mostly through borewells. The depth of borewells range from 50m to 150 m. In places like Banjara hills etc., the borewells have been drilled up to a depth of 150 m or more, it is found that the fracture system exists up to 150 m in parts of area. Therefore massive formation with fractures and joints form the potential aquifer system in the area. The discharge of these wells vary from a meagre quantity to 9 lps with an average of 2.5 lps.

In alluvial area ground water occurs mostly under water table conditions in the inter granual pore space. The aquifer system in alluvium is mostly of homogenous in nature. The ground water development is mostly through large diameter dug wells or dug cum bore wells with yield ranging from 3 to 5 m^3/day.

Ground Water Resource

Ground Water Resource Evaluation essentially involves the quantitative estimation of two components viz. input and output. In order to evolve proper management strategies and development of ground water resources, it is essential to quantify all the components of recharge and discharge to arrive at the ground water balance. The recharge (input) consists of rainfall infiltration, recharge from surface water bodies, return flow from irrigation, drains and other sources created from human activities and sub surface inflow. The discharge (output) consists of draft through ground water structures, natural discharge to streams and subsurface outflow. Determination of these components, especially draft is extremely difficult task because of non-availability of ground water structures census.

While estimating the recharge to ground water the following factors are considered:

Out of the total area of 1,000 sq. km only 600 sq. km of area is considered as suitable for ground water recharge, and 400 sq. kms is considered as not suitable to recharge because the area consists of hills, dwellings, roads etc.

Ground Water draft is arrived by multiplying the number of ground water abstraction structures with their unit draft. In the absence of reliable well census data for the city the draft figures are arrived by computation of ground water draft based on the total minimum demand and supply of water. The shortage of water is considered met through ground water, as there is no other source. The HMWSSB is supplying 660 MLD from surface water source. The total demand stands to be 1,040 MLD as per the prescribed norms of 135 LPCD.

The deficit is 380 MLD. However, in view of present living conditions of various sections of people in the area, the per capita consumption may exceed 100 litres/day. Thus deficit of water therefore works out to be 180 MLD (65.7 MCM/year).

The total annual ground water recharge is 73.3 MCM and the ground water draft is 65.7 MCM per annum. Thus, the amount of ground water balance is 7.6 MCM. However, since the computation of both ground water recharge and draft are based on many assumptions and not based on real statistical data, the ground water balance arrived at has limitations in its correctness.

Ground Water Quality

As ground water plays crucial role in meeting the domestic and industrial requirements of Hyderabad and environs, assessment of quality of ground water becomes essential. Calcium is more than 200 mg/l (max. permissible limit is 200 mg/l) in areas around Uppal, Golkonda, Yusuf guda, Sanath Nagar, Jeedimetla, Bolaram Industrial area, Attapur, Hyderguda village, Sanjeevayya Park, Kattedan, Bhulaxminagar, Nirala nagar, Doctor's Colony, Siri Nagar, LB Nagar.

The concentrations of nitrate in the Hyderabad ranges from 1.2 to 760 mg/l. the general range is 10 to 300 mg/l. The highest value is recorded in Siri Nagar in L.B. Nagar Municipality area. The highest value may be due to pollution of ground water by SIRI's Industry, which is located near the collection point.

The higher concentrations of nitrate (>100 mg/l) occurs in Industrial areas such as Bollaram, Alwal, Kattedan and residential areas such as Hyderguda, Athapur, Hayathnagar and Saroor nagar areas. Rest of the area falls in desirable to permissible limits (45 to 100 mg/l.)

Sum-up

Ground water developing in the area is mostly through borewells. The depth of the bore wells range from 50m to 150m. The average yield of bore wells ranges from 0.2 to 22.5 lps. During pre-monsoon period of 1998, in about 75 per cent of the area the depth to water levels ranged from 6 to 8 m. Around 15 per cent area had water levels of 8 to 10 m depth. During post-monsoon period of 1998, in about 80 per cent of the area the depth to water level ranged from 2 to 4 m. The water level fluctuation between May and November 1998 ranged from 2 to 4m in about 85 per cent of the area. Water table in the area varies from 460 m to 560 m AMSL. The water table gradient around Musi river is 12 m/km and it is around 7 m/km away from Musi river. Most part of Musi River is effluent in nature. The long-term water level records show a mixed trend of rise and fall. Out of 27 stations, 11 stations show raising trend of 0.44 to 3.49 m. The remaining 16 stations show declining trend of 0.11 to 4.85 m. Due to urbanization, the area available for ground water recharge from rainfall has considerably reduced.

Both surface and ground water sources are fully utilized to meet the water needs of Hyderabad and environs. The surface water sources are Himayat Sagar, Osman Sagar, Manjira barrage and Singur dam. The total storage capacity of these surface water sources is 38.367 TMC. The total water demand of Hyderabad (urban) and environs is 1,040 MLD. Out of which Hyderabad Metro Water Supply and Sewarage Board (HMWSSB) supply 660 MLD from surface water sources w.e.f. 10-9-98 and 380 MLD from ground water sources. The total water demand of Hyderabad and environs is likely to increase to 1,580 MLD and 2,000 MLD by 2011 and 2021 respectively. To meet the further water demand, HMWSSB is planning to draw 410 MLD from Nagarjuna Sagar reservoir in the first phase. The ground water resource evaluation indicates that the ground water recharge in the area is in the order of 73.3 MCM and ground water draft is estimated at 65.7 MCM. The ground water balance stands at 7.6 MCM.

Mumbai

The Greater Mumbai is located on the western most periphery of the Maharashtra state. In the past *i.e.* in 1885, it was elongated shaped group of seven islands, *i.e.* Mumbai, Mazgaon, Matunga, Mahim,

Worli, Soyster rode and Old Woman's Island but with the advent of development, narrow creeks were filled up and reclaimed and now it is connected to the mainland. The area of Mumbai was further extended by adding suburban areas in the north and along the border of Thane district. Now Greater Mumbai comprises south Salsette, Trombay and Mumbai Islands having a geographical area of 603 sq. km. (Mumbai city-69 sq. km. & Mumbai suburb-534 sq. km.) It is bounded by N. Latitude 18°53' and 19° 19' and East Longitude 72° 47' and 72° 58'. The Arabian sea lies on the southern and western sides while it borders Thane district in the northern and eastern sides. The population of the Greater Mumbai as per 1991 census was 1,34,53,848 which included population of Mumbai city (31.98%), its suburb (41.99%) and extended suburbs (26.03%). The population of the city is expected to be double the present population by 2010 A.D. It is observed that population has doubled every 20 years.

Water Supply and Demand

The entire Greater Mumbai is an urban area and hence there are no irrigation projects in the district. All the major surface water reservoirs are used for water supply. At present the water supply is mainly from surface reservoirs and river schemes. With the present available source, the Greater Mumbai area gets 2,700 mld of water.

Future Source of Water Supply

The projected demand of water for Greater Mumbai for the year 2021 is estimated to be 5,355 mld. To meet this additional requirement of water, four water supply projects are being under construction and by the completion of these projects the total supply will rise to 5,479 mld.

Ground Water Resource

Based on the available data, the ground water resources for Greater Mumbai are as follows:

1.	Total geographical area	603 Sq. Km.
2.	Area considered suitable for ground water recharge/development	300 Sq. Km.
3.	Ground water recharge	105 mcm

4. Ground water withdrawal 22 mcm
5. Balance ground water for development 83 mcm

The ground water is presently developed through dug wells as well as borewells. However, there is a wide variation in the yield from wells which is higher in post-monsoon months and much lower in summer months. If the average yield of the well is considered to be between 30 and 50 cubic metres/day, then the number of additional ground water structures feasible would be between 500 and 800 wells. In Greater Mumbai area, 3,950 dug wells and 2,514 borewells are existing and this development is wide spread in the area.

The utilization of ground water through bore-wells is more in the Andheri, Malad, Goregaon, Kandivali, Bhandup, Kurla, Chembur, and Ghatkoper area. In addition to that new bore-wells are being drilled at every upcoming society. Mumbai Municipal corporation is providing subsidy for the construction of the bore wells, so that this water can be utilized for the gardening, sanitation and other purposes. Further new bore-wells are taken up in the industrial area to cater their partial needs.

Sum-up

Available utilizable ground water and the draft shows that there is a plenty of scope to develop ground water for domestic as well as industrial purposes but the quality in many localities will be unsuitable for drinking due to high concentration of nitrates. Even though plenty of ground water is available in the area, more emphasis is given by authorities on creating surface water reservoirs which involves high expenditure, rather than developing ground water side by side in planned way. The conjunctive utilization of available surface water and ground water in systematic planned way will be the best solution for meeting the present and future demand of water supply.

Presently the ground water marketing is carried out in the entire Greater Mumbai area, for which the water is fetched through the existing dugwells and borewells or even new borewells are being drilled for this purpose. The ground Water marketed is used for the construction purposes, hotel use and sanitation purposes in the society. This way the utilization of ground water through borewells is more in Andheri, Malad, Goregaoan, Kandivali, Bhandup, Kurla, Chembur and Ghat-Koper area. There is no regulatory control in ground water marketing in the Mumbai Metropolitan. Excessive ground water development in the beach areas and coastal area can lead to saline water intrusion. Hence this needs to be regulated through suitable ground water regulation.

There are thousands of industrial units of various types in Greater Mumbai area. The industrial effluents disposed by many industries may cause pollution to ground water and to some extent even surface water supply also. The quality of ground water, in general, is suitable for drinking except for its nitrate content at many places, this maybe due to inadequate sanitation facilities for nearly 20 lakh population. The trace element analysis of water samples shows that in some of the water samples presence of Cu, Cr, Cd, As, Hg is observed which is harmful for human health.

ROOFTOP RAIN WATER HARVESTING

With the increasing scarcity of water in the urban areas, dependence on ground water has increased manifold. But the natural recharge to ground water has been decreasing following construction of buildings and paved areas. An urgent need is now being felt to counter the rapid depletion of ground water in the urban areas by artificial recharge of ground water aquifers. Nature has provided us the cheapest source of water through rains and this water can be harvested to supplement the groundwater recharge. As repeatedly mentioned, in urban areas, rain water available from roof tops of buildings, paved and unpaved areas goes waste. This water can be recharged to aquifer and can be utilized gainfully at the time of need. The rain water harvesting system needs to be designed in a way that it does not occupy large space for collection and recharge system. A few techniques of roof top rain water harvesting in urban areas have been designed by the Central Ground Water Board[19] which are indicated below:

(i) Rooftop rain water Harvesting through Recharge Pit

- In alluvial areas where permeable rocks are exposed on the land surface or at very shallow depth, roof top rain water harvesting can be done through recharge pits.
- The technique is suitable for building having a roof area of 100 sq. m. and are constructed for recharging the shallow aquifers.

- Recharge Pits may be of any shape and size and are generally constructed 1 to 2 m. wide and 2 to 3 m. deep which are back filled with boulders (5-20 cm), gravels (5-10 mm) and coarse sand (1.5-2 mm) in graded form—Boulders at the bottom, gravels in between and coarse sand at the top so that the silt content that will come with runoff will be deposited on the top of the coarse sand layer and can easily be removed. For smaller roof area, pit may be filled with broken bricks/cobbles.
- A mesh should be provided at the roof so that leaves or any other solid waste/debris is prevented from entering the pit and a desilting/collection chamber may also be provided at the ground to arrest the flow of finer particles to the recharge pit.
- The top layer of sand should be cleaned periodically to maintain the recharge rate.
- By-pass arrangement be provided before the collection chamber too reject the first showers.

(ii) Rooftop Rainwater Harvesting Through Recharge Trench

- Recharge trenches are suitable for buildings having roof area of 200-300 sq. m. and where permeable strata is available at shallow depths.
- Trench may be 0.5 to 1 m. wide, 1 to 1.5 m. deep and 10 to 20 m. long depending upon availability of water to be recharged.
- These are back filled with boulders (5-20 cm), gravels (5-10 mm) and coarse sand (1.5-2 mm) in graded form—boulders at the bottom, gravel in between and coarse sand at the top of the sand layer and can easily be removed.
- A mesh should be provided at the roof so that leaves or any other solid waste/debris is prevented from entering the trench and a desilting/collection chamber may also be provided on ground to arrest the flow of finer particles to the trench.
- By-pass arrangement be provided before the collection chamber to reject the first showers.
- The top layer of sand should be cleaned periodically to maintain the recharge rate.

(iii) Roof Top Rain Water Harvesting Through Trench with Rechargewell

- In areas where the surface soil is impervious and large quantities of roof water or surface runoff is available within a very short period of heavy rainfall, the use of trench/pits is made to store the water in a filter media and subsequently recharge to groundwater through specially constructed recharge wells.
- This technique is ideally suited for area where permeable horizon is within 3 m below ground level.
- Recharge well of 100-300 diameter is constructed to a depth of at least 3 to 5 m below the water level. Based on the lithology of the area well assembly is designed with slotted pipe against the shallow and deeper aquifer.
- A lateral trench of 1.5 to 3m width and 10 to 30 m length, depending upon the availability of water is constructed with the recharge well in the centre.
- The number of recharge wells in the trench can be decided on the basis of water availability and local vertical permeability of the rocks.
- The trench is backfilled with boulders, gravels and coarse sand to act as a filter media for the recharge wells.
- If the aquifer is available at greater depth say more than 20 m, a shallow shaft of 22 to 5 m diameter and 3-5 metres deep may be constructed depending upon availability of runoff. Inside the shaft a recharge well of 100-300 mm dia is constructed for recharging the available water to the deeper aquifers. At the bottom of the shaft a filter media is provided to avoid choking of recharge well.

(iv) Roof Top Rain Water Harvesting Through Existing Tube-wells

- In areas where the shallow aquifers have dried up and existing tube-wells are tapping deeper aquifer, roof top rain water harvesting through existing tubewell can be adopted to recharge the deeper aquifers.
- PVC pipes of 10 cm dia are connected to roof drains to collect rain water. The first roof runoff is let off through the bottom of drain pipe. After closing the bottom pipe, the rain water of subsequent rain showers is taken through a T to an online PVC filter. The filter may be provided

before water enters the tube-well. The filter is 1-1.2 m in length and is made up of PVC pipe. Its diameter should vary depending on the area of roof, 15 cm if roof area is less than 150 sq. m and 20 cm if the roof area is more. The filter is provided with a reducer of 6.25 cm on both the sides. Filter is divided into three chambers by PVC screens so that filter material is not mixed up. The first chamber is filled up with gravel (6-10 mm), middle chamber with pebbles (12-20 mm) and last chamber with bigger pebbles (20-40 mm).

- If the roof area is more, a filter pit may be provided. Rain water from roofs is taken to collection/desilting chambers located on ground. These collection chambers are interconnected as well as connected to the filter pit through pipes having a slope of 1:15. The filter pit may vary in shape and size depending upon available run off and are back-filled with graded material, boulder at the bottom, gravel in the middle and sand at the top with varying thickness (0.30-0.50m.) and may be separated by screen. The pit is divided into two chambers, filter material in one chamber and other chamber is kept empty to accommodate excess filtered water and to monitor the quality of filtered water. A connecting pipe with recharge well is provided at the bottom of the pit for recharging of filtered water through well.

PREVENTING URBAN WATER POLLUTION

A large number of people now live in cities. Suburbs that contain many acres of hard surfaces characterize these urban environments. Natural vegetation such as forests and fields slow rainwater down, allowing it to soak into the surface. In contrast, streets, roofs, parking lots and manicured lawns, all provide hard, impervious, surfaces that prohibit rain from soaking into the ground. Since the excess rainwater cannot soak into the ground, it accumulates and rushes into storm sewers and waterways.

There is a common misconception that water traveling into storm sewers ends up treated at the local wastewater treatment plant. Because of this misunderstanding, many citizens use storm drains and ditches as places to dispose of all kinds of pollutant. Thus, storm drains carry large amount of pollution away from urbanized areas mixed with the excess storm water.

Street litter, fertilizers, pesticides, herbicides, pet and yard waste, motor oil, household hazardous wastes, and paint are just a few of the pollutants that find their way into storm drains. This water travels from storm drains into local streams, ponds, and lakes and ultimately into local streams and rivers. Because storm drains can be a major source of pollution to our waterways, it is important to keep polluting materials out of them.

Following is an overview of the problem, and some suggestions on how one can help keep storm drains clean.

Household Hazardous Waste Pollution: Many household products we use daily contain toxic materials that can threaten public health and the environment. Drain and oven cleaners, paint thinners, and bathroom cleaners are just a few of the items we use that can cause serious health and environmental problems.

Solution:

- Unused household hazardous water may be disposed of at the local disposal site.

Fertilizer Pollution: Fertilizers contain large amounts of chemicals that are good for lawns and plants when used properly. The main chemicals contained in fertilizers are nitrogen and phosphorus. These two chemicals can cause accelerated growth of aquatic plants leading to oxygen depletion and large fish kills.

Solution:

- If fertilizers are spilled, those should be swept and not washed them into the streets and storm drains.
- Fertilizers should never be used before anticipation of a heavy rain.
- Fertilizers should not be over applied. More application does not mean a greener lawn; it means more watering and mowing.

Pesticides and Herbicides Pollution: Pesticides and herbicides contain toxic materials that pose both environmental and human health risks. Humans, animals, aquatic organisms, and plants can be severely threatened by these chemicals. The toxins found in pesticides and herbicides can runoff lawns and gardens into storm drains and streams whenever it rains.

Solution:

- Uses of pesticides and herbicides should be minimized and uses of alternatives, if possible, should be considered.
- All directions and warnings must be read before use.

Paint Pollution: When improperly used, paint can poison people and animals. Lead is an especially harmful content of some paints.

Solution:

- Old paint may be donated to community groups
- Oil-based lead paints may be disposed of at the local household hazardous waste collection site.

Motor Oil Pollution: Motor oil can damage or even kill aquatic vegetation and animal life. An oil slick contaminating two million gallons of drinking water can develop from one quart of oil.

Solution

- Used motor oil may be recycled.
- Used motor oil should never be poured down a storm drain or onto the grass or driveway.

Street Litter and Plastic Pollution: Large quantities of street litter end up floating in streams, rivers and lakes. Street litter is often made of plastics. Plastics take hundreds of years to biodegrade and can be harmful to birds and animals who mistake them for food.

Solution:

- Litter may be put into garbage cans. It should never be thrown into storm drain or on the street.
- Plastic and other materials may be recycled as much as possible.

Pet Waste Pollution: If left on yards, pet waste can release untreated bacteria and other harmful materials into stream.

Solution

- Pet wastes may be buried or flushed down the drain.

CONCLUSION

Rapid urbanization accompanied by faster population growth has already created enormous stresses on the natural environment throughout the world including the Indian subcontinent. These stresses extent far beyond the land that urban areas actually occupy to affect the land that provides the resources to sustain urban life. Urban areas claim the ecological output and life-support functions of both nearby areas and distant regions. For example, nearly 60 per cent of the water withdrawn for human use goes to the urban areas-about half of that to irrigate food crops for urban residents, roughly one-third for use by industry, and the remainder for drinking and sanitation[20]. While the links between population and freshwater resources are complex, there is no doubt that population growth increases the demand for freshwater, more particularly in the towns and cities. While new approaches to managing urban water supply and demand can help in the short term, reducing population growth in urban centers, particularly through migration from rural areas is essential to avoid imminent catastrophe in many metropolis, cities and towns.

Closely related to population growth is the changes in the transformation of demographic patterns. The most vital of these demographic characteristics, particularly in terms of projecting future water and food needs in a fast-growing economy like India, is the rate of urbanization. Urbanization accelerates changes in diet away from basic staples like sorghum, millet, maize and root crops, to cereals requiring less preparation (such as wheat), fruits, livestock products, and processed foods. More urbanization influences the rate of growth in domestic and industrial water demand, as well as agricultural water demand through changes in the food demand. Rural to urban migration-and its attendant significant effect on demand structures—increased quite rapidly over the past-few decades in the country and will continue to grow further in future.

India needs a Blue Revolution in water management, just as we need of another Green Revolution in agriculture. Time is of the essence. Dwindling freshwater supplies per capita are threatening the health and living standards of thousands of denizens of the country's cities and towns. Achieving a blue revolution will require coordinated policies and responses to problems at international, national, and local levels. For example, in India, currently there is no coordination among the six ministries, including, water resources, food and environment that manage water. It is necessary to bring the water policy exclusively under one ministry. In fact, many urban areas of the country are growing in population so fast that their economies, services and infrastructures cannot keep up. Further, there is also chronic shortage of resources to solve the growing water problem any time soon.

In most parts of the country the groundwater resources has been over-exploited causing alarming environmental problems. High urban usage of water requires steps to augment supply by ensuring roof—top rainwater harvesting compulsorily. Further, as urban water supply is heavily subsidized, there is little compulsion of conservation of scanty freshwater. Needless to say, proper pricing of water should be a weapon for conservation. Unless people are made to shell out appropriate rates for the water they use, conservation is sure to fail.

The unfinished tasks in water supply in urban areas, may be summed up as augmentation to attain the prescribed norms, higher degree of reliability, assurance of water quality, a high standard of operational and management efficiency, accountability to customers and in particular special arrangement to meet the needs of urban poor living in slums, constituting nearly 22 per cent of the total urban population, and levy and recovery of user charges to finance the maintenance functions as well as facilitate further investment in the sector.

Last but not the least, a vital part of a long-term solution to urban water supply problem lies on recognition of the links between rapidly growing populations and shrinking freshwater supplies. Recognition, knowledge, and concern can help build the political will to avert a crisis and develop the commitment needed to assure that humanity's apparently unquenchable thirst for freshwater does not exhaust the country's finite water supply.

REFERENCES

1. Population Reports, Series M, No. 16, May-June, 2003, Johns Hopkins Bloomberg School of Public Health, Baltimore, USA.
2. Food Insecurity Atlas of Urban India; M.S. Swaminathan Research Foundation and World Food Programme, Chennai, India. p. 145.
3. Hinrichsen, D., Robey B., and Upadhyaya, U.D., Solutions for a water-short world, Population Reports, series M, No. 14, Johns Hopkins School of Public Health, Population Information Programme, Sept. 1998. p. 33.
4. Hardoy, J., Mitlin, D., and Satterthwaite. D: Environmental Problems in an Urbanizing World—Finding solutions for cities in Africa, Asia and Latin America, London, Earthscan Punlications, 2001. p. 448.
5. Satterthwaite, D: Coping with rapid urban growth. London, Royal Institute

of Chartered Surveyors (RICS), Sept. 18, 2002 (Leading Edge Series). p. 35.

6. United Nations (UN) Population Division, World Urbanization Prospects: The 1996 revision, New York, UN, June 1998, p. 190.
7. WHO/Unicef Joint Monitoring Programme for Water Supply and Sanitation: Global water supply and sanitation assessment 2000 report. Geneva, World Health Organization (WHO) and United Nations Children's Fund (UNICEF) April 24, 2001. P. 79.
8. Aiyar, Shankar: Thirsty India; why water has become problem No. 1 in cities and villages; India Today, June 9, 2003. p. 39.
9. Tenth Five Year Plan 2002-07, Vol. II, Planning Commission, Government of India, New Delhi.
10. Ibid.
11. Ministry of Urban Development & Poverty Alleviation, Government of India, Annual Report 2002-03. p. 22.
12. WHO/UNICEF Joint Monitoring Programme for Water Supply and Sanitation: Global water supply and sanitation assessment 2000 report. Geneva, World Health Organization (WHO) and United Nations Children's Fund (UNICEF) April 24, 2001. p. 79.
13. World Health Organization, Health and Environment in Sustainable Development: Five years after the Earth Summit, Geneva, World Health Organization, 1997 (No. WHO/EHG/97.8) p. 245.
14. Ibid
15. WHO—Health: Creating healthy cities in the 21st century: Earthscan Publications, 1999; pp. 137-72.
16. Serageldin, I: Toward Sustainable Management of Water Resources, Washington, D.C., World Bank, 1995. p. 1-33.
17. Tenth Five Year Plan 2002-07, Vol. II, Planning Commission, Government of India, New Delhi. p. 640.
18. Ground Water in Urban Environment of India; Central Ground Water Board, Ministry of Water Resources, Faridabad, December 2000.
19. Rain Water Harvesting Techniques to Augment Ground Water; Ministry of Water Resources, Central Ground Water Board, Faridabad.
20. O' Meara,, M.: Reinventing cities for people and the planet. Worldwatch Paper, Washington, D.C. Worldwatch Institute, 1999. p. 68.

9

Quality of Potable Water

"All UN member states have pledged to reduce by half the proportion of people without sustainable access to safe drinking water by 2015"

—Millenium Development Goals

Introduction

Water is essential for growing food, for household water uses, including drinking, cooking and sanitation; as a critical input into industry; for tourism and cultural purposes; and for its role in sustaining the earth's ecosystem. Of the total 1,357.5 million cu kms. of water on earth, only 3 per cent *i.e.* 37.5 million cu kms. is fresh water. As populations grow and water use per person rises, demand for fresh water is fast increasing. Yet there is no more water on earth now then there was 2000 years ago, when the population was less than 3 per cent of its current size. Rising demands for water for irrigated agriculture, domestic consumption and industry are forcing stiff competition over the allocation of scarce water resources among both areas and types of use.

The earth was home to 6 billion people in 2000: 1.1 billion of them lacked safe water and 2.4 billion lacked adequate sanitation. Safe water in sufficient quantities is fundamental to human health. Safe water and sanitation shape health through potable water supply, safe food preparation, hygiene, better nutrition and relaxation. Environmental security depends on people's actions to control salinity, drainage and water pollution; manage droughts and floods; and manage land and water to guard those resources.

According to the latest United Nations report, India with 1,880m^3/capita per year, is ranked 133rd on the list of 180 countries in order of water availability per person per year. In terms of water quality it comes third bottom up, with an indicator value of 1.31; Morocco and Belgium complete the list with—1.36 and 2.25 values against them. India gets poor rating not only for the inferior quality of water but also for its ability and commitment (or lack of it) to improve the situation. The position is similar in the countries like Belgium, Morocco, Jordan, Sudan, Niger, Burkino Faso, Burundi, central African countries and Rwanda. On the other hand, the list of best quality water is led by Finland followed by Canada, New Zealand the United Kingdom, Japan, Norway, Russia France and Sweden.

Water-and sanitation related diseases are found to be widespread. Nearly 250 million cases are reported every year in the world, with more than 3 million death annually, about 10,000 a day. Diarrheal diseases impact children most severely killing more than 2 million young children a year in the developing world[1]. World Health Organization (WHO) estimates that almost one million Indian children die because of an unsafe drinking water. As many as 220 million Indians do not even have access to safe drinking water. About 86 per cent of the total diseases in the country are directly or indirectly related to the poor quality of drinking water. Poor quality of drinking water is a major cause of diseases including jaundice, typhoid and diarrhea. These diseases are major killers almost around the world. Many children have retarded growth due to poor quality of water. Inequitable distribution of water also worsens the problem as some people enjoy free access to water while others do not have access to even safe drinking water. Much emphasis is being placed on reducing biological contamination, but contamination from naturally occurring chemicals in groundwater and from industrial and agricultural waste is also becoming a serious problem all over the world, more particularly in developing countries.

Freshwater and Economic Development

It is axiomatic to say that a country's level of freshwater use reflect- and, in fact, is one of the key measures of its level of economic development. In developing countries including India, people use far less water per capita than in developed countries. Developing countries devote most of their supplies to agriculture. India, for instance, uses 90 per cent of all water for agricultural purposes, with just 7 per cent for industry, and 3 per cent for domestic use. The higher the development, the more water is used for domestic and industrial purposes and less for agriculture. Around the world freshwater demand per capita is rising substantially as countries develop economically. Withdrawals of water have grown in all three-major categories of use—to meet rising industrial demand, rising domestic demand, including municipal use, and increasing reliance on irrigation to produce food. As the freshwater availability is limited, increasing demand is being met with quality compromise.

The level of water use also reflects the level of urbanization in a country and urbanization has taken place appreciably in all regions in India. But low household water use in many of the cities and towns in India reflects difficulty in obtaining freshwater. Piped water systems are rare in rural areas. This is the position in other developing countries of the world too. In fact, two-thirds of the world's population, the majority in developing countries, get their water from public standpipes, community wells, rivers and lakes, or rainfall collected off roofs. Often rural people—usually women and girls—must walk many kilometers and spend many hours fetching water for their households. As the world becomes predominant by urban, while agriculture depends more and more on irrigation, it will be difficult for cities to meet the rising demand for freshwater. The situation is worse in developing countries, India is not an exception, where rapid urban growth often puts tremendous pressure on antiquated, inadequate water supply systems.

Further, years of rapid population growth and increasing water consumption for agriculture, industry and cities and towns have strained India's freshwater resources and in many areas the demand for water much exceeds nature's supply. Though population growth rate has slowed somewhat, the absolute number of people added to the population each year—the relevant figure considering the availability of and need for freshwater-remains near historic highs. In fact, in India with such a large population base, a modest population growth rate translates into large absolute numbers. The country's population growth rate, at about 1.8 per cent per year, means about 19 million people added each year to current population of 103.7 million (2002). Obviously, such population rate has ominous implications on per capita freshwater supply.

The Johns Hopkins School of Public Health, Baltimore, USA estimates that in 1995, 31 countries—mostly in Africa and the Near East-faced water stress or water scarcity and population growth alone will push an estimated 17 more countries, with a projected population of 2.1 billion, into these categories within the next 30 years. By the year 2025, 48 countries, with more than 2.8 billion people—35 per cent of the projected global population in 2025—will be approaching water crisis[2]. In India, water availability per capita which was earlier 5,000 cubic meters, has now dropped to 2,200 cubic meters[3]. In fact, the country is fast approaching a phase of extremely stressed water availability conditions. This is also affecting quality of water very seriously with concomitant adverse effect on the health of the people. As per the 2001 census report, only 74 million households in the country had a source of drinking water within their houses; 25 million had the source near the house while a whopping 32 million still had to travel long distance to fetch water. Only 36 per cent households (or about 70 million) have access to tap water and an equal percentage drink untreated water from hand-pump[4].

Water Stress and Scarcity

As their populations grow, more and more countries are facing water shortages. A country is said to experience water stress when annual water supplies drop below 1,700 cubic meters per person. At levels between 1,700 and 1,000 cubic meters per person, periodic or limited water shortages can be expected. When annual water supplies drop below 1,000 cubic meters per person, the country faces water scarcity. Once a country experiences water scarcity, it can expect chronic shortages of freshwater that threaten food production, hinder economic development, and damage ecosystems.

Malin Falkenmark developed the concepts of water stress and water scarcity based on an index of per capita freshwater needs. She estimated a minimum need of 100 liters per day per person for household use and from 5 to 20 times as much for agricultural and industrial uses[5,6]. These concepts have been widely accepted and used by hydrologists, the World Bank, and other organizations. For example, Population Action International (PAI) has relied on them to make projections of per capita water availability and to forecast water shortages in 2025 and 2050 (see Table 9.1)

Table 9.1: Growing Water Shortages

Country	*Population 1995 (millions)*	*Water Per Capita 1995**	*Population 2025 (millions)*	*Water Per Capita 2025**	*TFR 1998*	*% Growth Rate 1998*
1	*2*	*3*	*4*	*5*	*6*	*7*
			Water Scarcity in 1995 and/or 2025			
Algeria	28.1	527	47.3	313	4.4	2.4
Bahrain	0.6	161	0.9	104	3.2	2.0
Barbados	0.3	192	0.3	169	1.7	0.5
Burundi	6.1	594	12.3	292	6.6	2.5
Cape Verde	0.4	777	0.7	442	5.3	2.9
Comoros	0.6	1,667	1.3	760	5.1	2.7
Cyprus	0.7	1,208	1.0	947	2.1	0.7
Egypt	62.1	936	95.8	607	3.6	2.2
Ethiopia	56.4	1,950	136.3	807	7.0	2.5
Haiti	7.1	1,544	12.5	879	4.8	2.1
Iran	68.4	1,719	128.3	916	3.0	1.8
Israel	5.5	389	8.0	270	2.9	1.5
Jordan	5.4	318	11.9	144	4.4	2.5
Kenya	27.2	1,112	50.2	602	4.5	2.0
Kuwait	1.7	95	2.9	55	3.2	2.3
Libya	5.4	111	12.9	47	6.3	3.7
Malawi	9.7	1,933	20.4	917	5.9	1.7
Malta	0.4	82	0.4	71	2.1	0.6
Morocco	26.5	1,131	39.9	751	3.3	1.8
Oman	2.2	874	6.5	295	7.1	3.9
Qatar	0.5	91	0.8	64	4.1	1.7
Rwanda	5.2	1,215	13.0	485	6.0	2.1
Saudi Arabia	18.3	249	42.4	107	6.4	3.1
Singapore	3.3	180	4.2	142	1.7	1.1
Somalia	9.5	1,422	23.7	570	7.0	3.2
South Africa	41.5	1,206	71.6	698	3.3	1.6
Tunisia	9.0	434	13.5	288	3.2	1.9
United Arab Emirates	2.2	902	3.3	604	4.9	2.2
Yemen	15.0	346	39.6	131	7.3	3.3
			Water Stress in 1995 and/or 2025			
Afghanistan	19.7	2,543	45.3	1,105	6.1	2.5
Belgium	10.1	1,234	10.3	1,217	1.6	0.1
Burkina Faso	10.5	2,672	23.5	1,194	6.9	2.9
Eritrea	3.2	2,775	6.5	1,353	6.1	3.0
Ghana	17.3	3,068	36.3	1,464	5.5	2.9
India	929.0	2,244	1,330.2	1,567	3.4	1.9

(*Contd.*)

Table 9.1 (*Contd.*)

1	*2*	*3*	*4*	*5*	*6*	*7*
Lebanon	3.0	1,854	4.4	1,261	2.3	1.6
Lesotho	2.0	2,565	4.0	1,290	4.3	2.1
Mauritius	1.1	1,970	1.5	1,485	2.0	1.0
Niger	9.2	3,552	22.4	1,452	7.4	3.4
Nigeria	111.7	2,506	238.4	1,175	6.5	3.0
Peru	23.5	1,700	35.5	1,126	3.5	2.2
Poland	38.6	1,458	40.0	1,406	1.6	0.1
South Korea	44.9	1,472	52.5	1,258	1.7	1.0
Tanzania	30.0	2,964	62.4	1,425	5.7	2.5
Togo	4.1	2,938	8.8	1,370	6.8	3.6
Uganda	19.7	3,352	45.0	1,467	6.9	2.7
United Kingdom	58.1	1,222	59.5	1,193	1.7	0.2
Zimbabwe	11.2	1,787	19.3	1,034	4.4	1.5

Note: TFR = Total Fertility Rate * In cubic meters per year

Source: Garner-Outlaw & Engelman, Sustaining water, easing scarcity: A second update, Washington, D.C., Population Action International, 1997. Gardner-Outlaw and Engelman base their calculations on UN Population Division population estimates. The growth rate and TFR data come from; Population Reference Bureau, World Population Data Sheet, 1998, Washington, D.C., 1998.

Author's Source: Population Report, Series M, No. 14; The Johns Hopkins School of Public Health, Baltimore, USA.

Calculations of water stress and water scarcity are based on estimates of a country's renewable freshwater supplies and do not include water withdrawn from fossil groundwater. Fossil groundwater is essentially a non-renewable resource: It takes tens of thousands of years for these deep aquifers to replenish themselves. A country may temporarily avoid the effects of water stress by mining its nonrenewable water supplies, but this practice is not sustainable, particularly if the population continues to grow rapidly and per capita demand for freshwater increases.

We may highlight again that as of 1995, 31 countries, with a combined population of over 458 million, faced either water stress or water scarcity. This represents an addition of only 3 countries since 1990, when 28 countries, with a combined population of 335 million faced chronic water stress or water scarcity. The number of people estimated to live in water-scarce and water-stressed countries rose by nearly 125 million during these five years, however, largely reflecting population growth in water-short countries. However, India is so far neither in the category of water stress nor in water scarcity, though may be in the category of water stress by 2025[7]. But in many regions of the country acute scarcity of water is already prevalent. Thus, the average for the country does not depict the real ground level situation.

Need for 'Safe' and 'Wholesome' Water

River beds are the genesis of our ancient civilization, The entire green planet is very much dependent upon the wholesomeness of water. In our early religious rituals, water was worshipped as a God or Goddess and different religious activities sharply indicated about the necessity of maintenance of wholesomeness of water. Water courses have their own purifying system, but excess pollution load in the rivers or in any water course may not find their way to purify water in natural course.

After the Industrial Revolution pollution was generated day by day because of magnanimity of industrial use of water in absence of scientific purification of discharged waste water. In the last century serious water pollution caused throughout the world and the wholesomeness of surface water, as also the underground water was at stake. Some realization has, of late come and different Acts have also been enacted to prevent pollution in the water courses and also in the underground water reserve, now people could realize that a healthy society depends upon the maintenance of wholesomeness of water.

Water is regarded as man's best friend in the physical environment. It serves him in many ways. As an article of physiological necessity, it has a high place in the healthy maintenance of the body. It is required for life processes, for the proper functioning of the organs. Much of the ill health in the underdeveloped countries is largely due to lack of safe drinking water. There can be no stage of positive community health and well-being without safe water supply.

Water intended for human consumption should be not only "safe" but also "wholesome". A safe water is one that cannot harm the consumer, even when ingested over prolonged periods. Water may be safe, but if it has an unpleasant taste or appearance, it may drive consumer to other, less safe, sources. Drinking water therefore should be not only completely safe but agreeable to use and wholesome; such a supply may be termed "acceptable" or "potable". Safe and wholesome water must be:

- free from pathogenic agents;
- free from harmful chemical substances;
- pleasant to the taste; and
- usable for domestic purposes.

Water is said to be contaminated when it contains infective and parasitic agents, poisonous chemical substances, industrial or other wastes or sewage. The term polluted water is synonymous with contaminated water. Pollution and contamination are the result of human activity.

Water is required for many purposes. These may be stated broadly as follows: (1) Domestic uses: On the domestic front, water is required for drinking, cooking, washing and bathing (2) Public purposes: Water is required for public cleansing, fire-fighting, maintenance of public gardens and swimming pools and numerous other civic purposes. (3) Industrial purposes: Without water, there cannot be any industrial development. Some industries like the iron and steel and paper industry need quite a lot of water. (4) Agricultural purposes: The food and raw materials needed by the world cannot be raised without water. Water is therefore an essential factor in the economic, social and cultural development of a community.

From the stand point of public health, water should be provided in adequate volume. A limited supply will be of very little benefit, however superior the quality of water may be. A daily supply of 150 to 200 litres (35-40 gallons) per head is considered an adequate allowance. The consumption of water, however, depends upon climatic conditions, standard of living and habits of the people. The larger the quantity and the better the quality of water, the more rapid and extensive is the advance of public health.

Competition for Scarce Water Supplies

As the availability of fresh water is shrinking, in developed and developing countries alike, competition among water users is increasing. Quite natural, tensions are particularly high in water-short areas where population pressures, urbanization, and development needs combine to create intense demand for limited freshwater resources. Looking 30 years into the future, FAO estimates that feeding the world's population will require 60 per cent more food. Most of that increase will come from intensified agriculture supported by irrigation. But water is already scarce in many countries and the competition for water from industrial and domestic users continues to grow. So where will we find the water to feed the world?

Indian states have become embroiled in disputes over water rights and over dams that might provide more water for one state but at the expense of another. The country has more than 250 million city-dwellers even though the rate of urbanization is among the lowest in the world. The percentage of urban dwellers in India keeps increasing—from 10.8 per cent in 1901 to 17.3 per cent in 1951 and 27.8 per cent in 2001. Rural-Urban migration and high demographic natural increase in cities will further increase the proportion to more than 50 per cent of the total population by 2020. In most of the cities and other urban agglomerations, the water supply sector is faced with a number of problems and constraints. Fresh water sources are being heavily exploited to meet the demands of the urban populations. As water source fail to meet rising demands, groundwater reserves are tapped, often to unsustainable levels. Almost all cities depending on groundwater are faced with the rapid depletion of their water tables. In addition to quantity, the country now faces problems of water quality. Over extraction of groundwater in Chennai has led to salinity intrusion in the coastal aquifers. The mismanagement is clearly evidenced in the increasing inter-state (and local) conflicts over water resources. Also urban areas are in conflict with rural needs. Reportedly, Tirpur in Tamil Nadu suffers a water deficit of 22 million cu.m. a year—apart from serious degradation of water quality—and has begun to import water. The position in Chennai is also not different. Many farmers within 35 km. radius of the city have abandoned farming and instead sell their water to industrial and domestic users.

Consequences of Over Use and Pollution

Over use and pollution of the world's freshwater resources are a recent phenomenon. Some 200 million Indians do not have access to safe and clean water. An estimated 90 per cent of the country's water sources are polluted with untreated industrial and domestic waste, pesticides and fertilizers run-off. According to a publication of the Rajiv Gandhi Drinking Water Mission-'Introduction to Rural Water Supply and Sanitation Programme in India, Published in 1993'—about 1.5 million children under five years (as against WHO's estimate of 1 million) die each year from water borne diseases. The country also loses over 200 million workdays annually due to these diseases.

The consequences of water pollution are dire as in India people still use water directly from rivers and lakes for drinking and bathing. Groundwater pollution is particularly serious as some 80 per cent of the country's domestic needs are met from this source (UNICEF report). Diseases like hepatitis and diarrhea, and roundworm, hookworm and guinea worm injections amongst the population are commonplace and, as reported by the World Bank, nearly 21 per cent of all communicable disease in India are water borne. The Indira Gandhi Institute for Development Research in Mumbai estimates that water borne illness cost the country as much as Rs. 366 billion (approximately US$8.3 billion) per year.

Drinking water quality is evaluated on the basis of the physical, chemical, and biological parameters. However, parameters like fluorides (permissible limit 1-15 mg/l) and nitrates (permissible limit 10 mg/l) that have long-term health effects need to be analyzed too. The incidence of communicable diseases like gastroenteritis and diarrhoea in summer is related to the consumption of contaminated water. The source of water supply is very important. Municipal water supply is considered to be the best and least contaminated. Those who are traveling and those who live in slums, do not get water from municipal water supply; they have to rely on some poor source of water like still water from a drum or hand pump near a waste drain or seepage of waste water into the water supply or putrid water due to faulty joints. The supplies that withdraw water from the upper-ground water table, usually 30-60 feet from the ground level, do not have good quality water.

While water-related diseases vary substantially in their nature, transmission, effects, and management, adverse health effects related to water can be organized into three categories; water-borne diseases, including those caused by both fecal-oral organisms and those caused by toxic substances; water-based diseases; water-related vector diseases. Cause and route of transmission of these diseases, geographic extent, number of cases and deaths per year globally are shown in Table 9.2. Another category-Water-scarce (also called water-washed) diseases-consist of diseases that develop where clean freshwater is scarce.

Pollution is pervasive. Few countries, whether developing or industrialized, have adequately safeguarded water quality and

Table 9.2: The Major Water-Related Diseases

Disease	*Cause and Route of Transmission*	*Geographic Extent*	*Number of Cases* [a]	*Deaths per year*
		Major Water-Borne Diseases		
Amoebic dysentery	Protozoa travel the fecal-oral route via contaminated water, food, person-to-person contact.	Worldwide	500 million per year	*
Bacillary dysentery	Bacteria travel the fecal-oral route via contaminated water, food, person-to-person contact.	Worldwide	*	*
Diarrheal disease (including amaebic and bacillary dysentery)	Various bacteria, viruses, and protozoa travel the fecal-oral route via contaminated water, food, person-to-person contact.	Worldwide	4 billion currently	3-4 million
Cholera	Bacteria travel the fecal-oral route via contaminated water, food, person-to-person contact.	South America, Africa, Asia	384,000 per year	20,000
Hepatitis A	Virus travel the fecal-oral route via contaminated water, food, person-to-person contact.	World wide	600,000 to 3 million per year	2,400 to 12,000
Paratyphoid and typhoid	Bacteria travel the fecal-oral route via contaminated water, food, person-to-person contact.	80% in Asia; 20% in Latin America, Africa.	16 million currently	600,000
Polio	Virus travels the fecal-oral route via contaminated water, food, person-to-person contact.	66% in India; 34% in Near East, Asia Africa	82,000 currently	9,000

(*Contd.*)

Table 9.2 (*Contd.*)

1	*2*	*3*	*4*	*5*
		Major Water-Based Diseases		
Ascariasis	Fertilized eggs are passed out in human feces. Larvae in eggs develop in warm soil. Human ingests soil on food. Hatched larvae penetrate intestinal wall, where they mature.	Africa, Asia, Latin America	250 million currently	60,000
Clonorchiasis	Worms reproduce in gastropod snails, then are swallowed by freshwater fish or other snails. When humans eat raw or undercooked fish, the worms migrate to the bile ducts and lay eggs.	Southeast Asia	28 million currently	None reporte
Dracunculiasis (guinea worm)	Worm Dracunculus ingested by Cyclops (a crustacean). When humans digest Cyclops, worm larvae are released into the stomach. Larvae penetrate intestinal wall, then develop into worms, migrating through tissues. After a year, adult worm reaches skin surface of lower extremities. Female comes in contact with water, then discharges larvae into water.	78% in Sudan; 22% in other sub-Saharan Africa and a few cases in India and Yemen	153,000 per year	None reporte
Paragonimiasis	Worms living in pulmonary cysts lay eggs in human lungs, which are coughed up and then swallowed. Worm eggs are passed out in feces and break when in fresh water. Larvae find snail host in which to replicate, then move into freshwater crab or crayfish. Humans eat uncooked seafood. Worms migrate in pairs from stomach through intestinal wall and diaphragm to lungs, where they mate.	Far East, Latin America	5 million currently	None reporte

(*Contd.*)

Table 9.2 (*Contd.*)

1	*2*	*3*	*4*	*5*
Schistosomiasis (bilharzias)	Eggs of the schistosome worm are passed out in human feces. Eggs hatch on contact with water, releasing the parasite miracidium. Parasite moves into freshwater snail, where it replicates. It is released back into water, then penetrates human skin within seconds and moves into blood vessels. Within 30 to 45 days, miracidium grows into worm, which can lay 200 to 2,000 eggs per day for an average of 5 years.	Africa, Near, East, rain forest belt in Central Africa, Western Pacific, Kampuchea, Laos	200 million currently	20,000
Major Water-Related Vector Diseases				
Dengue	Virus is picked up by a mosquito from an infected human or animal. The virus incubates for 8 to 12 days and replicates. The next time the mosquito takes a blood meal, the virus is injected into a human's blood.	All tropical environments; concentrated in Asia, Central and South America	50-100 million per year	24,000
Filariasis (includes elephantiasis)	Worm larvae are ingested by a mosquito and develop. When the infected mosquito bites a human, the larvae penetrate the puncture and reach the lymphatics, where they reproduce.	Africa, Eastern Mediterranean, Asia South America	120 million currently	None reported
Malaria	Protozoa develop in gut of a mosquito and are passed on in its saliva each time it takes a new blood meal. Parasites are then carried by the blood to the human's liver, where they invade the cells and multiply.	Africa, Southeast Asia, India, South America	300-500 million per year (clinical)	2 million

(*Contd.*)

Table 9.2 (*Contd.*)

1	*2*	*3*	*4*	*5*
Onchocerciasis (river blindness)'	Worm embryos are ingested by black flies. Embryos then develop into larvae within the black flies, and the flies inject the larvae into humans when they bite	Sub-Saharan Africa, Latin America	18 million currently	None reported**
Rift Valley fever (RVF)	Virus usually exists in animal hosts. Virus is picked up by mosquitoes and other blood-sucking insects and injected into the blood of humans. Humans also are infected while working with body fluids of dead animals.	Sub-Saharan Africa	NA	1% of cases

a. Number of cases are reported as incidence ("per-year")—the number of new cases occurring in a year—or as prevalence ("currently")—the number of cases existing at a point in time.

* Included in diarrheal disease

** No deaths but causes 270,000 reported cases of blindness annually

NA = not available

Source: WHO 1995 except amoebic dysentery, bacillary dysentery, dracunculiasis, dengue, and RVF from WHO 1998 and clonorchiasis and paragonimiasis from Muller & Morera 1994

Author's Source: Population Reports, Series M, No. 14, February, 1999

controlled water pollution. Many countries do not have standards to control water pollution adequately, while others cannot enforce water quality standards. Increasingly, international development agencies are urging that developing countries devote more attention to protecting and improving water quality[8]. The developed world also must spend more and do more to clean up degraded waterways, or economic development will stall and the quality of life will fall.

The indiscriminate dumping of untreated industrial and domestic waste into rivers and lakes further exacerbates already stretched water situation. The pollution reduces the amount of water available for use, creating scarcity in regions which otherwise have abundant water resources. Further, the domestic sector produces about 75 per cent of the waste water but sewage treatment facilities are inadequate in most cities and almost absent in rural India. According to the Central Pollution Control Board, in 1995, of the 8,432 large and medium industries in the country, only 4,989 had installed appropriate measures to treat wastewater before discharge. Of the over two million small-scale industrial units, a number of which-like tanneries for instance—are extremely polluting, very few have any treatment facilities whatsoever and their untreated waste invariably find their way into the country's water systems.

In the North Arcot district of Tamil Nadu, for example, some 80,000 litres of effluent containing toxic chemicals like mercury and chromium are discharged untreated each day from the tanneries. With increasing groundwater exploitation, poisonous minerals like fluorides and arsenic present below ground also find their way into the extracted water. They cause flurosis and arsenic poisoning.

Agriculture is the biggest polluter, even more so than industries and municipalities. In virtually every country where fertilizers and pesticides are used, they have contaminated groundwater aquifers and surface waters. Animal wastes are another source of persistent pollution in some areas. The water that goes back into rivers and streams after being used for irrigation is often severely degraded by excess nutrients, salinity, pathogens, and sediments that often render it unfit for any further use, unless cleaned—typically at great expense-by water purification plants. In the US, agricultural chemicals, eroded sediment, and animal wastes have fouled over 173,000 miles of waterways. Farming is said to be responsible for 70 per cent of current water pollution in the US. In India, which depends on irrigated agriculture for food supplies, more than 4 million hectares of high-quality land have been abandoned because of salinization and waterlogging caused by too much irrigation.[9]

In localized areas the acquifer has also been polluted through leaching of agricultural chemicals and even industrial effluent. In Bichchri village in Rajasthan, for instance, some 80 wells have been polluted due to discharge of dangerous chemical into irrigation canals. Sample of vegetables available in some Delhi markets have been found to be contaminated with pesticide residues and heavy metals like zinc, lead and cadmium. The main reasons for heavy metal accumulation in vegetables were irrigation with sewage water and industrial effluents. Aerial deposition from thermal plants and vehicular exhaust were other sources of heavy metals.

Women and children are particularly susceptible to pollution. Toxins have a lethal effect on the reproductive health of women. Flourides in the water have also been known to affect the foetus. Not only are women biologically more susceptible to the impact of contaminated water, but their contact with polluted water is greater than that of men.

The world's tremendous output of pollutants challenges the capacity of waterways to assimilate or flush away pollution. Saying among water engineers is "the solution to pollution is dilution." This truism is taking on frightening dimensions. Each year roughly 450 cubic kilometers of waste water are discharged into rivers, streams, and lakes. To dilute and transport this dirty water before it can be used again, another 6,000 cubic kilometers of clean water are needed—an amount equal to about two-thirds of the world's total annual useable fresh water runoff. If current trends were to continue, the world's entire stable river flow would be needed just for pollutant transport and dilution by the middle of the this century, according to an estimate by the UN Food and Agriculture Organization.

Yamuna—An Extremely Polluted River

Yamuna river is as polluted as ever. A ten-year clean up effort, started in 1993, seems to have literally flowed down the drain. The river is dead after it crosses Delhi. A look at the two broad parameters of river pollution bear testimony to this. The biochemical oxygen demand (BOD) levels have been hovering between 15 and 30 micrograms per litre, while the normal BOD level should be just 3 mgs per litre. Higher BOD levels mean lesser chances of survival for plants, fish and other aquatic life in the river. Aquatic life has perished in the Yamuna river as the oxygen supply for organic matter has reduced. Another indicator of a highly polluted river is the high level of coliform bacteria, which thrives on human waste. The coliform count varies between 1 lakh and 10 lakh, while the normal count should be at 5,000.

Dead River

- 790 mgd sewage flows into Yamuna
- 552 mgd can be treated in DJB-run STPs
- 360 mgd is treated
- 50 per cent of the sewer lines are not connected to these STPs
- 30 per cent of STPs do not work due to lack of water to dilute sewage
  (mgd: million gallons a day)

Higher level of coliform means more risk of water-borne diseases. The pathogens in the river water kill the organic life in it. Experts say the pollution levels have remained the same for the past one decade. The reason is the large amounts of waste let out into the river by the city residents. The city generates about 2,700 million litres of sewage from households every day. Another 300 million litres are discharged into the river by the industries. There is no clear water that joins Yamuna when it flows 22 kms along Delhi. So there is no dilution of the densely polluted sewage in the river. Only half of the sewage generated by the river is treated by the inadequate mini sewage treatment plants (STPs), planned in the first phase of Yamuna action plan. There are 50 sub-drains which pollute the river the most. A Tata Energy Research Institute (TERI) study says that even if one controlled the pollution in 10 of them, half of the pollution will be curtailed. But the first phase of the plan has gone haywire with little planning and even lesser clean-up action. Centre for Science and Environment (CSE)'s director Sunita Narain says, "Nobody had ever thought of cleaning the river by attacking the problem at the roots. Only symbolic shramdaans won't do. A planning with a bigger picture in mind is the key".

The Dirty Truth

Pollution Levels in Yamuna

- *Biochemical oxygen demand*: 15 to 30 mgs per litre (normal 3 mg/litre)
- *Coliform becteria*: 1 lakh-10 lakh (normal is 5,000)

Main Pollutants

- *Domestic sewage*: 2,700 million litres per day (mid)
- *Industrial sewage*: 300 mid

Yamuna Action Plan (Phase I)

- *Duration*: 1993-2003
- *Budget*: Rs. 700 crore
- *Delhi's share*: Rs. 170 crore

Aims

- *Community toilets*
- *Sewage cleaning equipment*
- *Mini sewage treatment plants*
- *Public awareness and participation*

The story so far

- *Lack of amenities in most toilets*
- *Equipment non-functional*
- *Mini STPs inadequate to handle bulk sewage*

The existing plants do everything but treat the faecal coliform that kills all organic life in the river. This is despite the fact that the Supreme Court had directed the Delhi Jal Board (DJB) to bring down the bacteria level in the Yamuna. Upgrading the existing secondary sewage treatment plants to tertiary plants will cost about Rs. 1,500 crore. Even the western countries cannot afford it. There are no parameters to bring down the coliform level in the river water. But the

Central Pollution Control Board (CPCB)—that monitors all operations to clean the Yamuna—said it has already issued guidelines to the DJB on how to do it.

DJB is stated to have been directed to bring the coliform level to 22,500 per 100 ml. For that, DJB will have to install STPs that can do tertiary treatment of sewage. The existing plants do not have this facility. The board has built 17 sewage treatment plants that it had promised to the Supreme Court, but maintains that it has not been briefed properly by the CPCB on the technological requirements of the STPs. When the plants were being constructed, the CPCB gave DJB the permissible biological oxygen demand and the level of suspended solid in the treated sewage. Their secondary treatment plants are equipped to treat the sewage accordingly. The board plans to enhance its sewage treatment capacity to 724 million gallons a day by the end of the tenth five-year plan. These plants will only be equipped for secondary treatment of sewage.

Contamination of Rural Water Supplies

In most countries, policy documents on integrated management of water resources give highest priority to the drinking water supply in water allocation decisions. Rural drinking water may not seem to be problem because water used for domestic purposes is only a small fraction of a country's total fresh water consumption. In Pakistan, water withdrawn for domestic purposes was estimated at 26 cubic meters per person in 1990, compared to 1,226 cubic meters of water per person for irrigation[10]. The notion prevails that a small diversion from the irrigation sector could fulfill the demands of a growing population for domestic water. However, reallocation of water among sectors can be difficult, and a truly integrated water-management approach is constrained by traditional sectoral thinking and by priorities set by professionals in various disciplines.

Public health officials are mainly concerned with increasing deterioration of water quality due to industrial, agricultural, and urban waste and to insufficient investments in domestic water supply infrastructure. This global concern for water quality reflects the high standards imposed on drinking-water quality by institutions and professionals in the western world, where the quality of drinking water ranks above all and where the direct use of surface water from irrigation systems for drinking seems unacceptable.

On the other hand, those who manage water for agricultural production believe their main responsibility is to provide water to meet crop-water requirements. Although few irrigation managers see supplying water for domestic use as their mandate, many rural residents draw their domestic water supply, directly or indirectly, from irrigation systems. Without acknowledging this reality, efforts to make irrigation systems more efficient might adversely affect the availability of irrigation water for nonagricultural uses. For example, lining canals with concrete to reduce seepage losses can cause shallow drinking-water wells to dry up.

The main strategy for improving rural drinking water has been to install low-cost hand pumps that draw groundwater uncontaminated by disease-causing microbes. In Bangladesh alone, more than 4 million tubewells have been installed over the past 20 years to provide safe drinking water to 95 per cent of the population. This has reduced the incidence of diarrhea. Only recently have the high concentrations of arsenic in many of the tubewells become apparent. What seemed like a public-health success story became one of the biggest environmental health crises of the 21st century—chronic poisoning of more than 20 million people exposed to high arsenic concentrations in their drinking water. A similar crisis is developing in India, where an estimated 66 million people drink groundwater with too high a fluoride content. While arsenic is toxic and carcinogenic, fluoride is an essential element for development and protection of teeth and bones. In excess, however, fluoride leads to serious dental and skeletal deformities and other health problems. Installing filters or other devices at millions of tubewells to remove arsenic and fluoride is an almost impossible task. Therefore, alternative sources of drinking water must be found in affected areas.

In some regions, the availability of shallow groundwater for drinking is an increasing problem because of overexploitation for agricultural and industrial purposes. In some of the major breadbaskets of Asia, such as the Punjab in India and the North China Plain, water tables are falling by 2 to 3 meters a year. Wealthier farmers can continue to drill deeper tubewells with larger and more expensive pumps. But poor farmers are not able to do so. All stakeholders see falling groundwater levels as a threat to food security. Groundwater depletion also causes the shallow drinking-water wells of poor communities to run dry, a problem that has received less attention. Deepening these wells is costly and beyond the resources of the poor. Too little is known about how pumping groundwater for irrigation might affect the levels of arsenic and fluoride in drinking water. However, as stated earlier that over pumping in coastal areas causes saltwater to invade freshwater aquifers, making the water unsuitable for drinking. Poor water quality in tubewells and the reduced availability of water in shallow dug wells often make it necessary to look at other sources of drinking water. Harvesting rainwater is being explored, but it might not be feasible in very poor arid and semi-arid regions.

STATUTORY CONTROL OF WATER POLLUTION

Statutory control of environmental pollution in India is not of recent origin. We have had some 2200 legislations dealing with various aspects of environment protection. But the concerted legislative activity in the backdrop of the drawbacks of dissipated and piece-meal earlier legislations and inadequacy of such legislations to meet the evolving challenges of pollution, in fact, started after 1970 with the enactment of some specific legislations dealing exclusively with pollution problems. Statutory control of water pollution under the Indian legal system broadly falls under three heads: (i) Statutory rights of riparian owners under Indian Easements Act, 1882; (ii) Penal and Public Nuisance Actions Under Earlier Statutes; and (iii) Administrative Regulation under the recent Central Environmental Statutes.

Constitutional Remedies

A regulatory mechanism for the prevention of environmental degradation, through writ process is provided for in our Constitution. Under Articles 322 and 226 of the Constitution the Supreme Court and the High Court respectively, poses a wide latitude to grant relief and prevent environmental damage by issuing directions, order or writs.

Under Article 322, which itself is a fundamental right, any person whose fundamental right as conferred by Part III of the Constitution has been violated can invoke the Supreme Court's jurisdiction to enforce his right. Whereas, the writ jurisdiction of the High Court under Article 226 may be invoked not only for the enforcement of a fundamental right but for any other purpose as well. For that matter, the Supreme Court's jurisdiction under Article 322 is more limited than the jurisdiction of the High Courts under Article 226. As now, the Supreme Court has accorded judicial recognition to the right to a wholesome environment as being implicit on Article 221, a litigant may, accordingly assert his or her right to a wholesome environment against state, by a writ petition to either the Supreme Court or a High Court. Laws can be outlined as well as implemented; but it is far-fetched to think that laws can be 'enforced'. For a law to be established what is required is a consciousness for the good of all, civic sense for living in a society and judgment between good and bad.

Supreme Court Notice on Potable Water Quality

Recently (August 25, 2003) the Supreme Court (Coram Khare, CJ, Sinha, J) issued notices to several Union Ministries including the food and civil supplies, and the health ministries asking them to detail steps they had taken to draw up standards to check the quality of drinking water available across the country.

The Court order came on a Public Interest Litigation (PIL) on the level of pesticides in drinking water, including ground water. The Centre for Science and Environment (CSE), whose letter to the court on the issue was converted into a PIL, welcomed the Supreme Court's action on the "right to clean drinking water". CSE director Ms Sunita Narain also took the opportunity to demand "legally enforceable drinking water norms". In her letter to the Supreme Court written on February 5, 2003 Ms. Narain had pointed out that even after 55 years of Independence, India did not have quality standards that would help to clearly define clean and potable water.

POSSIBLE SOLUTIONS

Recycling Urban Wastewater

Poor surface-water quality is due mainly to disposal of untreated wastewater from urban areas in rivers, canals, and lakes. An alternative is to reuse wastewater for agriculture, an ancient practice being revisited because of growing water scarcity. Wastewater can be seen as a resource with high potential for reuse in agriculture, which presents an opportunity for increasing food security in rapidly growing urban areas. Definite environmental and economic benefits accrue from reuse that

- Avoids direct pollution of rivers, canals, and other surface water;

- Conserves water;
- Conserves nutrients, thereby reducing the need for chemical fertilizer;
- Disposes municipal wastewater at a low-cost, sanitary way; and
- Provides a reliable water supply to farmers.

However, a number of disadvantages cannot be ignored:

- Health risks for the irrigators and communities in prolonged contact with wastewater;
- Health risks for the consumers of produce irrigated with wastewater;
- Contamination of groundwater with nitrates;
- Build up of heavy metals and other chemical pollutants in the soil;
- Creation of habitats for mosquitoes and other disease vectors; and
- Possible limiting of marketing options (particularly for export) for agricultural produce.

To safeguard the health of irrigators and consumers, the World Health Organization (WHO) has formulated international guidelines on wastewater reuse in agriculture and aquaculture. The guidelines establish the number of fecal coliform bacteria and worm eggs allowed for unrestricted irrigation. Using wastewater for irrigation has the effect of localizing the health risk because the exposed group remains relatively small. Adequate health measures could be targeted at this exposed group. If untreated wastewater is dumped into surface-water sources, much larger populations of downstream water users could be exposed to certain health risks. This is especially relevant for arid and semi-arid regions such as Rajasthan, where irrigation canals are often the only open water bodies. These irrigation canals receive untreated wastewater from cities, which is used for washing, bathing, and even drinking.

The solution appears simple: treat the wastewater before use and disposal. Excellent treatment methods exist. But they are prohibitively expensive for most developing countries, which have the most need for this source of irrigation water. The reality today is that two-thirds of the urban wastewater generated in the world receives no treatment at all. The cost of providing wastewater treatment facilities for all cities is astronomical. Even if the resources were available, improved water quality is not guaranteed. Many of the existing wastewater treatment plants are not functioning properly because local authorities often prefer high-technology solutions to more appropriate, lower-cost alternatives.

Most conventional treatment methods remove the nutrients in wastewater, reducing economic benefits to farmers. Restricting the type of crops being cultivated with untreated wastewater to tree or nonfood crops that pass less contamination into the food chain is another option, but difficult to enforce in many developing countries. Crop restriction also reduces economic benefits from the use of wastewater, since the vegetables, most susceptible to contamination, are also the most profitable.

In the foreseeable future, many towns in developing countries will continue or expand the irrigation of high-value vegetable crops with untreated wastewater. Governments may wish to regulate reuse but are unable to offer practical solutions to the users. It is urgent, therefore, to develop a framework for evaluating different options and trade-offs so that governments and communities can make better-informed decision.

Low-Cost Cleaning of Water

An experimental mix of chemicals permits low-cost home treatment of highly contaminated water. The packet has been designed for use in developing countries, where some 5,000 children die each day from diarheal diseases—primarily because of poor sanitation and infected drinking water. The new treatment turns even dark, foul-smelling, germ-laden water into a drink as clean as most US tap water, says Stephen Luby of the Centers for Disease control and Prevention in Atlanta. The chemicals' cost should run about a penny per litre of treated water, according to Grey Allgood of Procter and Gamble's Health Sciences Institute in Cincinnati.

During tests in Guatemala, Kenya, Pakistan, and Bangladesh, residents were instructed to stir a four-gram packet of the chemicals into a 10 litre jug of river or other water for five minutes, until dirt and other suspended materials settled out. Villagers then filtered out the sediment by pouring the water through tightly woven cloth. Over the next 20 minutes, the water's residual chlorine bleach vanquished germs.

SOME RECENT STUDIES

Delhi Groundwater is Turning Toxic

According to a Central Ground Water Board publication, Ground Water in Urban Environment of India, Delhi's sub-soil aqua reserves have a problem of excessive nitrate content. In areas like IIT, NCERT campus, Naraina and parts of Shahdara, nitrate content is more than the permitted 10 mg per litre. Other parts of the city like Najafgarh and Kajhawla have more than the permitted 1.55 mg per litre of fluoride. Salinity in Water is a problem in large parts of Delhi. Some parts have heavy metals and fertilizer content in ground water as well. Different pollutants have different effects on people. High fluoride content leads to dental and skeletal fluorosis, bone deformation and other disorders. Excess nitrate is known to cause hole in chest, skin becoming slight blue in colour and digestive disorders. Saline water can cause an imbalance in the salt content of the body.

The state of groundwater in Delhi is gradually being poisoned with the release of pesticides, toxic chemicals and salts of heavy metals from the three landfills of Ghazipur, Okhla, and Bhalswa. These landfills receive over 75 per cent of the 5,000-5,500 metric tonnes of untreated waste collected from the city every day, of which 60 per cent is non-biodegradable. These landfills are not scientifically oriented as the pits are only used as dump sites. Landfills should have a layer of synthetic plastic or a special type of clay layer to stop seepage but these landfills have a layer made of malba topped with soil. Muncipal authorities blame financial status for the condition of the landfills. Presence of metal salts could lead to disease like flurosis. In certain areas, high traces of fluorides and nitrates have been reported which are much above the World Health Organization limits. Whenever it rains, soluble toxic elements like fluorides, zinc, copper sulphates, lead and pesticides seep down to the water strata from the landfills. The Central Groundwater Board has set up monitoring sites around the landfills to carry out tests on the state of toxicity of the waste.

A large number of Delhiities are either totally dependent on the sub-soil reserves or use it in addition to the meagre Delhi Jal Board

(DJB) supply. Moreover, all bottling units in Delhi are totally dependent on groundwater for filling up their bottles. In January 2000, the Central Ground Water Board (CGWB) and the Central Pollution Control Board (CPCB) had conducted a joint study to find out how bad the Delhi's groundwater was. The biggest problem was found to be that of excessive salinity, nitrate and fluoride content. Traces of hard metal and other contaminants were also found in some samples. Fluorides come mainly from sub-soil clays and rocks and nitrates from untreated sewage, leaking septic tanks, human faeces and leaching at the landfill sites.

Residues of pesticides were detected from samples drawn from places near agricultural areas like Alipur; Kanjhawala, Najafgarh and Mehrauli. But the contamination seems to have worsened in the past two years. The CSE study drew samples from the tubewells of bottling units located in places like Okhla. Shivaji Marg and Burai, and pesticides were found in all of them.

A Cocktail of Pesticides in Mineral Water Bottles

Mineral water bottles may contain a cocktail of pesticides, if the findings of a study conducted by the Center for Science and Environment (CSE) are to be believed. The CSE's Pollution Monitoring Laboratory (PML) analysed 17 brands in the national Capital and 13 brands sold in street—corner shops in Mumbai. The results showed that almost all of these contained five different pesticide residues harmful to health and far exceeding the specified standards for drinking water. The shocking report also mentions that the least contaminated bottled brands are ironically the ones that are least sold. The largest selling brand contained the third largest pesticide residues in the Capital. The concentration levels of pesticide residues ranged from 79 times to 104 times. The study was conducted between July 2002 and January 2003 on randomly selected sealed bottles bought from the markets.

Samples contained enough contaminants to cause cancer, liver and kidney damage, disorders of the nervous system, birth defects and disruption of immune system in the long run. Pesticides do not kill immediately, but can cause irreparable health disorders as they accumulate in body fat. The lab tests in Delhi showed that each sample had 36.4 times more pesticides than the stipulated levels while the Mumbai samples were a shade better. Most commonly found pesticide residues were lindane, DDT, malthion and chlopyrifos. It was also pointed out that the CSE had used European norms because the standards set by Bureau of Indian Standards (BIS) were "vague and undefined". The study reveals that the quality of mineral water bottles sold in Mumbai was better thatn that sold in Delhi but the worst offender in Mumbai had a pesticide contamination which was 16.7 times higher that the prescribed level but far lower than that in the brands sold in Delhi.

In a major judgment, the Delhi High Court (HC) has stripped the Municipal Corporation of Delhi (MCD) of its power to issue licence for manufacturing and processing "packaged drinking water" holding that the Bureau of Indian Standards (BIS) was the sole competent authority in this regard. The judgment said there is no purpose in two authorities scrutinizing the very norms for a premise where packaged drinking water had to be manufactured and the more stringent norms as prescribed by the BIS were liable to be followed.

WHO/UNICEF Study

Unicef sponsored Study reveals that as many as 200 districts in the country have a high fluoride concentration in ground water. About 25 million people, spread over 17 states, are affected, while another 66 million have been described as a population at risk. The country assessment report, on which the Planning Commission collaborated, is part of a joint monitoring programme for the sector. The report calls for enactment of a ground water legalization in all states. Aiming at equity and sustainability in access to ground water and its development needs. The study also calls for a river-basin or sub-basin based approach to water management in the country.

"The present approach restricts the issue only to political boundaries, involving a number of agencies and ministries with overlapping responsibilities" it notes. According to the report, a river—basin based approach will ensure that issues such as water allocation, controlling pollution and protection of water resources "are not dealt with in isolation and decisions on the overall development process and land-use planning flow from this" As for ground water contamination, the report notes that fluoride is not the only contaminant, with nearly 13.8 million people in 75 blocks in West Bengal also reported to be at risk owing to presence of arsenic in ground water. There are varying levels of iron, too, particularly in the north-eastern part of the country, and heavy metals like lead, nickel, chromium, zinc, copper and manganese, apart from nitrates and bacteriological contamination.

United Nations Study

A world water development report of the United Nations has categorized India among the worst countries for poor quality of water, as well as their ability and commitment to improve the situation. The Asian rivers are the most polluted in the world, with three times as many bacteria from human waste as the global average. These rivers also have 20 times more lead than those of the industrialized countries, says the report.

India ranks a poor 120 in a list of 122 countries ranked for their water quality as also their ability and commitment to improving its quality. In terms of water availability too, India has not fared too well, ranked a lowly 133 in a list of 180 countries. The poorest in terms of water availability is Kuwait where just 10 cubic metres of water is available per person per year, with the Gaza Strip, the Bahamas and Qatar a little above it. Among the top five water-rich countries are Greenland, Alaska, French Guyana, Iceland, and Guyana. Belgium is considered the worst basically because of the low quantity and quality of its groundwater, combined with heavy industrial pollution and poor treatment of waste water. It is followed by Niger, Burkina Faso, Burundi, Central African Republic and Rwanda.

India's neighbours, Bangladesh, Sri Lanka, Nepal and Pakistan, have fared better occupying the 40th, 64th,78th, and 80th slots. Attributing this to "inertia at leadership level" the report entitled "Water for People, Water for Life" observes that "the global water crisis will reach unprecedented levels in future with growing per capita scarcity of water in many parts of the developing world. The report had been compiled on the eve of the Third World Water Forum held at Kyoto, Japan, by 23 UN partners constituting the World Water Assessment Programme (WWAP) under the UNESCO. It further observes that water resources will steadily decline because of population growth, pollution and expected climate change.

Globally the challenge lies in raising the political will to implement water-related commitments. Water professionals need a better understanding of the broader social, economic and political context, while politicians need to be better informed about water resources issues. Otherwise, water will continue to be an area for political rhetoric and lofty promises instead of very much needed actions. The list of the countries with better quality is headed by Finland, followed by Canada, New Zealand, United Kingdom, Japan, Norway, Russian Federation,

Republic of Korea, Sweden and France. It ranks over 180 countries and territories in terms of the amount of renewable water resources available per capita.

The report adds that by the middle of this century at worst seven billion people in 60 countries will be faced with water scarcity, at best 2 billion in 48 countries, depending on factors such as population growth and policy-making. Climate change will account for an estimated 20 per cent of this increase in global water scarcity.

A FEW EXAMPLES OF WATER POLLUTION

Even as the government orders a high-level probe into the alleged presence of pesticide residues in bottled water, the industry is asking which safety/hygiene norms should Indian companies follow—Indian or European? While the two cola majors, Pepsi and Coke, say that their bottled water brands, Aquafina and Kinley, met the Indian BIS norms for pesticide residues. FICCI's Food Research Analysis Center (FRAC), which has been testing bottled water for the past two years, has also found that brands like Bisleri, Aquafina, Kinley and Purelife are meeting the Indian pesticide residue norms for bottled water, as laid down by the Bureau of Indian Standards (BIS). However, the National Accredited Board for Testing laboratories (NABL)-accredited FRAC has also, like the Center for Science and Environment (CSE), found that the Indian bottled water brands fall short of the European standards on pesticide residue. Ground water contamination in India was far higher than in the European countries, so it stood to reason that bottled water in this country would have more pesticides than European bottled water brands. The government on the other hand finds the presence of deadly pesticides residues in various brands of bottled drinking water, very disturbing and serious. Meanwhile, the two major cola companies stood firm on their grounds about the purity of their bottled water. According to a Pepsi spokesperson, " Aquafina bottled water is produced to meet WHO prescribed standards, and follows a stringent seven-step process of purification.

Analysts also say that the incremental costs involved in reducing the pesticide residue levels in bottled water to European standards may result in a spurt in the prices of bottled water. There is a need for cost benefit analysis on how much pesticide is acceptable at what costs. Cost-benefit analysis has to take into consideration the expenses on treating diseases caused by taking bottled water and resultant loss of time and wages.

The Arsenic Effect in West Bengal

Tubewell with arsenic contaminated water is the main source of water for the entire at Kamdev Kati village in Kolsur Panchayat of Deganga block (about 120 km from Kolkata) in North 24 parganas of West Bengal. The affect of consuming water with high levels of arsenic is physiologically and psychologically apparent amongst the villagers. A number of tubewells in the village contain arsenic way above permissible levels. According to the Bureau of Indian Standard, the permissible level of arsenic in drinking water is 0.05 milligram per litre. Victims of arsenic menace suffer from dispigmentation of the entire body, boils on the limbs, keratosis (which appears like cellulosed skin protrusions on palms and soles) and burning sensation through the body. This leads to being socially ostracized from ones own immediate family. Substantial number of people are suffering from the same ailment with differences in intensity.

In the early eighties, cases of arsenical dermatitis were reported from some districts of West Bengal. Analysis of water sample revealed that the groundwater was contaminated with arsenic in these villages. According to a latest survey, 1,312 mouzas (revenue villages), 15 non-municipla/outer growths in 65 blocks and 9 municipalities of Malda, Murshidabad, Burdwan, Hooghly, Nadia, North 24 Parganas, South 24 Parganas and Howrah districts have dangerous levels of arsenic in their groundwater. A total population of 5.36 million in these 8 districts is exposed to arsenic through groundwater use. This accounts for 7.87 per cent of the state's population. Due to continuous extraction of groundwater and peculiar geographical conditions, the extent of the problem may increase with the passage of time.

A revival of traditional water harvesting systems involving local communities could be just the solution to the arsenic menace in West Bengal. Efforts to provide sanitary dugwells in two identified arsenic affected villages-Kamadev Kati and Ambika Nagar in North 24 Parganas district of West Bengal have been started. If successful, this low cost option could be widely replicated in the region through non-government organizations or by the communities themselves. The rationale behind this project is simple. Arsenic content beyond permissible levels is usually found in water from deep tubewells, at a depth of about 30m. On the other hand, it is rarely found in phreatic or unconfined aquifers. It is this very property of phreatic aquifers that will be exploited. These shallow aquifers will be tapped by dugwells to provide arsenic free water, which can subsequently be recharged by harvesting rain.

In view of the seriousness of the problem, all available options must be explored and an appropriate technology then introduced. At this point of time the options available for dealing with arsenic included:

(a) Roof water harvesting and storage or artificial recharge;
(b) Conversion of the commonly occurring village ponds (pukur) into infiltration basins and installation of hand pump or shallow tubewell nearby; and
(c) Construction of a hygienic power well to tap the phreatic aquifer.

Polluted Water in Ludhiana District of Punjab

Lakhs of residents living in 40 villages of Ludhiana are yet to get a reprieve from the pollution caused to underground water by the Budha Nullah, a large sewerage drain that criss-crosses the district and carries toxic elements into the Sutlej. Reams of paper have been wasted by environmentalists and the media about the alarming destruction caused to the environment by the Budha Nullah. The Punjab State Human Rights Commission has gone through a year of litigation hearing the government agencies about the corrective measures they had taken to reduce the pollution level. Strong claims have been made before the commission that safe drinking water was being provided. But the stark reality is that the pollution levels have assumed alarming proportions in the Budha Nullah. People are forced to consume unsafe water and the crops and vegetables have been found with large traces of toxic substances.

The people live so close to the river that they get plenty of water for drinking and irrigation purposes. But the irony is that the water has been classified as unfit for human consumption. Groundwater is easily available at 30 feet, but thanks to the pollution on account of seepage of toxic effluents all along the nullah, safe water is available only at 350 to 400 feet. While 40 villages in Ludhiana district are most seriously affected and suffer from water-borne diseases, especially hepatitis-B and jaundice, the contaminated water is being served to lakhs of others in Faridkot, Ferozepur and Bathinda districts. It is this water which goes to the Harie barrage and is finally supplied through canals to those districts.

Water Pollution in Punjab

The Punjab Pollution Control Board Study (1995) states:

- The quality of the Sutlej at Nangal is good and water here is 'A' category (designated as best use drinking water source). The river water quality degrades to 'C' due to addition of pollutants as it flows downwards. It further degrades to 'D' as the Budha Nallah from Ludhiana disposes industrial waste waters and domestic sewage (about 10,296 kg/day) into the river. There is no fish in the river upto 20km downstream of Ludhiana.
- The quality of the Beas on entering Punjab is generally 'A' category. It falls to 'D' immediately after it receives the industrial and domestic waste of Mukerian and the Goindwal Industrial Complex.
- The quality of the Ravi is 'A' category due to less human and industrial activity.
- The quality of the Ghaggar is mostly 'D' category. The virtual absence of life in the river water indicates that the pollution load received by it is much more than its rejuvenating capacity.

Pollution in Haryana

To prevent the pollution of the Yamuna, the Haryana Pollution Control Board (HPCB) has ensured that all industrial units with 5KL and more of discharge into the Yamuna install ETPs under the Water (Prevention and Control of Pollution) Act, 1974. Haryana has 3,605 polluting units under the Water Act and 81 units under the Air Act, 1981. There are 1,060 hazardous waste generating units. Although not all units have been conducting regular mock drills to check disaster preparedness, NFL and Indian Oil Refinery, Panipat, conduct such drills frequently.

H.S. Bains, Chairman, HPCB, however admits that small-scale industries are not very inclined to follow pollution control norms. To ensure safe disposal of hazard waste, 32 acres have been earmarked in Faridabad.

Mewat Region (Haryana)

The Mewat area of Haryana, reels under acute water shortage. The water crisis is so severe that the early morning hours here see man and animal alike using village ponds. It is the same story in the other adjoining villages—Ghaera, Salambha, Firozpur Namak, Salaheri. Five blocks of Mewat-Nuh, Firozpur Jurka, Tauru, Nagina and Punhana—fall in Faridabad district. The 491 villages in these six blocks have population of around eight lakh. Over 80 per cent of the population is of Meo Muslims while the rest are made of Gujjar, Ahir and Jats. But the population that is the worst hit is staying in the foothills of the Aravallis.

In Mewat, the underground water level has fallen to 250-300 feet and people have been forced to drink brackish water from hand pumps and tubewells. Under the National Water Mission, the aim is to provide 40 litres of pure water to each citizen. But in Mewat, even a glass of pure drinking water is rare. The area has been reeling under acute drought from year to year. This has resulted in a poor rabi crop and ponds have dried up. The only pond in Ghasera village is Nuh block and here the water has turned green—a green, darker than the leaves. From six to seven in the morning, the male villagers use the water for cleaning and washing while the domestic animals and dogs drink the same water. From 8 am to 10 am, women can be seen washing clothes at the same pond and then bathing. The areas do get canal water, which is sweet, but compared to demand, this is negligible. The Mewat Development Agency (MDA) built 145 tubewells over the past five years; 17 of these are now providing only brackish water. Of the 491 villages, 139 are facing unprecedented water shortage. These villages do not even get potable drinking water and even the generators are useless as power supply is irregular.

The water shortage in the area has had its social repercussions also. The people are forced to illegally tap the canal water. There are some bitter fights over such tapping in these villages. The acrimony has increased with the water shortage becoming more acute. Rajaka, Karhera, Firozpur, Bhadas and Uleta villages are also facing the same situation. Sakras, Dohe, Ravali, Agon and Pathkhori of Firozpur Jhirka block, Ranyala Khurd, Uttarwar, Ransika, Bhagru Palwal and Firozpur Rajput—villages in Tauru block are also facing the same problem.

Pollution in Himachal Pradesh

With virtually nothing in the name of crisis management, Himachal may find itself in trouble in case of an accident of a severe order. Out of 2,000 units, there are 89 hazardous units in Himachal, major ones being cement-manufacturing plants, paper mills, bulk drug industries and leather units. Most units are located close to the residential areas, increasing the gravity of the problem. There is no off-site plan for any district even on paper, nor does one hear to any state crisis management committee in Himachal. None of the officials in the Himachal Pradesh Pollution Control Board (HPPCB) and the Labour Department seems to have heard of such a committee.

Another problem is the gross understaffing because of which the Labour Department finds it increasingly difficult to conduct inspections of units involving hazardous processes. There are just about 26 labour inspectors. No wonder the number of inspections conducted is less, while the number of prosecutions is even lesser. In 2000, there were just 665 inspections and nine prosecutions. In 2001 there were 772 inspections and four prosecutions, while 2002 had 3,577 inspections and just five prosecutions under the Factories Act. Sources disclose that understaffing and lack of mobility hampers efficiency of the department which, despite being aware of the danger, can do little about disaster control. Incidentally, the Labour Department in Himachal just has one clerk. The Joint Labour Commissioner, Himachal, S.C. Awasthi, however, maintains that the department is ensuring safety standards.

Out of the 4,000 units categorized under the Water Act, 14 are highly polluting. The main threat comes from the cement units operating in the Barmana and Rajbans areas. The already weak road network is being further destroyed by the numerous trucks these cement plants ply. A few larger plants are using millions of litres of drinking water from *khuds* and rivers in areas which have been experiencing acute water shortage.

Every summer, the people of Himachal Pradesh facing shortage of water are being supplied water contaminated with chemicals. Tankers of the Nangal Truck Union that ferry toxic chemicals including hydrochloric acid (HCL), liquid caustic, methanol and sulphuric acid from National Fertilizer Ltd. and Punjab Alkalies and Chemicals Ltd. are often being used by the Public Health Authorities of Himachal Pradesh to carry water to areas facing shortage of water. Normally two types of tankers are used to ferry chemicals from Nougat-based chemical plants. The tankers that ferry methanol are made of mild steel. The tankers that ferry HCL, sulphuric acid or liquid caustic and lined with rubber to protect these from being damaged by corrosive action of the chemicals. Though the tankers having simple mild steel bodies can be washed to some extent, it is very difficult to wash the rubber lined bodies of tankers. Chemicals carried in the vehicles is so toxic that if anybody enter the tanker body to clean it immediately faints. (Source: The Tribune, June 5, 2003).

Poisoned Water—Bhopal

On the 18th anniversary of the Bhopal gas tragedy, an NGO here has demanded the registration of a criminal case against the city's municipal corporation. Despite several health and environment reports stating that the water around the Union Carbide plant has been polluted, even poisoned by the toxic wastes of the factory for nearly two decades, the Bhopal Municipal Corporation has not provided safe drinking water to residents, the NGO alleged. It was only now that the municipal corporation recognized that water in several areas around the factory is damaging to human health. They have sealed six power wells and hand pumps and banned the use of water from these taps at Ayub Nagar, Kainchi Chhola, Dashhehra Maidan, Chandbari and Garib Nagar areas.

Between 1991 and 1996, two major environmental studies on the after effects of the gas leakages said that residents of the area were drinking highly contaminated water injurious to human health. The study conducted by the Public Health Engineering department stated: "Large areas of Bhopal city consisting of 10 wards were badly affected due to the disastrous methyl iso-cyanate gas leakage. This gas had badly polluted not only the environmental status of the locality, but also underground water reserves. The water samples of all these sources were analyzed and it was reported that the chemical oxygen demand (COD) is found to be higher than the permissible limits."

A study by the Boston Citizen Environmental Laboratory also claimed the water had been contaminated to a dangerous level. In 1999, a Greenpeace study said that the water was contaminated thousand times more than the average drinking water standards acceptable in developed countries.

While the assessment does take into account the efforts on to check this contamination, it also notes that in many instances, technology seems to have failed. The study notes that large section of the population in rural India continues to be dependent on unprotected sources of water.

OTHER RELEVANT ISSUES

The facts on deteriorating water quality and water pollution in the world as a whole as discussed above have been causing a great concern. In point of fact, pollution is pervasive. Few countries, whether developing or industrialized, have adequately safeguarded water quality and controlled water pollution. Many countries do not have standards to control water pollution adequately, while others cannot enforce water quality standards. As stated earlier, India is facing a serious problem of water pollution and there is now nation-wide debates on causes and the solutions of water pollution problem, and the scientists, environmentalists, economists and sociologists have been expressing their views on the matter through the printed media. It is found appropriate to reproduce (duly edited) one such article of N. Gopal Raj published in The Hindu of August 24, 2003.

"We have discussed in the previous pages problems relating to pollution of ground water which meets the needs of 60 per cent of India's households. So the quality of groundwater can potentially affect over 100 million households, rich and poor, urbanites and villagers alike. Contrary to what is often believed, groundwater is easily polluted. There are so many possible pollutants, both industrial and agricultural. The polluters, likewise, can be countless, from individual farmers and households to small garage-scale entrepreneurs and large industrial operations. Once polluted, restoring the quality of groundwater is impossibly expensive even when technically feasible. Filtering and other ways of removing or at least reducing the levels of pollutants when pumping out water is also costly and viable only for industrial scale operations. Even monitoring groundwater quality poses huge problems.

Payal Sampat, in her paper titled "Deep Trouble: The Hidden Treat of Ground water Pollution", published by the Washington, D.C.-based

World Watch Institute, writes graphically about the seriousness of the issue: We are now learning that the water buried beneath our feet is not only susceptible to pollution, it is in many ways more vulnerable than water above ground .... Because it is underground and slow moving, groundwater stores pollutants far longer than, say, rivers or air do. Hence, the very characteristic that makes aquifers ideal reservoirs of freshwater—their ability to accumulate and retain liquid for longer periods of time—also enables them to become long term sinks for contaminants. It's true that some aquifers recycle water back to the environment fairly quickly. But the average length of time groundwater remains in an aquifer is 1,400 years, as opposed to just 16 days for river water. Some aquifers contain water that fell as rain as much as 30 millennia ago. So instead of being flushed out to the sea. Or becoming diluted with constant additions of freshwater, as rivers, lakes and streams are, aquifers continue to accumulate pollutants, decade after decade—thus steadily diminishing the amount of clean water they can yield for human use".

Agriculture—A Major Cause of Pollution

Globally, agriculture is the biggest consumer of water. In India, agriculture consumes about 90 percent of the water (groundwater as well as surface water sources), compared to about three per cent for domestic use and less than five per cent by industry. The number of diesel and electric pumpsets has increased over 200 fold since 1950 and is currently estimated to number around 20 million. 1991 India's fertilser's as well as organic waste and sewage, has been implicated in the nitrate pollution of ground water. High nitrate levels are possibly the most widespread contamination of groundwater globally and can lead to health problems such as the blue baby syndrome, India uses over 11 million tonnes of nitrogenous fertilizer annually.

Even in the early 1990s, the agriculturally advanced States of Punjab and Haryana, where fertilizer is intensively used, had wells with nitrate levels well above the safe limits prescribed by the World Health Organization (WHO). A few years ago, the Central Ground water Board (CGWB) found that water from wells in over 300 sites scattered across Gujarat too had high nitrate levels. In 1997, Ashok Keshari of IIT Delhi found nitrate levels in the groundwater at many places in Varanasi and in Jaunpur town (Closed to Varanasi) that were several times higher than the WHO norm. In a published paper, he blamed the utilization of sewage effluents for agriculture, leakage from sewerage system, septic tanks and natural drains carrying municipal wastes, and application of fertilizers. Now wells with high nitrate levels were being found in the Faridabad area near Delhi. Recently, the environment magazine, Down to Earth, reported that the CGWB had found high nitrate levels in the ground water used in Bangalore and in many villages in Karnataka. In north Bangalore, the nitrate concentration was said to have doubled in a year.

The cadmium found in the sludge from the Coca-Cola bottling plants in Kerala and West Bengal too could have had its origins in the chemical fertilizer used. Phosphatic fertilizers are produced from phosphate rocks which vary in cadmium content from less than one to over 100mg/kg. Cadmium in phosphate fertiliser is of particular concern because its absorption by food plants provides the main exposure route for humans, according to Valeric Thomas of the Princeton University, U.S.

Harmful Pesticides

As the experience of the Rhine basin in Europe shows, cadmium can also be released into the air by the burning of coal and oil as well as other industrial activity, and then get deposited in the soil. Tight emission controls help reduce the cadmium load. Ever since the first synthetic pesticides were introduced in the 1940s, these chemicals have become an integral part of modern intensive farming practices. In India, pesticide use has doubled in the past three decades. Many of these chemicals are toxic to humans, and America and Europe set limits on the maximum pesticides residues permissible in foods. When applied to plants and the soil, these often harmful chemicals can readily leach into groundwater. Worse, they can persist in the groundwater long after the chemical has been banned. DDT is still found in groundwater in the U.S., three decades after it was banned there. Even when more short-lived pesticides are used, the environment of the aquifer can slow their breakdown. Scientists have found that the herbicide alachor, whose concentration halves in 20 days in the soil, can take up to four years to reach similar levels in groundwater. Moreover, the products created by the breakdown of some pesticides can be just as toxic.

Groundwater is Easily Polluted and Restoring its Quality is Impossibly Expensive

In India, there have been reports of high levels of DDT in groundwater. The March issue of Down to Earth pointed out that several pesticides banned in many Western nations can still be sold in India. Pesticide residues in bottled water and soft drinks are likely to be yet another indication that the groundwater is contaminated. The Union Government has taken steps to regulate pesticides residues in bottled water and, in the wake of the latest controversy, and has said in Parliament that it was considering making these standards applicable to soft drinks as well. Even though such regulatory measures are essential, they are nevertheless only treating the symptoms and not the root cause—the presence of pesticide residues in source water, observes the New Delhi-based Centre for Science and Environment (CSE), which has played a leading role in reporting the pesticide residues in these products. Pesticides are constantly seeping into groundwater as well as surface water due to the porosity of laws that govern the use of these toxic chemicals, it points out.

Excessive Groundwater Drawing

Some times, just excessive drawing of ground water, which is already a major problem in many parts of India, can create and worsen pollution. One obvious example is of seawater incursion into depleted aquifers near the sea. These aquifers would normally drain into the sea. But when they are depleted, the flow reverse and seawater seeps into the aquifers. Just two per cent seawater entering an aquifer is sufficient to make the groundwater too salty to drink or use for irrigation. Large-scale drawing of groundwater can create more serious hazards. The arsenic contamination of groundwater in West Bengal and neighbouring Bangladesh has affected millions and has been called the largest mass poisoning in history. Some research publications have suggested that the massive withdrawal of groundwater for irrigation could have played a part in releasing arsenic from the sediments which hold it. Now there are reports or arsenic-contaminated groundwater in parts of Bihar and Uttar Pradesh served by the Ganga, as well as in Chandigarh and parts of Haryana.

Moreover, such depletion of groundwater increased the flow within the aquifers. This, in turn, can accelerate the breakdown of sediments, thereby releasing polluting chemicals, as well as spread the pollution faster through the aquifer and other linked water channels. Such mechanisms could have aggravated the fluoride contamination of groundwater that is now being reported from different parts of the country, from Rajasthan and Gujarat to the Gangetic plains and the Deccan Plateau. As with the arsenic contamination, fluoride contamination too occurs through a natural process in which fluoride bearing rocks crumble and break down. The process can speed up if the chemistry of the aquifer is disturbed.

Industrial Pollutants

Industry, user of minerals, metals and fossil carbon fuels, as well as the primary producer of chemicals on a colossal scale, is a major source of pollution. Toxic chemicals in the industries' liquid effluents, solid wastes and even their atmospheric releases can ultimately find their way to the groundwater and rivers causing untold damage. Some 4.4 million tonnes of hazardous wastes are being generated by 13,011 units spread over 373 districts of the country, according to the latest annual report of the Union Ministry of Environment and Forests. Just three States—Maharashtra, Gujarat and Tamil Nadu—account for over 63 per cent of the total hazardous wastes generated in the country.

> *The falling water table has only added to the problem*

The Hazardous Wastes (Management and Handling) Rules of 1989 were intended to plug loopholes in the way industries handle their toxic waste. But these Rules have not yet been seriously implemented nationwide, points out Ramesh Ramaswamy, a Bangalore-based consultant industrial ecologist. Proper sanitary land fills which can handle the toxic wastes produced by industries, have still to be set up. As a result, these dangerous wastes continue to be disposed of haphazardly. "There are only 2 or 3 such landfills in Gujarat and Maharashtra and event these are reported to be inadequate," he says. The system for collecting and handling municipal waste and sewage, which too can contain toxic material, is also inadequate.

Household Wastages

Households too can add to the pollution. The Ministry of Environment and Forests estimates that 22,900 million litres a day (mld) of domestic wastewater is generated from urban centers, almost twice as much as the 13,000 mld of industrial wastewater. The treatment capacity available for domestic wastewater is only for 6,000 mld and for industrial wastewater 8,000 mld, according to the Ministry. Nor, according to experts, is there a proper system in India for handling municipal solid wastes. An added problem in India is that the small and unorganized sector industries dump their wastes into the sewage and municipal refuse system, making them more toxic and dangerous.

At a time when the sustainability of India's groundwater is already under strain from excessive withdrawal, pollution from industry, agriculture and households jeopardizes what remains. Even the scattered reports of groundwater pollution from different parts of the country are alarming. If action to reduce pollution on all fronts is not initiated rapidly, the source of water for the majority of Indians will rapidly dwindle. A crisis is in the making.

Minimal Monitoring

State Governments are responsible for ensuring safe drinking water in our taps but individual tapping of groundwater—a common practice even in urban areas—is far from safe as there is no system in place for regular monitoring. Even though aquifers are located all over the country, the Central Pollution Control Board (CPCB)—entrusted with the task of monitoring water quality under the Water (Prevention and Control of Pollution) Act, 1974—has barely 24 stations countrywide, with an additional 181 stations operated by the State Pollution Control Boards.

The Government's Regulatory System is Inadequate

While the Central Ground Water Board with a network of 15,163 locations monitors the quality of groundwater, its parameters do not include bacteriological and pesticides contaminants, both of which are included in the Bureau of Indian Standards (BIS) and the World Health Organization standards for drinking water. While there are standards for bottled water and drinking water, there are no standards for ground water since it is treated as a source, rather than a direct supply.

Some of the 181 stations established sometime in 1999-2000, are not even fully operational yet, sources say. The Government agencies, as usual play the "blame-game." While the CPCB maintains that the Central Ground Water Authority (CGWA), structurally the same as the CGWB, is the "custodian of ground water," the CGWA argues that pollution is the concern of the former. The CGWA—which draws on the large network of the CGWB—has over 15, 163 locations that are monitored four times a year, In addition, there are 30,000 locations under the State Departments that are monitored every six months. "They have a larger network, so there is a need for coordination. We should pool our resources. The technical know-how can be given by us while the monitoring can be carried out by them", says the former chairperson, CPCB, Dilip Biswas.

The Water Quality Assessment Authority, aimed at facilitating coordination was created two years ago but is yet to become effective. And while the monitoring situation continues to be bleak, the prevention mechanism appears to be failing as well. "Industries are meant to dispose of the waste through effluent treatment plants but most do not. Poor system and creation of cesspools is another problem with urban areas. And finally, we have been advocating for bio-pesticides but the Ministry of Agriculture is not happy with that suggestion," says the CPCB, Additional Director, R.C. Trivedi.

Prevention the Better Cure

The Big industries, which are more easily noticed and monitored, are not always the biggest polluters. By sheer numbers, small industries can create major problems. A classic example is Tirupur in Tamil Nadu, which exports some 71,000 tonnes of knitwear annually and produces another 50,000 tonnes for domestic use. Some 4,000 small-scale units are involved in various segments of this activity. Together, these units consume over 90 million litres of water a day, a quantity sufficient for a town of two million people.

The industrial process produce wastewater containing about one tonne of dyes daily. Although central effluent treatment plants have been set up, the salinity is not removed and it is not known what sort of toxic chemicals remain, and in what quantities, when the wastewater is discharged into a nearby dry river, points out Ramesh Ramaswamy, an industrial ecologist who has published a study on the Tirupur problem. The ground water around has been polluted to such an extent that more than half the industries' water requirement is met by tankers from over 50 km away.

The large number of small-scale industries and informal sector units pose major problems for pollution control, pointed out by the "India Development Report 2002" published by the Indira Gandhi Institute of Development Research. Since such units use outmoded and inefficient technologies, the amount of waste they generate is disproportionate to their size. They lack the knowledge, funds, technology, space and skills to treat their effluents. There are 3.2 lakh small-scale industries, producing about 8.500 products, some of which are highly polluting. The net impact of these units nearly equals that of all the large and medium industries put together, the report says. Enforcing the pollution regulations in these units is "painfully difficult" and aggravates the problems of the regulators, already constrained by understaffing, meagre resources, and, often, a lack authority.

Although technologies do exist to remove pollutants from groundwater after it is pumped out most of these methods are very expensive to implement. The least expensive way to prevent humans and other non-target species (both terrestrial and aquatic) from being exposed to harmful chemicals is simply to prevent the chemicals from entering the hydrological system in the first place. Prevention requires regular monitoring and understanding of the quantity and quality of pollutants entering the groundwater. That is a mammoth task in itself. It requires analyzing the groundwater for a wide range of potentially toxic chemicals, including many inorganic and organic compounds, heavy metals, and pesticide residues. The equipment for such analysis is expensive and each test, such as for pesticide residues, can also be quite costly.

It is not sufficient that dangerous wastes from large industries are monitored. As an FAO—sponsored study remarked: "The dispersed nature of source of pollutants is a core challenge facing both monitoring and control of groundwater pollution related to agriculture. Unlike industry or municipal sewage systems, agricultural pollutants are dispersed over large land areas. While return flows in drainage canals can be monitored it is difficult to determine the extent of direct seepage of pollutants through soils and into the groundwater until contaminant concentrations in groundwater become significant".

A nationwide system of monitoring and surveillance is essential, says Sunita Narain, director of the Centre for Science and Environment (CSE). In order to reduce the magnitude of the task, a sampling method, with water from a few randomly chosen wells being tested, could be thought of.

In the end, monitoring and surveillance can only be a prelude to action. Agricultural chemicals, such as fertilizers and pesticides, can be applied more effectively, thereby reducing the quantities which are used. The CSE, has been demanding an effective pesticide policy to ensure that chemicals registered and sold are 'safe' for use. The permissible limits for contaminants in food products need review, and this could lead to a wider discussion regarding permissible levels of pesticide and fertilizer use in agriculture and horticulture, Instead of subsidizing chemical pesticides, the Government should be promoting bio pesticides.

The method of industrial ecology will be to analyse material flows in a region and, if required, certain industrial sectors of concern, according to Mr. Ramaswamy. Such analysis will serve as a diagnostic sheet, showing the major sources of pollution that need priority action. Legal provisions, such as the Hazardous Waste Rules, needed to be implemented fully. In the case of small industries, many of which are clustered together in industrial estates, the India Development Report suggests delegating some of the responsibility of monitoring and enforcing pollution controls to the industry associations. The associations would have an incentive to regulate and discipline their members as otherwise all units in the estate could suffer. The experience at the Ankleshwar Industrial Estate in Gujarat, the largest estate of chemical industries in Asia, shows that such an approach, though not without problems, has potential. It cannot be assumed that sewage, household garbage and other forms of municipal solid wastes are harmless. These too need to be carefully handled and properly disposed of. It is necessary to act now and stop the contamination of groundwater, warns Ms. Narain. Otherwise, it would only worsen and become more difficult and costly to handle.

Conclusions

Contamination of groundwater with arsenic and fluoride and increasing pollution of surface water with waste from urban areas are major water quality problems. Since the same water serves multiple uses and users, solutions should be found within the framework of sustainable water management and in optional recycling of water. This requires that policymakers, governments, donors, international organizations, and the research community appreciate the close links between water used for food production and water used for drinking. The growing worldwide scarcity of good-quality fresh water makes it essential to bridge the gap between the different sectors involved in water resource management.

REFERENCES

1. Damme, Hans Van: Domestic Water Supply, Hygiene and Sanitation; Overcoming Water Scarcity and Quality constraint, *Focus* 9; 2020 Vision for Food, Agriculture and the Environment; IFPRI.
2. Population Reports, series M, Number 14; Population Information Program, The Johns Hopkins School of Public Health, Baltimore, U.S.A.
3. Kumar Dhirendra: Catching Raindrop; *Development Alternatives*, Vol. 12, No. 06, New Delhi.
4. *The Statesman*, New Delhi, Dated April 18, 2003.
5. Falkanmark, M. and Uidstrand, C: Population and Water Resources: A delicate balance. *Population Bulletin* 47(3): 1-36, November 1992.
6. Gardner-Outlaw, T. and Engleman, R. Sustaining Water, Easing Scarcity: A second update. Washinngton, D.C., Population Action International, 1997, p.2-19.
7. Engelman R. and Leroy, P. Sustaining water: Population and the future of renewable water supplies. Washington D.C. Population Action International, 1995.
8. Serageldin, I: Water Resources Management: New Policies for Sustainable future. *Stockholm Water Front*, June-1, 1995. p.2-3.
9. Serageldin, I: Toward Sustainable Management of Water Resources. Washington D.C. World Bank, 1995, p. 1-33.
10. Hoek, Wim Van Der: Emerging Water Quality Problems in Developing Countries; Overcoming Water Scarcity and Quality Constraints, IFPRI 2020 vision, *Focus* 9.

Part III

POSSIBLE SOLUTIONS

10

Water Resources Planning

Waters! you who are health-giving, give us energy, so that we may look on great delight

—Rig Veda

Introduction

Access to enough water of sufficient quality is fundamental for all human, animal and plant life as well as for most economic activity. At the global level, plenty of water is available. But to meet the demand, water has to be supplied where and when it is needed. However, with two-thirds of its surface covered with water, Earth is often called planet of water. Of all the Earth's water 97.5 per cent is seawater, 1.75 per cent is captured in the polar ice caps and the remaining 0.75 per cent is potable fresh water, which is found in rivers, lakes, swamps and underground. More than 1.2 billion of the planet's total population of 6 billion people lack access to safe water and 2.3 billion have inadequate sanitation. One child dies from lack of water or from unsafe water every eight seconds, according to UNICEF. These spatial, temporal and qualitative characteristics of water pose the greatest challenge to meeting the rising demand on all sectors. The fact is that almost all uses lead to degradation of water for its other users as they put something back into the water. Water quality and competition between uses are therefore two critical issues for the future of water use. There is no single formula by which these complex and inter-related problems can be solved. Increases in water supplies, and especially storage, are needed, but so is demand management, including not only economic instruments but also education and other efforts to change behaviours. Appropriate technologies and institutions must both play a role.

Throughout history, farmers and nations have depended on irrigation to produce sufficient, stable food supplies. In fact, water for agriculture is critical for food security. Presently, an estimated 40 per cent of agricultural products and 60 per cent of the world's grain are grown on irrigated land. Irrigation accounts for around 70 per cent of water withdrawals world-wide and over 90 per cent in low-income developing countries. But water for irrigation may be threatened by rapidly increasing non-agricultural uses in industry, households, and the environment. New investments in irrigation and water supply systems and improved water management can however meet part of the demand. But in many arid or semi-arid areas and seasonally in wetter areas—water is no longer abundant. The fresh water resources are getting exhausted day by day and 26 countries around the globe are now considered to be water scarce. A burgeoning population, misuse of water resources and poor management practices have been causing depleted supplies, falling water tables, shrinking in land lakes and stream flows diminished to ecologically unsafe levels, leading most part of the world to the threshold of water famine.

There are about 260 international rivers covering almost one half of the land surface of the globe and an un-accounted number of shared aquifers providing fresh water to the people settled within these basins. Unfortunately, individual perceptions on water differ as each one sees water only by its use, thus leading to a fragmented approach in managing water. Most of the beneficiaries are obsessed by the use factor—drinking, industry, irrigation, environment, etc.—as if all are competing and not part of the whole ecological system complementing each other for sustainable development. However, as the water is a non-substitutable resource and which is becoming scarce with population growth and developmental activities, it is of uttermost importance for a judicious water resource balancing in India.

The present mindset of a majority of the population is that water is government's business. This needs to be replaced by a model in which all major stakeholders participate at all levels. Without this change, technology solutions will be of no use. The change could be brought about by public awareness, education, identification and dissemination of best practices and incentives for action, thereby facilitating conservation of the precious resource and equitable distribution among the needy people.

The Concept

"Water Resources Planning" is a scientific tool aimed at not only optimising the utilization and management of water resources but also to make the developmentalists aware of environmental and social equity considerations as well as to make environmentalists aware of economic and development needs. Very precise and pragmatic planning being the need of the hour, water resources planning has assumed critical importance, particularly in the country where only two of the 20 major river basins are contained in a single state.

When we look at our total water resources, we receive an average annual precipitation of nearly 4.000 cubic kilometres, the average flow in the river system of the country has been estimated at 1,880 cubic kilometers and the actual quantum of water utilisable through surface structures is only about 690 cubic kilometres. Besides, the quantum of utilisable ground water is about 450 cubic kilometres. As against this, the water demand for irrigation which was about 321.3 cubic kilometres is expected to go up to 331.7 cubic kilometres by 2025 A.D.[1]

With a view to optimally utilising the available water resources of the country by storage and inter-basin transfer from surplus to deficit and drought prone areas, a national perspective for water resources development has to be prepared. Water is to be treated as a national asset than state control. This decision shall check inter-State water disputes and diversion of water from downpour to drought prone areas.

Strategy

- Integrated Development and Management of Water Resources
- Development of New Resources
- Optimal Utilization of Developed Water Resources
- Preserving the Resources
- Augmenting the Available Water Resources
- Stakeholders Participation in Water Resources Management.
- Demand Side Management Through Mass Awareness
- Dissemination and Application of Technology and Research

Strategy for Adoption

We have to recognize the potential for water use efficiency and conservation improvements to save enough water to meet new human demands and to satisfy environmental needs for water. The strategy for overall development of water resources should be a judicious mix of all the options available. Storage reservoir, ground water development and watershed development, each has its own place and none is superior to the other. All the three options have to be pursued simultaneously. Rational use of water should take full account of the efficiency for the system as a whole which is a product of (a) conveyance, (b) distribution and (c) field application. But unfortunately, a scientific approach to water management has not yet received the attention it deserves over a major part of the globe and India is no exception to it. Basic components of an efficient use of water involve:

1. Improving communication between suppliers of irrigation and those receiving it.
2. Effective educational programmes to minimize downstream health impact.
3. Since reservoirs cannot be easily replaced, prevention of siltation through catchment protection must get high priority.
4. Necessary protective steps to avoid problems of water logging and salination.

Rationalized Use—The Need

Rationalisation can be brought about through proper canal irrigation water management which unfortunately is not at all that good in our public canal irrigation. Farmers located at the head end of a canal system, say, a canal outlet, tend to appropriate more of irrigation supplies, to the disadvantage of their brother farmers at the tail end of this outlet. Besides reduction in maximum total output from canal irrigation, this unequal sharing of water leads to water logging and soil salinity problems which is more visible in head-end reaches of canal system. This brings us to the question of environmental degradation from canal irrigation development. That is why, environmentalists insist that agricultural development should take place on a durable and sustainable basis.

Similarly tube-well irrigation has also created major environmental problem. Let us take the case of Punjab where total irrigated area has exceeded the ultimate irrigation potential. A study related to Punjab clearly shows that a major portion of ground water is being exploited with the help of electricity and diesel operated shallow tubewells and it accounts for 94.8 per cent in the total draft. The level of exploitation there is so high that there is an annual decline of 20 cm in the groundwater table. We must confess that the better management of our water, soil and forest resources is too serious a matter to be left to the tender mercies of Government organizations alone.

The position is not very different in Haryana where largescale drilling of tubewells by private dealers lacking technical knowhow, has seriously affected 32 of the 108 development blocks in the State. The average water level in the State has declined by 1.90 metres, whereas in fresh belt areas of the State its average fall is 4.38 metres. It varies from 0.94 metre in Yamunanager district to 12.09 metres in Mahendergarh district. On the other hand, the water table in saline areas has risen during this period, ranging from 0.41 metre in Rohtak district to 6.02 metres in Sirsa district. Exploitation of ground water resources has been taking place primarily through digging of shallow tubewells in the State. The number of these tubewells has now risen to more than 500,000 from merely 28,000 when the State was carved out of erstwhile Punjab in November, 1966. The density of shallow tubewells varied from five to 25 units per square kilometre in the State[2].

According to an official study in Haryana, the State has a total quantity of 8.423 million cubic meters (mcm) of available ground water, of which nearly 7.348 mcm has already been exploited, mainly by irrigation tubewells. The Estimates Committee chairman of the Vidhan Sabha, Mr Om Prakash Beri, MLA, once launched an agitation seeking judicious apportionment of the existing canal water among all the districts. The exploitation of ground water resources in certain districts is as high as 365 per cent against the state-average of 78 per cent and the all-India average of 42 per cent. NABARD, on its part, has stopped financing tubewells. The Haryana SEB, too, has been issuing instructions to field engineers against releasing new tubewell connections where ground water has been over-exploited.

Another issue is the excessive use of canal water, which poses a serious threat to soil health of areas served by the Bhakra Canal system. The water table in these areas has risen by 20 to 40 feet during the last 25 years. According to experts, if the rising water table in these areas is not effectively and immediately checked, the cotton tract of the state may turn into a big marsh and the grain bowl into an arid and infertile pocket. The threat becomes all the more serious in view of the reports that about 4,000 sq km (almost 1/10th of Haryana's total area) has already been affected by salinity. There is likely to be a major water scarcity in another 30 to 35 years. Water rather than land use would determine agricultural operations in a situation like that.

The indiscriminate drawing of water through tubewells and the excessive watering of soil for crops like sugarcane has resulted in unsustainable agriculture on both counts. Wells could dry up there in the next few years if excessive drawing of water continues. At the same time excessive watering has lead to the soil becoming saline, and thus unusable for agriculture. Needless to say that these practices are no longer sustainable. Henceforth, biological and ecological issues should dictate agriculture, not technology and economics, as has been the practice for so long. In Maharashtra, a water scarce region, 40 per cent of rain water flows down the western ghats, into the sea. The narrow Konkan strip, with its laterite soil, receives 40 per cent of the state's precipitation but has only two per cent of the state's irrigation. The central part of the state is largely drought prone where 13 per cent of the land is under irrigation, of which 55 per cent uses water drawn from the ground water resources.[3] Of the total of 6,229 Blocks as of January 1992 in the country, as many as 12 per cent had been declared as dark or grey where further exploitation of underground water is prohibited. (Table 10.1)

Downstream Regulation Mode of Irrigation Systems

At present in our irrigation system, 'Up-stream' mode of water regulation is practised, in which water in quantity available, taking into consideration the land holding and is delivered at fixed intervals (wara-bandi turn), and the farmers are required to adjust their crops according to the availability. But in the "Down stream" mode of regulation, water is made available according to the 'demand' of the farmers; and this is more flexible mode and is considered more suitable to HYV crops now grown. The latter mode is being adopted in Egypt, and the average yield of the various crops is much greater than that obtained from other areas.

So the 'down stream' mode should be introduced in the command areas of the systems. Reduction of evaporation losses from water bodies is also essential. Evaporation from water bodies can be reduced by application of materials that spread 'mono-molecular film' on the water surface. This has been found feasible on small man-made water-bodies, like 'Fish Ponds', and can be used to effect considerable water savings. It may be said that in the days to come, every drop of water is going to become the most valuable possession of a region, so it must be used to maximize the productivity resulting from each drop.

Table 10.1: Categorization of Blocks/Talukas/Watersheds According to Dark, Grey and White (As of January 1992)

Sl. No.	*State*	*Number of Blocks*		
		Dark	*Grey*	*White*
(1)	(2)	(3)	(4)	(5)
1.	Andhra Pradesh	28	25	256
2.	Arunachal Pradesh	—	—	48
3.	Assam	—	—	134
4.	Bihar	—	55	533
5.	Goa	—	—	12
6.	Gujarat*	18	14	151
7.	Haryana	24	20	54
8.	Himachal Pradesh	—	—	69
9.	Jammu & Kashmir	—	—	123
10.	Karnataka	9	9	157
11.	Kerala	—	—	151
12.	Madhya Pradesh	3	14	442
13.	Maharashtra**	34	57	1,390
14.	Manipur	—	—	26
15.	Meghalaya	—	—	29
16.	Mizoram	—	—	20
17.	Nagaland	—	—	21
18.	Orissa	—	—	314
19.	Punjab	69	21	28
20.	Rajasthan	63	29	144
21.	Sikkim	—	—	4
22.	Tamil Nadu	43	76	259
23.	Tripura	—	—	17
24.	Uttar Pradesh	17	77	801
25.	West Bengal	1	35	305
	Union Territories	—	—	3
	No. of Blocks (Except in Gujarat & Maharashtra)	257	361	3,947
	No. Talukas in Gujarat	18	14	151
	No. of Watersheds in Maharashtra	34	57	1,390
	Total	**309**	**432**	**5,488**

Source: Ground Water Statistics, 1992, Central Ground Water Board.

* Gujarat 183 Talukas/ 218 Blocks ** Maharashtra 141 Watersheds/366 Blocks.

Environment Management Needs Special Cells

In the present era of rapid urbanization, over-population and unabated industrial growth all around the world, a constant need to shift focus from "development" to sustainable development" has repeatedly been felt. Last decade saw the holding of Earth Summit at Rio de Janeiro and the world realized the importance of preserving the environment and maintaining the eco-system in which it prospers. New ozone holes, global warming, polluted waters, barren lands, lifeless forests, greenhouse effect and toxic atmosphere were noted to be the visible negative impacts of man's relentless march without caring for the environment. It became evident that for the survival of both the human beings and the animal world, efforts to minimize environmental degradation were needed to be accelerated by every country, every industry and every individual.

Today, rules and regulation demand safe disposal of effluents by the industries without causing any adverse effect upon air, water or soil. Setting up an environment management cell has become essential for every industry. Monitoring the pollution control aspects is a full fledged job, needing a strong organizational setup and a testing infrastructure. There may be instances when some industrial units may have escaped the eye of the law and continued to discharge the effluents without their treatment but such an approach may result in heavy losses in the long run. With the passage of time, the rules, the control and the penalties are going to be heavier only. It is therefore a better and wiser step for every industry to set up its own unit for environment management, treat its effluents, discharge them safely and be free of the stress that will always keep lurking if these measures are not taken. Once an industry accepts the fact that saving the environment is an integral part of its production process and whatever little or large expenditure it needs is unavoidable, the double-mindedness and the perplexity come to an end.

Any industry, may be a production unit or a generation unit has to ensure that its operations do not affect:

1. Air Quality
2. Water Quality
3. Soil Quality
4. Noise Levels.

Every industry releases air and water effluents during its operation and produces noise. Air effluents affect air quality, water effluent affects water and soil quality and noise affects physiological and psychological states of living beings. To keep the pollutants contained by the effluents within permissible limits, air and water need, to be sampled, analyzed and treated while the noise levels are to be kept low.

In addition, every industry has to evolve a plan to add greenery and plantation to the area in and around that industry. Since the industries are run on a continuous basis, time intervals at which the effluents are to be sampled and analyzed are to be decided keeping in view the quantum of discharge and an environment management cell needs to be created accordingly to carry out the job.

Chart. 10.1 shows setting up of an environment management cell in an industry. The cell should work out the number of stations required to draw the samples of air, water and soild and the number of stations to measure the noise levels. It should also draw a landscaping and arboriculture plan to add greenery to the area occupied by the industry to monitor the progress of the new plants and to keep a record of effect of pollutants on the plantation. This cell should further assess the number of samples to be drawn and the intervals at which the air, water and soil samples are to be collected at the decided stations. Thereafter, the process of collection of samples, their analysis, the comparison of percentage of pollutants contained in the samples with the corresponding permissible limits and the conveyance of results to the concerned units of the industry for corrective measures should be a regular process carried out in an immaculate manner without allowing the laxity to set in over the passage of time.

An Environment Management Cell should fully equip itself with the necessary testing equipment, samplers, chemicals for testing equipment to record climatic conditions and noise level meters. High Volume Samplers to suck air through a glass fibre filter paper, respirable particular samplers, electronic balance for air quality; pH meter, conductivity meter, BOD incubator, COD apparatus, Turbidity Meter and Magnetic Stirrer for Water Quality and Precision Integration Sound Level meters to measure sound pressure levels should form essential part of the equipment handy to the environment management cell.

Chart 10.1

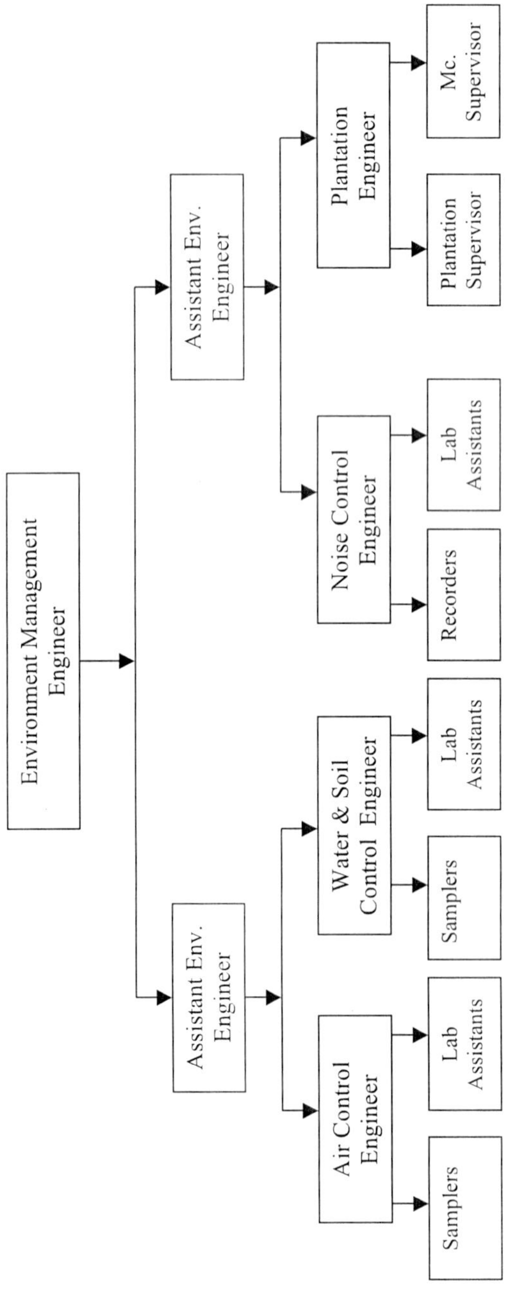
Environment Management Engineer
Assistant Env. Engineer
Assistant Env. Engineer
Air Control Engineer
Water & Soil Control Engineer
Noise Control Engineer
Plantation Engineer
Samplers
Lab Assistants
Samplers
Lab Assistants
Recorders
Lab Assistants
Plantation Supervisor
Mc. Supervisor

Participatory Irrigation Management (PIM)

Current weaknesses

Why have governments typically assumed responsibility for the overall management of water? Water's special characteristics make it difficult to use unregulated markets to deliver water efficiently or to allocate it equitably. The large capital requirements and economies of scale in water infrastructure tend to create monopolies, warranting regulation to prevent overpricing. Public investments in water infrastructure are often necessary given the large size and long time horizon of projects; underdeveloped capital markets in many lower-income countries and the potential for political interference reduce incentives for private investments. Water's natural supply is subject to extreme variability over time and space. Certain aspects of water-related activities, such as flood control, are public properties, which cannot easily be charged for on the basis of use. Finally, water moves through an intricate cycle within the ecosystem that makes water activities highly interdependent and results in numerous spillover effects, many of which affect the environment. These effects cannot be reflected in market prices. Yet there are many weaknesses in the ways governments manage water, in developed and developing countries alike.

A Conceptual View

The World Bank defines "participatory development as a process through which stakeholders influence and share control over development initiatives and the decisions and resources which affect them". Participation is intended to improve the performance of irrigation, to improve planning and water distribution and to develop a sense of ownership of the newly created infrastructure. WUAs are the means and embodiment of participation in irrigation management. The WUAs have been given a legal status through registration under the Societies' Registration Act. The legal status is expected to gain recognition for them from various agencies. The objectives of PIM are to ensure distribution of the available canal water equally and uniformly among all shareholders, to upkeep the field channels in running condition through deweeding and desilting, to attend to repair and maintenance of field channels, to ensure efficient and economic use of canal water and to determine crops to be sown for optimum utilization of available water. The basic aim is the welfare of the shareholders and increasing agricultural production. Formation of farmers' associations is not an object in itself, but a means for a new joint venture in development. The PIM is expected to foster a sense of ownership, equitable right to water, self-management, institutional access to irrigation information, choice of crops, improved water distribution system, better system maintenance and more disciplined and efficient use of water.

History and Progress

As per the vision of the National Water Policy for progressively involving the farmers in the management of water, Command Area Development (CAD) Authority created all over the country pioneered the formation of Water Users Associations (WUA) at the field channel/watercourse/tertiary canal level. Under the CAD programme, a provision of one-time functional amount to farmers associations of Rs. 500 per hectare, of which Rs. 225 per hectare each to be provided by the Central Government and the state government respectively and Rs. 50 per hectare to be contributed by the farmer associations, was made. This functional grant is to get the WUAs kick-start and to set them on course to embracing the self generated operation and management responsibilities. Ninth Five Year Plan recognized Participatory Irrigation Management (PIM) as a thrust area and it asserted that participation of beneficiaries would help in the proper upkeep of irrigation system and the optimal utilization of irrigation water. So far, the farmer user groups formed under these programmes have been voluntary and have been formed to take care of the O and M needs of the system at that level. Of late, a number of World Bank funded water resources consolidation projects, particularly in Haryana, Tamil Nadu and Orissa have also formed WUAs, and are attempting for farmers' organization and turn over (FOT) for the proper upkeep of the system. Here, the emphasis is on the complete turn over of the system to the farmer groups. Andhra Pradesh has forged ahead on the national scene by enacting "The Andhra Pradesh Farmers' Management of Irrigation Systems Act, 1997" and is trying to institutionalize participatory irrigation management.The State-wise coverage of irrigated area under WUAs is shown in Table 10.2.

Table 10.2: The State-wise Coverage of Irrigated area under WUAs

States	*Area Covered under WUAs (Lakh Ha.)*
Andhra Pradesh	45.7
Madhya Pradesh	26.3
Karnataka	3.0
Haryana	1.9
Rajasthan	1.9
Kerla	1.5
Bihar	1.1
Punjab	1.1
Maharashtra	0.9
Gujarat	0.8
Orissa	0.7
Assam	0.3

Source: Planning Commission,
Author's Source: The Economic Times, 27-11-2002.

Limitations

With all the noble objectives expressed above, PIM movement in India has forged ahead. The one-time success of the agencies involved in propagating the cause of PIM has been noted by all and has become a part of our development history. However, the sustainability of the experiments undertaken in different states leaves a question mark as the WUAs are a product of specific time and space in which the system improvement works have been undertaken with either external or internal finance. With the donor agency/government funding, lakhs of rupees on improvement, modernization/rehabilitation of the system have been spent and there is natural exuberance and excitement amongst the farmers which has helped in organizing them into a group. However, with the task at hand getting accomplished, the WUAs have tended to lose their sheen. Studies point out that many of these user groups have become dysfunctional and defunct over the years. Therefore, there is a definite need for institutional intervention in order to make these groups lively and vibrant for all times to come. The State of Andhra Pradesh has set a model which aims at making water user groups functional, self-sustaining and also at expanding the range and expanse of WUAs.

Model

Now coming to a workable model for making PIM more vibrant, some of the steps which need to be undertaken are listed below:

(i) The WUAs should be made financially viable and sustainable on a long-term basis, for which a definite recurring source of revenue needs be located. In other words, a definite contribution by the users on a crop or annual basis per acre can be made to the accounts of WUAs so as to sustain and facilitate the discharge of its responsibility towards operation and maintenance of the watercourses/field channels. The collection of the same can be undertaken by the beneficiaries themselves or by the revenue agencies as in Andhra Pradesh.

(ii) The WUAs should be given a better legal foundation by a suitable legislation and should be constituted through the mode of secret ballot. The office-bearers ought to be elected through an election which may be conducted under the eyes of the District Collector/District Returning Officer. The process has to be transparent for legitimacy and acceptance.

(iii) The higher hydrological formations may be created at the secondary and tertiary canal levels to involve farmers in management of water. Suitable devolution of authority and responsibility must be made to make the participation more meaningful. Proper strengthening of these committees with incorporation of suitable kind of control mechanism vis-à-vis the officials of the irrigation agencies can be made to make the arrangement effective. These formations also need to be vested with authority under suitable legislation for: (a) exercising control over the lower formations, (b) getting a formal contribution for O & M of the system; and (c) performing specific roles and discharging due responsibilities.

(iv) In the states of Punjab and Haryana, there is a system of warabandi prevailing under their respective Canal and Drainage Acts which provide for equitable distribution of water. However, the issue of equity remains unaddressed in many other states. Therefore, there is a need for adopting provisions elsewhere as well similar to those in the Canal and Drainage Acts of these states or there is a need for providing a conflict resolution mechanism in the WUAs at different levels. The WUAs need to be empowered to undertake equitable allocation of water in the chak and in case of their failure, a corrective alternative needs to be provided.

(v) By bringing more area under irrigation over the years, the agriculture has been intensified, but in certain cases there is a greater need for diversification of agriculture as well. For that, deputationists from agricultural universities, state agriculture departments and other specialist agencies can be sent to WUAs and their emoluments can be borne by the government while they can work in close conjunction with the community for bringing about agricultural diversification.

(vi) There is also a need for training the farmers in the art of self management and there is a greater need for training programmes to move to the farmer fields.

Decentralizing water service delivery.

With a foundation of appropriate legal and regulatory procedures and given an overall coordinating framework, the management and delivery of water services can be decentralized, increasing local and community control and improving efficiency. Many countries are encouraging users' associations to take more control of water management. In an effort to both decentralize and involve beneficiaries, Mexico is transferring management of 78 irrigation districts (covering more than 1.8 million hectares) to water users' associations, which will be responsible for operating and maintaining all canals and water distribution.

Prescribing and encouraging the participation of stakeholders

Experience has shown that the participation of "stakeholders"-individuals and institutions that would be affected by decisions about water resources management-in formulating strategies, planning, designing, implementing, and managing water activities are not only good management practice but also helps to build necessary political and social consensus. Water users are willing to pay for services that are designed to cater to their needs, which improve the ability to financially sustain water service activities. In projects in Bangladesh and Kenya, water users not only participate in establishing rural water and sanitation systems but also manage them.

India can certainly learn from the experience of such associations in Maxico, Bangladesh, Kenya as quoted above as well as of Indonesia, and Philipines where the farmers are involved not only in the planning and construction of the system but its actual utilization. As soon as irrigation scheme has been built, social scientists from research institutions provide them additional guidance. This type of working in partnership has added a new dimension of accountability downward to farmer clients, as well as upward in the bureaucratic hierarchy. In short, water users must be involved in the governance of water resources.

Water selling by volume is another suggestion. A system by which such Associations can be involved in buying surplus water in bulk and distribute the same to deficit farming members according to their need. This type of control is possible only through farmers' own institutions which can also take the additional responsibility of improving the knowledge of the farmer about irrigation water management. Farmers could then devise any system to allocate costs among individual users, though their expected concern with transmission cost minimisation would, in all likelihood, lead them away from volumetric pricing at this level. This suggestion takes the Vaidyanathan Committee's argument for effective devolution of power a little further, suggesting that primary collection of revenue, in effect, be privatized. Farmers' participation in system management thus emerges as a prime pre-requisite for a smooth and successful transit to a more efficient system. Once sufficiently large organized user-groups come into being, they will themselves be interested in seeing to it that projects are implemented economically and that the quality of water management improves.

The public in general and local people in particular must take a keen interest in a matter which concerns their future. Water policy decisions during the coming decade will force us to consider how we are going to achieve some of our most fundamental goals. But it is hoped that we will not opt for policies which will result in more polarization and in stalemate. Rather, we need to use our genius to develop approaches that will permit the simultaneous accomplishment of our deeply held values. The general choice must not be between economic growth and environmental quality. Rather, the appropriate policy is one that will permit us to choose a combination of market and non-market goods that is to our liking and chosen by means that will not damage us in the process. The choice need not be between extreme individualism and collective paralysis. Rather, the appropriate policy is one that recognizes that certain problems require concerted action but that also tries to create an environment for individual decision making that will promote the common good. Normative water requirements of crops in many parts of the country exceed the available supplies. The problem is further aggravated by the wrong practice of flooding the fields as a result of flow irrigation.

The farmers co-operatives should take the initiative to participate in the management of irrigation water with the government. If the farmers come forward to complete the unfinished or ongoing irrigation works, the government should welcome such a move. Besides, the government should be happy if they take over the management of irrigation completely. In many countries, the agriculturists shoulder 30 per cent cost of irrigation projects and the government after completion, hands over the entire management such as maintenance. But, the government looks after the safety and security of the dam. The Aga khan Trust had motivated the peasants of the Chandravadi village in Mendarda taluk of Junagadh district, to partly share the cost of the canal in their village. The farmers contributed Rs. 4 lakh while the Trust donated Rs. 18 lakh for the construction of the canal.

Lessons Learnt from Successful Water Users Organizations

A study of Panchayats from Indian experience in Gujarat, Maharashtra, and Andhra Pradesh and in the Philippines and Indonesia shows that the key factors of success for water user organizations are: (1) strong government supports; (2) Sense of ownership by members; (3) strong leadership; (4) strong linkages with other organizations; (5) adequate levels of incentives; (6) well developed and operated organizations based on hydrological units; (7) regular capability programme of the farmers.[4]

1. Strong Government Support

This lesson included an improved or new system which provided farmers with adequate and dependable supply of water. All the successful systems had these attributes. Three of the five associations were also given three year management grants especially as members took over maintenance and management for the first time. The Government provided strong policy support.

Officers of Agriculture and Irrigation Departments attended the meetings of all these successful organizations. At four of the successful water users associations (WUAs), Government also provided technical assistance. Government supported their decisions and developed good rapport with the organization. In Phillippines no water was supplied to an area until functional water organizations were established.

2. Sense of Ownership

At communal systems in Indonesia such as a Subak in Bali, the question of ownership does not arise. The farmers are the owners and the Government does not indicate or interfere with the business of the association.

3. Strong Leadership

It was found in all except one case that leadership was dedicated, knowledgeable and skilled. In all cases leaders are elected by members to whom they are accountable. Leaders typically have good personal linkages with other influential persons and officers of various organisations. In case of Hyderabad, Philippines and Indonesia, leaders are given special training. Leadership is a critical factor in forming or strengthening a water users association.

4. Strong linkage with other organizations

Successful water users associations are linked to farmers, Government, local markets and cooperatives for selling their agricultural produce. In four cases there were linkages with banks for credit provided on the guarantee of the associations. Also, in these cases water groups were closely linked to agricultural and irrigation departments. Experience indicated that initially it is important to establish only a single purpose as against a multipurpose organisation. Too many farmer organizations fail because members and government officials load them up with too many tasks. Irrigation management itself is complex enough for a farmers organisation to be handled effectively initially.

5. Adequate Level of Incentives

At all sites visited inside India and in other countries, an assured water supply was viewed to be priority incentive. Management grants were made available to water users associations in four out of the five cases. Only in the case of Philippines, maintenance was carried out by the organisation and then paid by the National Irrigation Administration (NIA). Typically the members in the five cases received water at a lower price than non-members. Loans from banks and co-operatives in three cases were guaranteed by the WUA. In Gujarat, in two cases, the successful WUAs were registered under the Co-operative Society Act and received tax and other benefits. An interesting "commission method" for collecting water charges exists in Philippines. NIA pays a 35 per cent commission to WUAS for 50 per cent collection rate and more incentives for higher collection rates.

In Philippines, three of the successful WUAs provide better management incentives to their membership by hiring local persons to patrol the system. Also a water manager opens and closes the gates, identifies damages and performs minor repairs and maintenance. In four successful cases, the Agriculture Department assured that all inputs would be supplied at the right time at right price and right place. Another interesting incentives provided by NIA in the Philippines is that the water rates can be paid in kind after harvest. Finally, in all but one case, members of the associations were involved in planning, implementing, distributing and allocating water, maintenance and rehabilitation programmes.

6. Capability Development Programme

Regular Training Programme for members is a feature of organisations in Hyderabad State (Pochampud) Project and the projects in the Philippines and Indonesia. In other projects also there is a provision of regular training to the farmers and lower functionaries.

7. Organizational Set-Up

All the successful WUA (five cases) used the hydrological unit versus the village or residential unit as the basis for organisation. Typically, the basic unit was the chak or outlet. In four of the five successful cases, the minor level meetings were held monthly or whenever it was needed. An annual meeting is held for members in which many senior officials and other influential people participate. In all but one of the five successful cases of WUAs, officials were not to vote. The WUAs in the Philippines have an interesting approach. They constitute committees to handle separate functions such as fee collection, maintenance, repairs and other activities. As stated earlier, the associations in Gujarat, Philippines and Indonesia hire their own canal or water patrols. In four successful cases, WUA's resolve conflicts and collect fees and fines quickly. At these successful WUAs, a decision is made each year with the irrigation system managers about how water will be supplied. In three of the five cases, written rules and sanctions were generally established prompt and swift sanctions and other appropriate measures. In all the cases the officers are elected by the members and do not have to be approved by the Government.

Finally, in only one case the WUA is linked to a temple or principles of one or more religions. This is the Subchak of Bali. Indonesia has long history of religious rituals, customs and practices which are completely integrated into the culture. Farmers associations in Philippines, Srilanka and Indonesia have hired community organizers for accelerating the community activities of the association and working as catalytic agents. These have proved to be of great use.

Haryana Case

The Government of Haryana (GOH) has given an undertaking to the World Bank that in order to promote participatory irrigation management, it would be introducing water users' associations (WUA) of farmers styled after Maharashtra. These associations would purchase water in bulk from the government and distribute it among farmers, using the Warabandi method, that is, in proportion to the holding size. The WUAS would assess water charges in the same way as is being done by the GOH-by charging a separate rate per unit area for each of the crops raised.

These rates are however not related to the cost of water consumed, and are decided by some thumb rule, always to the farmers' advantage. Also, the choice of crop is solely decided by the farmer. Thus, at no stage is the farmer under compulsion to make optimum use of a scarce input like water or ensure a minimum return for the same to the supplier. This may not have mattered much to the GOH. But this would certainly imperil the working of a small body like WUA, as it would have no control over its income. The income would be fluctuating according to the cropping pattern of the farmers. The only way to ensure financial stability of the WUAs is to resort to volumetric assessment, so that the units for the purchasing and selling of water become the same. This can be done by an innovative method, Warimetric. This was developed in Haryana in 1976 and had been successfully tried on six distributaries for eight long years, but is now lying dormant.

The principal advantages of warimetric are:

- It makes assessment on a volumetric basis, as stipulated by the National Water Policy. This makes the farmer quantity conscious;
- Being automatic, it does not need collection of field data and appointment of field staff like patwaris. This makes the WUAs fool-proof against corruption and leakage of revenue;
- It enables every farmer to himself check the accuracy of his water bill; This would liberate a farmer from the age-old bondage in which he is being held by the Patwari;
- It does not require capital investment or working capital, and is hence a zero-budget activity;
- It eliminates the risk of a WUA ending up in losses in the bargain of purchasing and selling water; and
- It would be a significant social revolution and a real gain for participatory irrigation management.

The GOH and the farmers are paying a heavy price for not having switched to warimetric. A zero-budget method like warimetric is ideally suited for a non-technical and non-professional body like the WUA. Haryana can become a pioneer in this field.

Water Users' Project—Swajaldhara

Having achieved only about 30 per cent utilization in people's participation in water sector reforms, the Center recently announced changes in guidelines. As part of reforms in the sector, the Government wants people in rural areas to pay 10 per cent towards the cost of construction of a project and then take the responsibility of operation and maintenance. Admitting that utilization in this scheme had been only Rs. 178.66 crores of an allocation of Rs. 590.30 crores since its inception in 1999, the scheme named "Swajaldhara", had been extended to 67 districts of 21 states during 2002-03.

The key changes in the scheme are instead of being top-down it would now be demand-driven and with a community participation approach. Secondly, the administrative unit for the implementation would now be the blocks/gram panchayats/ village/ habitation instead of the whole district. Panchayats and communities would plan, implement, operate, maintain and manage all drinking water projects.

The partial capital cost sharing would be 10 per cent in cash payable by communities. There would be a 50 per cent concession in gram panchayats and villages where half the population comprises Scheduled Castes and Tribes. The full ownership of drinking water assets would be with gram panchayats or community or the beneficiary groups who would be given funds directly for the project.

At present, about 15,000 habitation in the country have no potable drinking water, while two lakhs are partially covered and 2.17 lakh villages are identified with problems with the quality of water. The gram panchayats and blocks adopting the reforms principle would be eligible for the "swajaldhara" projects.

Integrated Water Resources Management (IWRM)

World Water Commission, while settling the framework for action resolved that "every human being, now and in the future, should have enough clean water for drinking, appropriate sanitation, and enough food and energy at a reasonable cost. Providing adequate water to meet these basic needs must be done in a manner that works in harmony with nature".

Integrated Water Resource Management (IWRM) is a process to promote co-ordinated development and management of water to maximize economic and social welfare in an equitable manner, without compromising the sustainability of vital eco-system. The idea behind integrated water resource management is to generate water wisdom, create knowledge and awareness about water, protect and restore water resources, achieve water and food security, extend sanitation coverage, meet the challenge of rapid urbanization and manage floods. It calls for a holistic approach as human interventions (such as afforestation, deforestation, urbanization, damming) do affect the productivity of water resources and to treat water as a natural capital asset that must be maintained. Therefore, the Integrated Water Resource Management shifts the focus from sub-sectoral to cross-sectoral water management.

The Dublin Principles enunciated in 1992 set a paradigm for integrated water resource management and lay down that:

(a) fresh water is a finite and vulnerable resource, essential to sustain life, development and the environment,
(b) water development and management should be based on a participatory approach involving users, planners and policy-

makers at all levels (including management at the lowest appropriate level),

(c) women play a central part in the provision, management and safeguarding of water, and

(d) water has an economic value in all its competing uses and should be recognized as an economic good.

The traditional sectoral and fragmented approach to water resource management has resulted in creating government bodies representing conflicting interests. Policy objectives have been defined without involvement of water users and without consulting the cross-sectoral and institutional bodies. Therefore, the limited resource has not been employed to maximize social welfare.

As the overall quantity of water cannot be altered significantly by human actions, fresh water resource has to be regarded as a natural capital asset and hence to be used in a judicious manner to optimize the benefits for the community. In a situation of competition for scarce water resource, there is a possibility for water being allocated to low value uses and no incentives are provided to treat it as a limited resource. Therefore, there is an urgent need to recognize water as an economic good. In any case, management of its demand through economic instruments has been long recognized and occasionally practiced as well.

There is a definite case for integrating the green water flows and blue water flows. Similarly, integration of surface and groundwater management, integration of upstream and downstream water related interests and integration of quantity and quality in water resource management are the challenges before us. However, to operationalise the concepts which have been discussed above, following steps are needed:

(i) An apex body at the state level should be set up for IWRM and it should be equipped with appropriate authority and capacity for monitoring and regulating water resources management. It should also create and manage the central database for efficient coordination, monitoring and evaluation to support decision-making. This body should harness the need for different sectors and should ensure monitoring and evaluation with each implementing agency.

(ii) Judicious assessment of sectoral water requirements, promoting less water consuming crops, adoption of efficient technology to obtain maximum output and consideration of water charges based on volumetric basis should be the concern of this body.

(iii) To manage and maintain the water quality, it needs to review the standards and norms for safe use, enactment of proper legislation and establishment of effective mechanism for enforcing of norms and standards and undertake measures for awareness through training and media campaign amongst users.

(iv) Need for activating the efforts of state agriculture departments, rural development departments, area development boards and to involve NGO in watershed management.

(v) State Environmental Plans may integrate and activate efforts of various agencies to establish an environmental database and suitable implementation programme.

Conjunctive Use

It may be mentioned in this context that much espoused, but seldom applied concept is the conjunctive planning, management and regulation of surface and ground water. In most of the cases, the conjuctive use of ground water is wrongly identified with the use of ground water as a supplement to surface water irrigation. The concept of conjuctive use essentially envisages the simultaneous and balanced use of both surface and ground water, keeping in view the agro-climatic conditions as well as ecological integrity of a command.[5]

This calls for an integrated approach to the development and use of complete water system in the area in the shape of major, medium and minor irrigation projects on a water shed basis. Such an approach would also put at rest existing controversies between major, medium and minor irrigation projects. The issue of developing this or that irrigation resource loses significance in an ultimate analysis, when the exploitable water resources, taken as a whole, happen to be far from bountiful. These would hardly provide irrigation cover to two thirds of our crop area even on full exploitation say, by the year 2010 A.D.- eventually we have per force to develop all categories of irrigation works: mega dams, big dams, small dams, check dams, tanks, dugwells, tubewells, borewells, etc. And once we begin to show due appreciation of the inherent complementarities or interlinkages between irrigation sources, the issue of choice among irrigation sources becomes really a non-issue in a pragmatic sense. If at all priority to groundwater irrigation is to be accorded, it should be in the high rainfall areas of the east Gangetic plains and Madhya Pradesh where much of the groundwater resource is unexploited.

The farmers in Canal Commands should be encouraged to acquire wells of their own. Water from these wells, may be used in conjunction or combination with canal supplies. The dependability in irrigation gives the farmer much needed confidence to intensify his farming resulting in intensive use of farming inputs. Conjunctive irrigation also results in balanced use of water resources.

In point of fact, the issue relating to complementarity of each of the sub system is not fully appreciated. A little insight into the system will show that it is large dams alone which substantially augment groundwater regime in low and medium rainfall regions of the country, thereby sustaining a thriving groundwater-based farming in such regions. For instance, two-thirds of the groundwater recharge in Punjab originates from canals fed by waters of Bhakra and Pong reservoirs. Without such recharge, groundwater-based farming in this state would be a fraction of what it is today. Similarly the seeped-in waters are found to quadruple the rate of return of investment in well irrigation in Ahmednagar district of Western Maharashtra. In a specific case of the Mula Canal in this very district, these waters have boosted the ground water based farming, both in terms of number of sustainable wells and crop output per well. The eightfold increase in output from wells is an indirect output contribution of the Mula project, and is no less than the direct output resulting from the canal irrigated acreage of this project.

In Tamil Nadu, both canal and tanks together make a substantial augmentation to its natural groundwater availability. Somewhere between one-fourth and two-fifth of the total output of the order of 7 million tonnes from wells established in the state may be traced to seeped-in waters of canal /tank origin. Whatever be this proportion, bulk of the seepage into groundwater table is from the state canal network. The Minor Irrigation Census place the extent of supplementary irrigation by wells at nearly 2.9 million ha representing nearly 10 per cent of the estimated potential of ground water works. This distinction between sole and supplementary irrigation by wells is unfortunately not observed by any state other than Tamil Nadu.

Allied with this issue is the question relating to big and small dams. There are advantages /disadvantages and limitations in each case. But a controversy of this type is completely misplaced under Indian conditions. There are different situations in the country where only a big or small dam is possible. In view of the vagaries of Indian monsoons, the imperative need for irrigation development for protecting crops during drought time is the prime issue. Under these circumstances we should assume that controversies of this nature are only a futile exercise and rather a hindrance to irrigation development and for that matter progress of agriculture in the country.

COMMAND AREA DEVELOPMENT PROGRAMME (CAD)

Greater emphasis has to be placed on improving water management in the irrigation commands with a view to augmenting agricultural production. Command Area Development programmes initiated in 1974-75 must receive much higher priority than is being accorded to them. The CAD programme aims mainly at reducing the gap between the potential created and utilisation thereof. This is to be achieved through the integrated development of command areas by constituting a multi-disciplinary team under the overall control of the CAD Authorities.

Components of the Programme

1. On-Farm Development (OFD) works *i.e.* development of field channels and field drains within the command of each outlet; land leveling on an outlet command basis; reclamation of waterlogged areas; enforcement of a proper system of "Warabandi" (rotational water supply) and fair distribution of water to individual fields; realignment of field boundaries, wherever necessary (where possible, consolidation of holding are also combined); supply of all inputs and services-including credit; strengthening of extension services; and, encouraging farmers for Participatory Irrigation Management (PIM).
2. Selection and introduction of suitable cropping patterns.
3. Development of ground water to supplement surface irrigation (conjunctive use under Minor Irrigation sector).
4. Development and maintenance of the main and intermediate

drainage system (irrigation sector).

5. Modernisation, maintenance and efficient operation of the irrigation system upto the outlet of one cusec capacity (irrigation sector).

Programme Coverage

Beginning with 60 Major and Medium Irrigation Projects in 1974-75, the programme now covers 234 projects with a culturable command area of 22.78 million hectares spread over 28 States and 2 Union Territories.

Programme Implementation

The Command Area Development Wing of the Ministry of Water Resources coordinates and monitors the implementation of the Command Area Development Programme at the national level. Proposals received from the States for inclusion of new projects under the Programme are examined and, if found techno-economically feasible, are included under the Programme. Progress is measured through physical and financial progress reports of the programme as received from the States. The quality of works is ensured through monitoring, including field visits. Moreover, technical guidelines and manuals have been circulated to the States in this regard. Functionaries are trained on specific subjects from time to time, besides holding various meetings, workshops, seminars on different technical and managerial aspects.

The programme is being implemented by the State Government through Command Area Development Authorities (CADAs) set up by them. However, in some States, namely Arunachal Pradesh, Himachal Pradesh, Meghalaya, Nagaland, Tamil Nadu and Tripura, CAD Authorities have not been constituted and the Programme is being administered through the concerned line (Agriculture/Irrigation) departments.

Financing Pattern

The financing of the activities carried out under the Programme comes from the following three sources viz; State outlays; central assistance on matching basis for certain identified activities, and Institutional finance.

The financing pattern for providing Central Assistance to the States keeps on changing from Plan to Plan as per the past experience. The financing pattern in force from April, 1996 is however, as follows:

1. Grants are admissible on matching basis to the State Governments for establishment, topographical and soil surveys, planning and design of OFD works, supervision of OFD works, construction of field channels and field drains, enforcement of Warabandi, adaptive trials, demonstration and training, crop compensation, subsidy to small and marginal farmers on identified items, evaluation studies sponsored by the States, reclamation of waterlogged areas and one-time functional grants to the Water Users' Associations.
2. Hundred percent grant from the Central Government is given for orientation training for senior level officers and also for evaluation studies, sponsored by the Central Government.
3. Subsidy is admissible for land leveling and shaping, groundwater development and sprinkler and drip irrigation to small and marginal farmers, on the pattern followed under the Integrated Rural Development Programme (IRDP), now Swarnajayanti Gram Swarozgar Yojana (SGSY).
4. Loan for the purchase of equipment and machinery is provided to the States on a matching basis.

Physical Achievements

The core components of physical works are construction of field channels and field drains, implementation of warabandi (rotational water supply) and land leveling and shaping. Cumulative achievements upto March, 2001 since inception are given below:

Sl. No.	*Item of Work*	*Cumulative Achievement till March, 2001*	*2001-2002*	
			Target	*Anticipated Achievement*
1.	Field channels	153.42	2.35	2.35
2.	Field drains	10.46	0.35	0.35
3.	Warabandi	99.18	6.00	6.00
4.	Land levelling and shaping	21.72	0.22	0.22

Source: Annual Report 2001-02, Ministry of Water Resources, Government of India.

Finally in order to achieve improved water management, proper linkages have also to be developed in the implementation of the programme in such a way that the irrigation engineers, agricultural scientists and those engaged in managing irrigated agriculture could properly coordinate their activities in accomplishing this difficult task. It is important that the lab-to-field flow is accelerated with the help of appropriate and specific trial and pilot projects. With the technologies and methods now available, even modest expenditures on conservation and efficiency improvement could make unnecessary many of the inordinately expensive, ecologically disruptive water projects that have dominated water planning agendas for decades.

Watershed Management as the Integrated Package for Water Resources Development

Out of total about 329 m ha of the country's geographical area, it is estimated that only 206 million ha have any potential for production which includes 143 m ha agricultural lands. Out of total agricultural lands, about 85 m ha reported to be suffering from varying degree of soil degradation. Of the balance 123 m ha having no potential for production, 40 m ha is completely unproductive and remaining 83 m ha is classified as forest lands of which over half is denuded to various degrees. India has 480 million heads of livestock (as per 1997 census) which use 13 m ha of lands, most of which is over-grazed. As per the Soil Status Report, about 175 m ha (*i.e.*, 85 per cent of total land area having productivity potential) are degraded to varying degrees, out of which 150 m ha are degraded on account of water and wind erosion. Presently, rural and urban settlements, roads, railways, water bodies, mines, defence and industrial installations use about an estimated area of 21 m ha which may increase to 27-28 m ha by 2010 A.D. All these lead to the conclusion that there is no scope now to expand cultivable area except to reclaim the waste lands including those suffering from salinity, alkalinity and waterlogging on one hand and increase irrigation cover together with optimization of water use efficiency for the available cultivable area on the other hand.

Out of total annual run-off of 1,880 cu km only 60 per cent is estimated to be total utilised water resource (surface and ground water put together) in the country and thus remaining 40 per cent, *i.e.* about 752 cu km flows away without being utilised at all. In this connection, it would be worth while to recall the sound advice given by Parakrama

Bahu the Great, the 12th Century King of Sri Lanka, who said "let not even a small quantity of water obtained by rain go the sea, without benefiting man". This is precisely the aim of watershed management. The run-off water is normally collected in ponds/tanks and according to Irrigation Census of 1987, the total tanks in the country are put at little over 5 lakh with concentration in the States of Madhya Pradesh, Andhra Pradesh, Kerala, Karnataka, West Bengal & Tamil Nadu. Besides providing much needed water for irrigation, these tanks have also beneficial impact in the form of ground water recharge. In earlier times, tanks used to be properly maintained jointly by the rulers and beneficiaries. In post-independence era, people's participation in maintaining these tanks have been greatly diminished to the point of being almost non-existent now. Consequently, between 1962 and 1986, total command area under tank irrigation shrunk from 4.8 to 3.1 m ha *i.e.*, by about 1.7 m ha, mainly due to silting and thereby loss of storage capacity and water carrier system going into disuse.

This trend has to be revised partly by dovetailing governmental employment generation programmes like JGSY and partly by motivating and involving beneficiaries in regular maintenance of tanks. In this connection, it may be stated that under Sardar Sarovar Project in Gujarat, it has been decided that for the purpose of construction of branch canals, distributories and distribution system, maximum use of earth should be made by obtaining from nearby tanks or by creating new tanks in Government wasteland instead of obtaining the earth only from borrowed pit excavations. This approach will not only reduce the land requirement for canal system works by about 10 to 20 per cent but will also create permanent assets including improving the existing assets, *i.e.*, village tanks/ponds. It has been estimated that as a result of this innovative approach, about 5,000 tanks spread over entire command of Sardar Sarovar Project will be deepened, benefiting about 3,400 villages.

Realizing the fact that soil and water conservation practices alone are insufficient to encourage for adoption, the strategies of water conservation and crop production are harmonized on a catchment basis rather than on individual farmer's holdings. A judicious blend of rain water conservation and alternate land use with modern methods of crop husbandry has definitely changed the fate of rainfed agriculture. The impact is visible in terms of improvement in productivity over the base year and increased recharge of the ground water in dug wells. Impressive success stories at a few sites will definitely attract the farmer's attention provided his participation in the change is ensured and his conventional wisdom on ancient water conservation measure is adequately utilised to build up the present day programmes.

Reforming Government Water Services Agencies

The water services agencies also need to be reoriented to rise to the occasion and to deliver more to the community, and for the same to happen, certain steps, as mentioned below need to be attempted.

(i) Right-sizing of the state irrigation departments/agencies to a manageable level, introduction of VRS/golden handshade, allowing option to work in private sector and return to the employer after a fixed period are some of the immediate action points.

(ii) Water charges collected by WUAs and state irrigation departments should be self-sufficient for O & M system. Annual review of rates, establishment of an independent water tariffs regulatory authority; keeping of two rates, one for culturable command areas and the other for irrigated areas, can be considered.

(iii) The oversized irrigation department should give way to new corporate entities which may be based on river-basin criterion and should be more accountable. Diffusion and delegation of decision-making to the front-line managers can set better performance standards.

(iv) The emergence of a new power equation with the arrival on the scene of water user entities vested with rights and responsibilities situated in the legal enactment makes out a definite case for orienting the agency staff and for that, apart from some form of institutional control of the WUAs on them, training and orientation programmes will have to be launched in a big way.

(v) To manage the problem of over-drafting, unplanned and unmanaged exploitation of groundwater, a comprehensive groundwater legislation envisaging a single agency for monitoring and regulation of groundwater be made in each state. Metering and higher power rates in rural areas with improved electricity services be incorporated in the management of groundwater.

Adoption of New Techniques and Technology

In water management, over the years, the focus has been on the supply side leading to massive creation of major, medium and minor irrigation projects. However, owing to the limited possibility of expansion in capacity and also the enormous cost involved, there is a greater need for demand side management as well. The conservationists point out that minor change in water usage habits can save up to 30 per cent of water for use of others. Needless to add, suitable awareness and education programmes can contribute a great deal to this. Similarly, use of new technology such as sprinkler irrigation system, drip irrigation system, laser assisted land leveling, underground water conveyance system, zero tillage technology, furrow irrigated raised bed planting can make substantial savings. Similarly, irrigation at critical stages of crop growth can restrict waterings and save water. Use of bio-fertilisers and control of nitrate pollution will help improve the quality of water. Similarly, the phasing out of un-metered flat tariff for agriculture, a ban on over installation of tubewells in 'dark blocks' (where groundwater development is more than 85 per cent), mandatory harvesting of roof-top rain water can be some of the new policy options.

Conclusion

Despite the earth being called planet of water, the fresh water availability is extremely limited. Access to clean water for drinking, cooking, bathing, and other households needs is fundamental. Water for agriculture is critical for food security. But water for irrigation may be threatened by rapidly increasing nonagricultural uses in industry, households, and the environment. Investment and policy reforms in water management will be significant determinants of future food production, demand, prices and trade. Rapidly growing domestic and industrial water demand in India may increase water scarcity for agriculture, and with a continued slowdown in water investments, could be serious threat to future growth in food production and in meeting other demands of water. There is therefore an urgent need of water resource planning which will aim not only optimizing the utilization and management of water resources, but also to make the developmentalists aware of environmental and social equity considerations as well as to make environmentalists aware of economic and development needs. The constraints for improvement in these spheres are often neither financial nor technical—they are political, social, and managerial. The fundamental premise is that people's initiative and management of their own quality of life must be at the centre of planning and action. Visions and plans articulated at local level are the building blocks at the next levels of regional, national and even global action.

REFERENCES

1. Rosegrant Mark W., Cai Xining, et.al: World Water and Food to 2025—Dealing with Scarcity, International Food Policy Research Institute, Washington, D.C.
2. *Financial Express*, 6-10-1994.
3. Saranji Mr. Agriculture Commissioner, Maharashtra, Lecture delivered at an International Seminar and published in *Economic Times*, 18-12-1995
4. Irrigation Management and Farmers' Organization, A case study of M.P. Water and Land Management Institute, November, 1991.
5. Navalwala, B.N. Water Resource Development and Water for Life, FAO (1994).

11

Water Conservation: Resource Management

"On whom the flowing waters, ever the same, course without cease or failure night and day. May she yield milk, this Earth of many streams, And shed on us her splendour copiously!"

—Atharva Veda 12.1.9

Introduction

The valuation of water today is becoming synonymous with 'blue gold' because of the significant economic ramification the terminology conjures. Therefore the term 'blue elephant' was coined as just another 'phrase' to signify a search for coherent strategy for global water policy instruments to assist in managing this valuable resource. Although water covers three quarters of the planet, only a small fraction is accessible as fresh water. If the total amount of water withdrawn, almost 70 per cent is needed to produce the food that fuels human activity, the question is now uppermost in the minds of the experts all over the world is thus "will there be enough water to grow food for the almost 8 billion people expected to populate the earth by 2025?" According to International Food Policy Research Institute (IFPRI, Washington), more than 80 per cent of this population will live in developing countries and 58 per cent in rapidly growing urban areas. In response to population growth and rising incomes, IFPRI projects that worldwide cereal demand will grow by 46 per cent between 1995 and 2025, and in developing countries by 65 per cent; meat demand by 56 per cent worldwide and more than double in the developing world; and calorie requirements and dietary trends will translate to even greater water demand if the food produced is to supply adequate nutrition.

As a matter of fact, water, as an input to agriculture, is critical for sustaining the food security. India faces the daunting task of increasing its foodgrain production by over 50 per cent in the next two decades[1]. increasing competition for water in agriculture, industry, domestic and environment uses is already manifested in inter-and intra-sector, basin, state, district and village level conflicts.

These will escalate further as India's annual per capita water availability goes below water scarce threshold level of 1,700 cubic meter within the next two decades. In six of the country's 20 major river basins (with less than 1,000 cubic meter of annual per capita availability), water resources are under stress and depleting. By the year 2025, five more basins will become water scarce and by 2050, only three basins in India will remain water sufficient[2]. Supply expansion, to meet expanding needs, is constrained by availability and rising economic and environmental costs associated with its development and use.

Needless to say, most of the increase in foodgrains will have to come from intensified agriculture supported by irrigation. The success of irrigation in ensuring food security and improving rural welfare has been impressive hitherto, but past experiences also indicate that inappropriate management of irrigation has contributed to environmental problems including excessive water depletion, water quality reduction, water-logging, and salinization. Long-term hydrological records have shown a significant reduction in the annual discharge of some of the world's major rivers. In some basins, excessive diversion of river water has led to environmental and ecological disasters for downstream areas, and pumping groundwater at unsustainable rates has contributed to the lowering of groundwater tables and to salt water intrusion in some coastal areas.

Further, continued rapid growth of domestic and industrial water uses, growing recognition of environmental demands for water, and the high cost of developing new water resources threaten the availability of irrigation water to meet growing food demand. Water is also wasted in domestic and industrial uses. "Unaccounted-for-Water" (UFW) typically constitutes a high proportion of urban water supply. Moreover, growing number of the world's rivers, lakes, and groundwater aquifers are being severely contaminated by human, industrial, and agricultural wastes. High withdrawals of water and heavy pollution loads have already caused widespread harm to a number of ecosystems. This has resulted in a wide range of health effects in which humans have been harmed by waterborne diseases and contaminated food. Rising human demands only increase pressure on ecosystem and intensify the need to maintain an adequate water supply to wet lands, lakes, rivers and coastal areas to insure the healthy functioning of ecosystems.

Agriculture: Primary User of Water

There are no two opinions that out of all the requirements of water the domestic requirement is the most critical one for human survival. But it is also true that this domestic need of water forms only a very small part of our total water requirements and availability. Even today when India's population is over a billion, the total domestic need for water per year of the whole country is hardly about 33 cubic kilometre (cu km.) that is to say, only 3 per cent of the annual utilizable fresh water availability of 1,150 cu. Km., which is comprised of 700 cu. km. of surface water and 450 cu km. of groundwater. And yet crores of people and cattle suffer regularly for want of drinking water. Surely the government have failed miserably in their duty to ensure regular supply of potable water for all. It is the mismanagement, apathy, flawed models of development, neglect of traditional methods of water harvesting, etc. that have led to the present state of water scarcity even for drinking.

Nevertheless while worrying about the non-availability of drinking water which is indeed very critical for life, we must not, however, lose sight of other requirements of water, e.g. for agriculture, power, etc. as these too are very important for the nation's survival and also because the availability, requirement and management of water for non-domestic needs have significant bearings on the availability of water for the domestic purposes.

An old nemesis to the environmental health of river basins is irrigated agriculture. An important component of food production for thousands of years, irrigation continues to be practised in some form or the other in every country of the world today. Chile and Peru derive 50-55 per cent of their agricultural production from irrigated land, China 70 per cent, Pakistan 80 per cent, and in Egypt 98.6 per cent of the croplands are irrigated. Irrigation is extremely water-intensive, and most irrigation practices are grossly inefficient. According to one estimate, as much as 2,500 tonnes of water may be required to produce one tonne of cereals. If so, there is certainly an urgent need for efficiency in the use of irrigation water. More so when competing uses like industry or domestic water supply system is economically much more productive and gets high priority. Worldwide, only 37 per cent of irrigation water is consumed by crops, while most of the rest is wasted, contributing to salinization and waterlogging where drainage is inadequate. Salinization causes 'sterilization' of some of the world's best and most productive farmlands. Salinity now seriously affects productivity on about 20-30 million hectares (about 7 per cent) of irrigated land around the world. And it is a persistent problem in the plains of eastern and western China, the Indian subcontinent, the Central Asian Republics, South Eastern Europe, the Middle East, North and West Africa, and Australia. In other words, salinization has become commonplace wherever irrigation is practised on a large scale.

It was estimated that in 2000 AD the water requirements for agriculture was as high as 630 cu km per year. This requirement is estimated to rise to a staggering 770 cu km per year by the year 2025. Thus while meeting the domestic requirement of water ought not to be a problem any time, it is bound to become exceedingly difficult to meet the growing need of water with every passing year even under normal monsoon condition. The need for agricultural water is going to increase every year as we need more and more foodgrains and other commercial crops to sustain our rapidly growing population increasing by about 1.7 crore per year. So, the crux of the problem which constraints us is how to meet our ever increasing needs of water for agriculture. Modernisation of canal water delivery system and command area development to raise the irrigation efficiency is the need of the hour. It may be pointed out that in many regions in the country the efficiency continues to be around 40 per cent or even less as compared to 75 per cent in Israel.

Increased Pollution of Water Resources

A recently published report, The World Water Resources Report (1994-95), compiled by the World Resources Institute, in collaboraton with UNEP (United Nations Environment Programme), states that nearly 70 per cent of the water resources of India are polluted and in large stretches of major rivers, water is no longer fit for bathing, leaving alone drinking needs. The river Yamuna receives about 200 million litres of untreated sewage every day.

The story of the river Sutlej as, it flows past the industrial town Ludhiana is similar, and plan has been framed to clean it at a cost of Rs. 164 crores. Hardly, a few cities care to give even primary treatment to sewage before discharging into the natural watercourse or river, with the result that both the surface and ground water resources are getting badly polluted. With such a 'bleak water scenario' the position is going to become more serious in decades to come, and the 'water crunch' is quite a disconcerting prospects indeed. This calls for maximum conservation of the available resources, and rational management and efficient use. Every drop of water is going to become the most valued possession of a region. In fact, what is needed in the water management are 'crop per drop', *i.e.* irrigation efficiency and thereby productivity, 'job per drop', *i.e.* giving utmost importance to economic value of water and 'drop per drop', implying and indicating an ecosystem approach to river basin management.

Silt and Sedimentation

Full utilisation of the potential created also involves another important area. All our big dams and water storage tanks are silting at an alarmingly high rate due to indiscriminate felling of forests in the catchment areas, threatening to reduce the total life span of all of them drastically. Our costly and irreplaceable reservoirs, which represent not only valuable irrigation potential but very often hydel and flood control potential also, must be saved from the threat of premature siltation. Although soil conservation measures in river valley catchments have been undertaken under successive plans to reduce the erosion of soil cover and the corresponding inflow of silt into reservoirs, there has been large-scale denudation of vegetation and forests in the catchment areas of reservoirs, leading to increasing silt load in rivers.

The sedimentation rates in the reservoirs is thus much higher than the one assumed in the design, thus reducing useful life of the reservoirs. On a rough estimate we are losing a staggering 2 million acre feet (MAF) live storage capacity annually in our major and medium dams, corresponding to a loss of nearly 300 thousand hectares of irrigation potential every year. It costs a minimum of Rs. 50,000 for irrigating a hectare of land. We are, thus losing Rs. 5.000 crore annually in the form of capital assets alone.

The country has incurred a double loss, in not getting the projects completed resulting in cost escalations, and in importing agricultural goods and spending additional thousand crore of its scant and precious foreign exchange. Of course, the loss to consumers who has to pay a higher price for imported agricultural goods and the greater loss in terms of income, employment and savings of the country is another important side of the story.

Wastage Through Seepage

Huge wastage of water and water-logging of vast cultivable areas caused by the seepage from badly managed canals are no less a serious problem. Experts have estimated that 38 per cent of canal water is wasted on account of seepage and evaporation between dam gate and the field. Of the remaining 62 per cent of what reaches the field, only 15 per cent is actually utilised by the plants because of the wasteful flood irrigation method. According to International Land Reclamation Institute, losses in the use of water in some north Indian canals are estimated as follows: loss in main canal and branches 15 to 20 per cent, loss in major and minor distributors six to eight per cent, losses in field channels 20 to 22 per cent, losses in application and evapo-transpiration 25 to 30 per cent.

Another very disturbing phenomenon is the fast depleting underground water and vast areas are being declared as 'grey' and 'dark' blocks. Undoing of traditional water recharge systems and non-creation of new ones are adding to the already acute situation. Though the spread of irrigation has fostered a tremendous surge in crop production over the last generation, more than 80 percent of the world's cropland is still watered only by rainfall. These lands produce two-third of the global harvest and provide the subsistence diet for many growing third world populations.

Some of the land can and will, in time be brought under irrigation. But given the high cost of new irrigation projects expanding irrigation may in many cases not be a feasible near-term solution to raising food production. Effective water management is as crucial to increasing crop yields in non-irrigated agriculture as it is in irrigated areas. Yet with irrigation occupying center stage in recent decades, the potential to improve water productivity on rain-fed farmland has been largely neglected. it is true that rain-fed farming (often called dryland farming in arid and semi-arid regions) is inherently a risky enterprise. Regardless of what historical records may show, dryland farmers have no guarantee that rain will sufficiently water their crops throughout the growing season.

The problem in India is threefold. Firstly, irrigation potential already created at such a heavy cost is not utilised fully. Secondly, most of the projects have been started without proper studies and have spilled over the plans, leading to cost escalation on the one hand and delays in the availability of water on the other hand (Table 11.1).

Thirdly, the most disturbing feature is the huge wastage of available water resources due to wrong practices and policies.

Table 11.1: Status of Plan Projects

Status of Projects	*Number of Projects*		
	Major	*Medium*	*Total*
Completed in Pre-Plan period	74	143	217
Projects taken up during plan era till the end of IX Plan	301	966	1267
Projects completed during Plan era till the end of IX Plan (1951-2002)	142	724	866
Projects Spilling over to X Plan (2002-2007)	159	242	401

The reduction of run-off and soil losses through a greater emphasis on afforestation and better land management must be taken up on a nation-wide scale because such a strategy also constitutes our best insurance against the recurring droughts and floods.

There is also a continuously declining water level in Punjab and Haryana. Similarly, large-scale exploitation of groundwater in Tamil Nadu and Gujarat caused intrusion of sea water into the coastal area aquifers thereby rendering these usable waters to unusable saline waters. It is only through the integrated approach that it will be possible to maintain hydrological balance which is vital for sustainable agriculture in any region.

Unfinished Projects

A number of inefficiencies arose during the course of irrigation project implementation, Of these, time and cost overruns in the case of major works, as revealed by audit reports, have caused much dismay. The cost overruns in many projects appear truly mind boggling till one takes pains to correct the actual costs incurred for the continual price inflation. Some perceptive irrigation planners have tried to bring out that the price rise in respect of inputs used in major irrigation projects has been about 3 per cent higher than that in the case of foodgrains. A better measure of the execution inefficiency is the time overrun, which can in turn be largely ascribed to inadequate project funding by the State Government. The question is: why this project under-funding?

Too many projects start simultaneously, some of them even not approved by the Planning Commission. Too much leakage of resources in the shape of kickbacks. Pre-emption of limited funds in executing World Bank-aided projects which are, acre for acre to be irrigated, far more expensive than other schemes. The diversion of invisible funds into providing subsidies to existing irrigation projects working at a huge loss.

According to irrigation experts, avoidable delays occur due to several reasons. Approvals are made at random without considering available resources as a result of which the work is stalled. Some projects are announced before considering their economic feasibility and viability by way of promises in the election manifestoes. There is usually a huge time lag for the work to commence. Also, if the project is seen to be environmentally damaging, it hampers the progress resulting in time and cost overruns. According to the Eighth plan document, the "biggest single malady in the major and medium irrigation sector right from the first plan has been the continued tendency to start more and more projects resulting in a thin spread of scarce resources and consequent time and cost overruns. Though all the plans, without exception, declared their intention to give priority to complete the ongoing schemes, the addition of new schemes continued unabated. The scarce rupee resources are being spread too thinly over a large number of projects. Thus new projects are taken up even before old ones have fructified and the resultant delays push up the costs".

Many a time, the Planning Commission is fed with wrong data to get a project cleared and then once the project is started the data is changed to acquire more funds. Complementarities of the entire irrigation systems are not completed in time and costs go on escalating, allegedly for very substantial cuts to contractors, engineers and others. Input complementarities with water may also be lacking. These issues need to be tackled to improve the working of this crucial sector.

The financial management of public irrigation works is unquestionably dismal. These works (with the sole exception of Uttar Pradesh) do not generate sufficient revenue to cover even their working costs, not to speak of recouping the high investment costs. A major factor for this is the policy of gross under-pricing of canal water by every state government with resource costs becoming more and more unfavourable, thanks to the gathering momentum of farmers' agitation in recent years. But do politicians realize the myopia of continued under-pricing of farm resources? In the case of irrigation, preventive maintenance gets poorer attention having its toll in two ways: (1) poorer irrigation service to the farmers and (2) mounting up of the replacement investments which have to be provided for at an utmost priority basis-this further erodes the pool of invisible funds for new and on-going projects.

A disturbing feature of the irrigation scene relates to the unfinished major and medium projects. P.K. Thungon, then Minister of Water Resources (MWR), stated in the Lok Sabha on March 1, 1993, that there were 105 major and 165 medium irrigation projects in varying degrees of completion which included the Tungabhadra (1945) and the Bhadra (1947) both in Karnataka. It may be of interest to mention here that according to the annual report of MWR 1981, major projects had spilled over into the Seventh Plan with the cost spillover being Rs. 22,236 crores. The Seventh Plan provided only Rs. 11,555 crores, or just about half the amount required for their completion. And considering the rate at which cost escalations have taken place, it will not come as a surprise if this allocation actually meets not more than a quarter of the spillover costs. In other words, it is quite likely that some of the ongoing projects will be finished only in the 21st century and by that time the costs per hectare would have shot through the roof.

Wasteful Practices

A FAO study (titled: 'Agriculture Towards 2,000') covering 90 developing countries (except China), found that about 20 to 35 per cent of water released from reservoirs actually reaches the crops being irrigated. As per the World Bank Study, as much as 50 per cent of all water stored and carried for municipal purposes is being lost.

In industrial sector also many products can be manufactured with significantly less water than currently being done. For example it has been found that water required for manufacture of one tonne of 'paper board' varies between 62,000 to 376,000 tonnes, the upper figure being six times the lower value.

The solution would then lie to aim at reducing the present volume of water use, through increase of conveyance, distribution and application efficiencies, at the same time conserving the quality of water resource. For improved water use efficiency we should adopt better on-farm management and command area management programmes, covering rational overall system management.

Use of non-conventional sources of water, such as recycling of municipal and industrial waste water, after appropriate treatment, use of 'marginal' or saline water for irrigation combined with improved irrigation like drip irrigation should be attempted wherever feasible. Also research should be carried out in the field of genetic engineering to develop those types of crops, which ate more tolerant to the saline waters. Sprinkler irrigation water can be used gainfully with salinity index less than 1,000 ppm, and when water has salinity index above 1,500 ppm, considerable burning of leaves of plants is noticed with sprinkler system.

Through 'Mulching practice', by the use of LDPE (low-density polythene sheets) as much as 25 per cent of the water supplied to crops can be saved by reducing the evaporation losses from the soil around the plants. Through advanced methods of irrigation like 'drip,' sprinkler', advanced surface or sub-surface piped systems, considerable saving of water is possible, as will be clear from Table 11.2.

Table 11.2: Efficiencies of Different Irrigation Systems

System of Irrigation	*Efficiency*
* Drip and Sprinkler	Above 85 per cent
* Furrow System	75 per cent
* Border with PLL (precision land leveling)	65 per cent
* Flooding	below 40 per cent

Source: Facts For You, March 1996.

Sustainable Water Resource Management

Developing countries like India need to work on sustainable ways to manage their natural resources. Moreover, there is widespread pollution and resultant degradation of water resources. Increasingly, the basic and prior need of fresh water is threatened like never before. The inception of Green Systems was in the very spirit to provide sustainable solutions to the challenges that face India. Green Systems believe that there is a better way, one that handles natural/fresh water resources benefiting and preserving the natural process with a sustainable water resource management approach.

Green Systems India has worked towards alleviating these problems in various sectors, in the form of integrated environmental management that well and truly puts the beneficiaries on the path of sustainability. Green Systems has undertaken umpteen water resource management projects throughout the country. One of them has been in Chattisgarh region, a rocky and water deficient area. Green Systems provided solutions for harvesting and managing water resources to the tune of 11 lac meter^3 per year. This entails effective management of this invaluable natural resource.

WATER PRICING: POTENTIAL AND PROBLEMS

Water pricing, policy has the potential to mitigate water scarcity. Because of its key role in managing water demand and augmenting water supply, water pricing is an important policy instrument for creating incentives to conserve and allocate water efficiently. By providing financial justification for developing additional supplies from conventional and unconventional sources, pricing policy can make more water available to users. Unfortunately, the water pricing policies being pursued in most countries fail to perform these vital roles due to faulty approaches and inappropriate institutions, both of which have their roots in political economy.

The state provides many irrigation and domestic water supply systems at subsidized rates because doing so delivers public goods in the form of secure food supplies, public health, or legitimacy for the government. But low water charges and poor cost recovery lead to declining funds available for investment in water infrastructure, poor maintenance of existing systems, inefficient water allocation, and growing conflicts between those with and without access to water. Economic and political pressures stemming from these problems have promoted a renewed debate on water pricing policy.

Role of Water Pricing Policy

Financially, water pricing is the main mechanism for cost recovery. Economically, it signals the scarcity value and opportunity cost of water and guides allocation decisions within and across water sub-sectors. The financial function requires water rates to cover the cost of supplying water to users. The supply cost is usually calculated by adding the operation and maintenance costs and the capital costs of constructing the system. But full cost recovery also requires water rates to reflect the long-term marginal cost (the cost of supplying an additional unit of water including the social cost of externalities).

The economic and allocative role of water pricing requires water rates to capture the scarcity value (or the marginal productivity/utility) and to equalize the opportunity costs (the value of water in its next best use) of the resource across uses. As water moves from least productive to most productive uses, places and time points for efficient allocation, there will be a convergence of the scarcity value, opportunity cost, and longterm marginal cost of the resource. Unfortunately, such a convergence is rarely seen in practice. Unrealized opportunities still exist and water rates can be designed to capture at least a part of these opportunity costs.

Cost Recovery and Economic Incentives

The effectiveness of the financial and economic roles of water pricing policy depends on pricing methods, sectors, and countries. Water pricing may be either volumetric (based on the quantity of water used) or flat rate (based on area irrigated or households benefited). Volumetric pricing is conducive to creating incentives for efficient allocation and use, but the cost of establishing volumetric water delivery structures is often prohibitive, especially in large and spatially spread surface irrigation systems serving many smallholders. As a result, area-based fixed rates are dominant in most irrigation systems. However, volumetric water rates are widely used in many urban water supply systems.

Full cost recovery is legitimate both in economic and equity terms only if the users gain all the benefits from investment. On this count, some argue that since food prices fall when irrigation increases food production, consumers benefit, and that higher water rates for full cost recovery could adversely affect farmers incomes. But this need not be the case because farmers benefit from irrigation partly from higher output and partly from increased land prices. In any case, the kind of full cost recovery being contemplated in many contexts does not cover even 10 per cent of the additional income from irrigation, let alone the land value appreciation.

Water rates are still subsidized even in countries with a relatively mature water economy such as Australia, Israel, and the United States. This is rooted in the political economy of water, as the powerful state and user interests often oppose charging the full cost of water. As a result, the gap is vast between the observed water rates and the ideal economic prices of water, as reflected by its scarcity value and opportunity cost. This illustrates the real magnitude of the task of designing pricing policies effective enough to play, simultaneously, the financial and economic roles.

Groundwater Extraction and its Pricing

Watershed projects focus on surface water basins. The aim is to harvest rainwater by reducing the volume and velocity of surface runoff. But watershed projects suffer from a supply side preoccupation and neglect of groundwater, which is today our primary source of drinking water and irrigation. Over-extraction of groundwater and its unsustainable use are the major factors intensifying droughts suffered by the hard rock regions of India. In these regions, comprising nearly 70 per cent of India's landmass, natural rates of groundwater recharge are low. Great care needs to be exercised here to preserve the balance between rates of groundwater extraction and recharge. This balance has been repeatedly violated in recent years.

What is required now is to take a three dimensional view of ground water and see each aquifer as a common property resource. Wells and tubewells are to be viewed as the means used by farmers to extract water from this aquifer. This extraction needs to be carefully and collectively regulated. To place this regulation on a solid scientific basis, each aquifer needs to be mapped and its storage and transmission characteristics carefully studied. Such studies are best done with the close involvement of the farmers of the area.

They would help the community to decide on the intensity and sequencing of use of water from different wells in the aquifer. To give a rough indication of a possible scenario, wells with both low storage and low transmissivity would be preserved for drinking water. Wells with a low storage of water and high rates of transmission out of it, would be used in the first part of the season as water would not last for long in these wells. Wells which have high storage of water and lose it also slowly, would be used only at the fag end of the season.

In the present era of market mentality, there are those who suggest that the best way to regulate groundwater and prevent its over-extraction is to develop groundwater markets. For them the solution as always lies in "getting prices right". They forget that we are dealing with a common property resource with significant externalities—a classic market failure scenario. They fail to understand that as water becomes scarce, prices will tend to rise, gradually putting it beyond the means of the poorest farmers. And those users, such as large corporates, who can afford to pay and bid the highest, will enjoy a virtual monopoly over water and could indulge in its unbridled exploitation.

Without proper collective arrangements for groundwater use, there tends to be an infinite regress of competitive extraction, with farmers outbidding each other in depths of drilling. Competitive extraction of groundwater leads to disastrous outcomes, the worst of which are, for example observable in coastal areas of Gujarat and Tamil Nadu. Here, saline ingress of seawater poses a virtually irreversible environmental hazard for farmers who have engaged in competitive pumping of groundwater. There is no alternative to a vigilant community, mobilized in favour of women, the poor, dalits and adivasis, which understands the characteristics of its groundwater aquifers, if we are to move towards sustainable groundwater use regimes. Of course, these local-level bodies will need legal recognition and administrative backing of the state.

Under Pricing of Water and Lack of Cost Recovery

The other problem relates to misuse or over exploitation of water which is due to the wrong pricing policy for irrigation water and lack of education of the farmer in proper water management practices. Pricing water well below its economic value is prevalent throughout the world, as stated earlier, sometimes for cultural and religious reasons. Underpricing often results in misallocation and inefficient use. In many countries, expanding the supply is politically more expedient than raising fees; therefore, pricing and demand management have received much less attention. Farmers in both industrial and developing countries often pay little for their publicly supplied irrigation water and so have few incentives to refrain from growing water intensive crops or to conserve water. This has invariably led to gross misallocations, whereby agriculture, which absorbs the lion's share, often includes low-value uses compared with higher-value domestic and industrial uses.

According to the National Water Policy (1987), water rates should be such as to convey its scarcity value to the users and motivate them in favour of efficient water uses, and at the same time, being adequate to cover annual maintenance and operation charges and recover a part of the fixed cost. Agricultural productivity per unit of water needs to be progressively increased in order to be able to compete with higher value uses of water.

Most of the States have at present very low irrigation water rates at substantially varying level and have not revised for the last 2-3 decades. Most of the North-Eastern States (except Assam and Manipur), do not even charge any irrigation water rates. Maharashtra is the only State where the irrigation water rates are announced for a 5 year period at a time with a provision for 10 per cent increase per annum so as to cover the full O & M cost as well as interest payable on the public deposits raised through irrigation bonds. The State Governments of Andhra Pradesh, Maharashtra, Haryana and Orissa have revised the water rates recently.

In London, the Overseas Development Institute (ODI)-an independent body launched a book, 'Managing Water as an Economic Resources', by one of its Research Fellows, Economist, James Winpenny. It draws on evidence from developing countries to show high elasticity of demand in response to price changes, and gives examples of successful water markets being a commodity. It will largely eliminate problems of waste and failure to maintain systems. This is a concept they would like to export worldwide.[3]

A field study has estimated that 60 per cent of the farmers irrigating rice and wheat fields, apply excessive water which is encouraged by electricity charges at highly subsidised flat rates. It has been established that if India continues to use land and water resources in the manner and with the efficiency as we do at present, we will simply not be able to meet our food and energy needs for an ever growing population.

One aim should be to discourage farmers from producing crops that give a poor economic return for water. In cases where farmers own the water but crops could not use it efficiently, they might be convinced that they could pull off a better deal by selling the water instead. An advantage of such a method for financing positive water projects in developing countries is that the economic pricing of water can transform loss-making operations into self-supporting ones and generate cash surpluses. Concern at the possible effect of increased prices on poor communities is met by this argument.

In developing countries the poor are among the worst victims of existing systems. They regularly pay prices per unit for their water (from private vendors) many times higher than those paid by wealthier people with their own connections. Any reform that raises charges, improves cost recovery and generates funds for expanding and improving the system, promises to be socially equitable even if charges to piped consumers are raised. The structure of tariffs can further promote distributional goals by offering low 'lifeline' rates for minimum levels of consumption.

Water charges, on an average, typically amounting to 2-3 per cent of the crop's value in most states have no influence on farmers' management decisions. The cost of irrigation water on the basis of investment, on an average, works out to 91 paise per thousand litres against which at prevailing water charges, the highest water rates for rice and sugarcane are 1.2 paise and 11.6 paise per thousand litres respectively. Most of the States have not revised their irrigation water rates for the last 15-20 years. The direct fall-out of such low irrigation water rates with added problem of low recovery is that today the water charges collected do not cover even working expenses, not to speak of depreciation charges and contributing even a moderate return on the investment. It is worth noting that just prior to independence (1945-46), public irrigation schemes showed a surplus of Rs. 7.92 crores after meeting working expenses, interest charges and writing off losses on unproductive works. After independence, the position has deteriorated rapidly and consequently it is estimated that financial loss (difference between working expenses and revenue collected) as incurred by the irrigation sector during the Seventh Plan period (1985-90) was Rs. 3,600 crores (without considering interest) and Rs. 8,900 crores (with interest on investment). On an average, gross revenue collection per hectare of irrigated command was found to be about Rs. 68 against which the estimated working expenses were Rs. 217 per ha and thereby resulting in financial loss of Rs. 149 per hectare per year. The Vaidyanathan Committee on Pricing of Irrigation Water worked out that in order to cover the cost of operations and maintenance (O&M) charges, the increase in revenue in various states should range from 1.5 times in Punjab and Orissa to over 14 times in West Bengal.

The Irrigation Commission, 1972, had recommended irrigation rates in cereal crops as five per cent of the gross income of irrigated crops and 12 per cent for cash crops. There are instances of farmers at the same time paying as much as a third of gross produce for obtaining irrigaton from private wells and tubewells. In Mehsana district of Gujarat, a private tubewell owner charges as high as Rs. 60 per hour for supply of irrigation water. This works out to about Rs 5,040 per hectare for wheat and Rs. 2,880 per hectare for mustard. Against this, the maximum rate being presently charged for canal irrigation is only Rs. 150 per hectare for wheat and Rs 118 per hectare for mustard.

The World Bank has also called for hiking water and power tariffs in the agriculture sector to optimise returns on projects funded by it. At the time of appraisal, World Bank funded irrigation projects show an economic rate of return (ERR) of 15-30 per cent. But despite the thorough project preparation and quicker financial flows, on completion such projects typically account for ERRs below 10 per cent. The fact is that irrigation is highly capital intensive and the sector accounts for between 24 and 38 per cent of most state outlays. But there is no correspondence between the capital costs and the pricing of irrigated water. With zero charges in some places, far from covering capital costs, low tariffs have led to a criminal neglect of maintenance of irrigation infrastructure and thus massive under-utilisation of the potential already created. (Tables 11.3 and 11.4).

Table 11.3: Financial Results of Irrigation and Multipurpose Projects

Sl. No.	*Year*	*Capital Outlay at the End of the Year*	*Gross Receipts (G.R.)*	*Working Expenses (W.E.)*	*Interest on capital outlay at End of the Year (I)*
(1)	(2)	(3)	(4)	(5)	(6)
1.	1974-75	3847.8	60.7	94.6	137.9
2.	1975-76	4454.3	86.9	95.4	157.7
3.	1976-77	5137.8	104.7	112.8	174.9
4.	1977-78	5996.3	96.9	127.2	215.5
5.	1978-79	6966.4	108.1	155.2	255.5
6.	1979-80	8090.0	100.7	140.5	292.3
7.	1980-81	9346.7	103.4	225.7	301.5
8.	1981-82	10790.5	120.2	265.3	415.6
9.	1982-83	12342.3	117.1	127.7	872.7
10.	1983-84	14062.1	165.1	273.9	562.8
11.	1984-85	15929.7	129.7	334.0	635.7
12.	1985-86	17971.2	223.8	486.9	681.7
13.	1986-87	20234.3	166.7	489.6	867.3
14.	1987-88	22311.2	138.7	1400.3 *	—
15.	1988-89	24608.0	166.4	2128.0 *	—

Source: Combined finance & Revenue Accounts of the Union and State Government in India.

Remarks: '*' Inclusive of Interest on Capital Outlay at the end of the year which is not separately available.

Table 11.4: Statewise and Yearwise Percentage Recovery of working Expenses Through Gross Receipts in Multipurpose and River Valley Projects

Sl. No.	*Name of the States*	*1974-75*	*1980-81*	*1984-85*	*1988-89*
1.	Andhra Pradesh	15.5	7.0	21.8	1.1
2.	Bihar	72.5	63.2	29.8	12.1
3.	Gujarat	71.1	34.4	35.4	5.7
4.	Haryana	55.3	64.4	37.7	16.6
5.	Jammu & Kashmir	6.0	1.8	5.8	7.1
6.	Karnataka	97.6	7.6	29.7	10.3
7.	Kerala	51.2	49.5	37.8	11.8
8.	Madhya Pradesh	129.3	45.5	30.2	35.1
9.	Maharashtra	166.0	93.8	47.3	5.2
10.	Orissa	30.7	46.8	64.3	46.2
11.	Punjab	63.2	73.8	81.5	25.5
12.	Rajasthan	49.6	55.9	43.6	9.7
13.	Tamil Nadu	22.5	15.6	7.7	2.0
14.	Uttar Pradesh	123.3	134.4	48.4	10.0
15.	West Bengal	38.2	11.0	8.8	3.6
	ALL INDIA	64.2	45.8	38.8	7.8

Source: Combined Finance and Revenue Accounts of Union and State Governments in India.

Remarks: Working expenses for the years 1987-88 and 1988-89 are inclusive

Realizing that water rates have not been changed for decades by most of the States and that revenues fall drastically short of even the working expenses covering barely wages of the work force, the Planning Commission has recommended a 30 per cent increase in irrigation charges spread over 3 years.The inter-ministerial group, set up by the Planning Commission, recommended:

(a) 1 per cent rate of return on total investment in the irrigation sector.

(b) States should set aside 10 per cent of their annual plan outlay for the projects in this sector.

(c) The quality of service be improved to justify the proposed increase in irrigation tariff.

(d) The age-old system of fixing irrigation rates on crop be replaced by a volumetric system.

On 1 per cent rate of return on investment, the states have represented that the present investment works out to Rs. 25,000 per hectare, and even this rate of return would mean increasing the tariff by another Rs. 250 per hectare. Thus, the total hike will work out to Rs. 560 per hectare. Their suggestion is that water tariff should be increased every year by 10 per cent to recover operational costs. The states have fully endorsed the recommendation that 10 per cent of the total plan outlay should be earmarked for irrigation, with the rider that this amount should be given as Central assistance.

The Planning Commission has recommended a 30 per cent increase in irrigation charges spread over the next three years. This was long overdue. According to irrigation experts, millions of hectares of irrigated land have been lost due to lack of funds for operations and maintenance. In fact, some irrigation projects undertaken during the second five year plan are yet to see the light of the day. What brought about this abysmal state of affairs ? Like other infrastructural inputs, water is treated as a public good to be subsidized by the government. Most states have not revised water charges for decades. And even in cases where they have, the rates are much lower than the increase in costs. As a result, revenues from irrigation projects fall drastically short of even the working expenses. The revenue proceeds barely cover the wages of the workforce. This has led to an all round deterioration of financial performance of all irrigation projects. But the government is in no position to sustain the mounting subsidies on irrigation. And in any case, there is little reason why government should subsidize a user-oriented, capital intensive infrastructure like irrigation.

The States which are reluctant to revise water rates will sooner or later realize the adverse impact of underpricing on the availability of water. Under pricing has not fulfilled the socialist goal of ensuring equity in resource use. On the contrary, it has led to inefficient resource use and distorted the distribution pattern in the agricultural sector. The relatively well-off farmers merrily lift water by using powerful pumpsets. This depletes the water table faster and makes it difficult for small farmers who cannot afford to hire or own pumpsets. Already some states, like Punjab, are grappling with the problem of a falling water table. If the trend continues, agricultural growth will be severely affected.

In the present hurry to convert problems of water scarcity into inter-state disputes, the causes of this scarcity are not being adequately explored. Even many of those who recognize that the limited water resources are not being efficiently utilized do not trace this deficiency to the failure of government controlled water management. In several states, the centuries old tank irrigation network has decayed under government control. While government control destroyed traditional tank management practices, the bureaucracy did not have a better alternative to put in place. The government's ability to enforce the desired cropping pattern is also weak, and the pricing of water is too low to yield funds to invest in measures that would improve the efficiency of water utilization, such as the lining of canals.

A National Water Policy that seeks to cover not only the distribution of water but also its efficient utilization cannot, therefore, continue to assume that government control improves water management. It would have to countenance alternative systems to manage water. These systems would have to be able to generate the financial and labour resources needed to maintain and expand the irrigation network. And for that, the water must be priced at levels that not only take care of maintenance costs and future investment, but also contribute a surplus to make the effort of managing water worthwhile. And a system that can charge an economic price for water and monitor it effectively enough to prevent its theft would need the support of the farmers. Since the precise form of such water management bodies will necessarily vary quite substantially in a country as diverse as India, the National Water Policy may be able to do no more than present some broad guidelines. But it must eschew exclusive reliance on management by government[4].

Given the diverse agro-climatic and cropping pattern, different states have to evolve their own sustainable irrigation system. The basic principle is to convert each project into an independent self-financing system. This can be achieved in two ways. Either the irrigation systems can be run in the form of corporations like any other public utility service or be owned by local farmer groups. The second alternative is doubtless a more viable one. The farmers, who are the actual users, must take the responsibility of maintenance, allocation and collection of water rates. User participation will impart accountability to the system. While it is desirable to move from a flat rate to a volumetric rate, the task is not simple. As the Vaidyanathan Committee (1992) has rightly suggested, the recommendations have to be implemented in stages. While some states have belatedly realised the problem, the majority have chosen the path of populism. But can we afford few more million hectares of irrigated land to vanish?

WATER MARKET IN RAJASTHAN

The Government of Rajasthan have taken a decision to abolish subsidies on irrigation water, putting the state well ahead on the road to reform. The need for power reforms is well known. That for water is less understood. Water is a resource that should be allocated according to price. City dwellers, in particular, value water greatly, and if the market could charge them, we would not have the kind of shortages as we see today. Water would be readily available in cities—albeit at a price. This would also make water utilities in the cities economically viable, capable of generating profits, and hence self-sustaining. Unfortunately, the state has messed up the water market by fooling around with prices. Irrigation subsidies mean that a lot of water that could have found urban markets gets wasted in rural areas: waterlogging and salination are common problems where free irrigation water is inefficiently used. If irrigation water is priced, then farmers in water scarce areas like Rajasthan would economise on water use and would choose their crops accordingly. With water pricing, and water markets, the people of Rajasthan can be sure of one thing: There will always be an abundance of water. They should realize that, just as they have understood that paying for power will ensure reliable power supply, so too will payments for water ensure regular water supply. Indeed, 70 per cent of the planet is made up of water. Very little of the planet is made up of crude oil. If diesel and petrol and kerosene are available aplenty in Rajasthan, why should a more abundant resource like water be scarce? Two reasons: one is pricing. If water is not priced, the market cannot work, and no businessman will step in to supply it. The second is property rights. One can own an underground oil field. Why can there not be property rights over ground water (even shared rights) and river water? With pricing and property rights, water markets will evolve and permanently rid us to the shortage economy bequeathed by statism.

Finally, there is a need to have ecological development of water resources to contain adverse environmental impacts at regional and local levels. The National Water Policy, 1987 and 2002, recognized quality of environment as a primary consideration in the development of water resources. Conjunctive use of canal and ground water could minimize the resource degradation problems to a large extent. Lining of canal irrigation network, surface and sub-surface drainage, bio-drainage and proper on-farm water management are called upon to check the rise in water-table. The declining trend, in groundwater level and consequent desertification can be checked by encouraging groundwater recharge measures.

Conclusion

It has been pointed out time and again that although water covers three quarters of the planet, only a small fraction is accessible as freshwater. The time when water could be freely used without being subjected to legal rights and obligations is past. Because of phenomenal population growth in the country, industrial and agricultural development, water has become scarce. As competition has become more acute, the tendency to appropriate water and exclude others from using it has increased. This has created pressures on the regulation and appropriation of water resources, and rights for different purposes, such as drinking, agriculture, industry and hydropower, and rational management of this non-renewable resource has become imperative. Unfortunately, the present mindset of the majority of the population is that water is government's business. Water continues to be heavily subsidized in most sectors resulting in colossal misuse and wastage of this precious resource. However, there is now a growing recognition of water pricing as a key policy instrument for cost recovery and demand management. Water markets can help immensely in improving allocation and use. A gradual approach to introducing water markets may be the most appropriate, putting into place water management systems, defined water rights, and water user participation.

REFERENCES

1. Kumar P.: Demand and Supply Projection for India, Agricultural Economics Policy Paper 98-01, New Delhi: IARI, 1998.
2. River Basin management: Issues and Options, Paper for Water Resources Day, Indian Water Resources Society, 1997.
3. Raja Arthwi, The price to pay for scarcity; *Water and Environment International*, Vol—3, No. 28, May 1994.
4. *Economic Times*, January 8, 1996.

12

Rain Water Harvesting

"Sovereigns over precious things,and Rulers over men, Waters! We seek healing balm of you"

—***Rig Veda***

Introduction

The over exploitation of our natural resources is creating problems all over the world. One of these resources, without which, it is not possible for us to sustain our lives, is water. Access to enough water of required quality is fundamental for all human, animal and plant life as well for most economic activity. Availability of clean water for drinking, cooling, bathing, and other household needs is a must. In fact, water is the source of life and livelihoods and it also sustains natural ecosystem. A country's level of freshwater use reflects—and, in fact, is one of the key measures, of—its level of economic development. In developing regions of the world people use far less water per capita than in developed regions. Developing countries devote most of their water supplies to agriculture. India, for instance, uses 90 per cent of all water for agricultural purposes, with just 7 per cent for industry, and 3 per cent for domestic use[1]. However, this essential resource is under threat. The challenges of growing water scarcity are heightened by the increasing costs of developing new water source, degradation of soil in irrigated areas, depletion of groundwater, water pollution and degradation of water-related ecosystems, and wasteful use of already developed water supplies, often encouraged by the subsidies and distorted incentives t hat influence water use.

Needless to say, the world's fresh water shortages result more from uneven distribution over time and space than from absolute scarcity. Even in some of the world's highest rainfall areas, rain falls in concentrated periods followed by prolonged dry spells. Then it is necessary that users must store rainwater when it is abundant to use later when it is scarce. Similarly, even in some of the arid and semi-arid areas, occasional intense rainfall generates runoff that can be captured and stored for subsequent use. For example, harvesting only 100 millimeters of rainfall over a one-hectare plot would yield 1 million liters of water. Ancient water harvesting systems in India, nay, throughout the world took advantage of this principle to smooth inter-temporal variations in water availability, making human settlement possible in a wide variety of ecosystems. As technology was then limited, people depended less on large river systems and underground water resources. Less than 150 years ago, no government anywhere in the world provided water to its people. They depended mainly on the rain that fell in their villages and towns or on the floodwaters of nearby streams. Local water harvesting declined with the development of large systems in which water is transported to a long distance through canals and pipes or pumped from great depths below the ground. However, growing scarcity and inter-sectoral competition for water, along with rapid depletion of groundwater and the problems associated with major surface—water control systems, have raised interest in revitalizing water-harvesting systems that capture rainwater wherever it falls.

Water harvesting is useful for such areas which receive frequent storms of a large magnitude to cause surface runoff. Semi-arid areas with an annual rainfall in the range of 400 to 750 mm having rolling terrain, including a hard plateau region, can benefit from the water harvesting technique by making ponds, dykes at slope bottom and a series of small dams or seasonal streams to intercept surface runoff. These will be beneficial for short term supplements in a drought-like condition. The forage around such water bodies will increase with an improved moisture regime despite a still high evaporation of about 2,000 to 2,500 mm. A medium rainfall in the range of 750 to 1,000 mm and also high rainfall areas are ideally suited for water harvesting. The evaporation in these areas is about 500 to 1,000 mm and 2,000 mm respectively. While meeting the dry period demand, water will also percolate down to improve the aquifer. Hilly areas where surface runoff quickly drains away leaving the upland high and dry would be immensely benefited by such measures.

The drastically increasing population and urbanization is one of the major causes of water crisis. India has more than 250 million city dwellers, even though the urbanization is among the lowest in the world.

To meet the rising demands which would not be satisfied by surface water sources alone, ground water is being extracted. The increasing use of ground water has led to substantial decrease in water table in several parts of the country. Another problem associated with urbanization is that, the concrete structures and tarred roads prevent water to seep into the ground. The water flows on the ground and into the drains and is finally wasted or contributes to pollution of surface waters. This shows our ignorance to realize the potential of rainwater to meet the growing water demands.

EARLY EXPERIENCE

As early as the third millennium BC, farming communities in Baluchistan impounded rainwater and used it for irrigation. Dams built of stone rubble have been found in Baluchistan and Kuch. Dholvira, a major site of the Indus Valley Civilization had several reservoirs to collect monsoon run-off and also excellent drainage systems. Some of the answers to this current crisis may lie in the lack of reliance on our own traditions by our communities. Indians, over centuries developed a range of techniques to harvest every possible form of water, from rainwater, stream and river water as well as flood water, They have tapped water from hill streams or springs known as kuhls carrying a discharge of 15-100 litres per second. In Meghalaya, a 200-year old system of tapping stream and spring water for irrigating plants by using bamboos still exists.

Gujarat—A Living Example

There is a living example in Gujarat of how rain water harvesting has helped the citizens of Dwarka. It is not as if Lord Krishna has performed a miracle in this city, but in times like that of the drought of 2002, it could pass off as a miracle. While thousands in Gujarat were suffering from drought, people in Dwarka could afford to sit back and relax. Women did not have to queue up for their quota of water, nor did they need to buy it. This, despite the fact that the area received only three inches of rainfall. It was the centuries-old tradition of water conservation that had given Dwarka reason to rejoice. Almost every house in the city has an underground tank which is used to collect rainwater every year. Sealed from all sides, the tank has a small opening at the top, through which a bag of lime (which acts as a disinfectant) is dropped into it.

Residents here have had water tanks from stone water cellars to cement-plastered ones—for as long as they can remember. Even today, buildings are not planned without the provision for an underground tank. It is this conserved water that had come in handy when Saurashtra was hit by one of the worst water crises in 100 years. Realizing the importance of rainwater conservation, the Dwarka nagarpalika made the provision of a tank almost compulsory during the construction of houses.

According to Prof Rameshbhai Bhayani, who has conducted a study of these tanks as a source of water conservation, Dwarka never faces water shortage because people carefully conserve water. Saurashtra may be hit by a drought, but one will never hear of water shortage in Dwarka. Since this region receives very little rainfall, it is a very old tradition of conserving rainwater. About 98 per cent of the houses here have these tanks full of water. All they do is to connect drainage pipe from rooftops and terraces to the tank below. If the rainfall is good, some tanks, which are as big as 15 by 15 feet, collect enough water. Since it is used strictly for drinking and cooking it sometimes lasts up to two years.

Even the Dwarka Peeth, the seat of the Shankaracharya, emphasizes on this method of rainwater conservation. There are many such tanks in the Peeth and the water in some tanks remains for nearly three years. This is the reason why despite an irregular supply of municipal water in Dwarka, few people fetch water from elsewhere. People there use municipal water for bathing and other purposes only. For drinking and cooking they use water from such tanks. They know it is pure rainwater and since they built a tank 10 years ago they have never faced any shortage. Before every monsoon, people of Dwarka begin preparations to trap all rainwater. Sometimes, they block terrace pipelines leading to the tank, let the first rains do the cleaning and then start filling the tank.

Equally heartening are reports that people themselves beginning to pomote rainwater harvesting, especially in Gujarat. A voluntary organization, Lakhota Jal Sanchay Abhiyan Samiti has collected money from the residents of Jamnagar town to desilt the huge Lakhota Lake constructed by the former royal family. Under the leadership of the Swaminarayan Gurukul in Rajkot, foundation stones were laid for check dams in more than 15 villages in just five days around the middle of April 2002. In Ahmedabad city itself, the Khadia Itihas Samiti has completed a survey of the underground water tanks (tanka) that were traditionally built in each house to harvest roof-top rainwater and is planning a campaign to revive them. These developments show that community-based rainwater harvesting may well become a widely adopted paradigm in the years to come both in the urban and rural areas.

Rajasthan

Credit must go to the people of the villages of Rajasthan and particularly Jodhpur, where old water system still exists and where the traditional system is maintained even after the advent of piped water. Villagers which neglected their traditional systems and relied solely on piped water sources faced scarcity under drought conditions. An ironic contrast of water management is that between Jaisalmer and Cherrapunji which get 100 mm and 15,000 mm of rainfall respectively. Jaisalmer had enough water for itself until recent years, while Cherrapunji, the wettest place on earth faces a drinking water shortage. Alwar district, also in Rajasthan, has been successful in harvesting water thereby bringing prosperity to its villages.

Although we have also made dams in the past, no significant dam construction has come about in the last 25 years. While there should be urgent efforts to complete big and small dams, compulsory water harvesting schemes with the participation of villagers also need to be implemented.

Finally, it must not be forgotten that a drought in 1987 almost brought down the union government. Therefore, if the country and the government need to focus on a single meaningful issue for now, it has to undertake the completion of all water-related projects without any further delays and also emphasise the need to revert to our traditional methods of water harvesting.

DEFINITION AND TYPES OF WATER-HARVESTING SYSTEM

Water harvesting is an ancient technique of not letting precious rain water flow away, but storing it. It includes all efforts made to conserve and effectively utilize the limited water resources we have. The other component of harvesting is artificial recharge of ground water in which the stored water is filtered and sent to aquifers in the soil or rock formations in the earth that hold water.

Treating of Household Sewage of a Village or Small Colony Where Municipal Drains are Provided

It should be quite easy to treat and re-cycle household sewage of a small colony or village before this is used for agriculture or pisci-culture. A mini Eco-Park with small natural filtration arrangements like duck—weed ponds and fish ponds are enough for such treatment. After testing of water from the fish pond, the same could be used for domestic purposes, other than drinking. Small size trees should be planted around the mini 'Eco-Park' This refers to the small scale concentration, collection, storage, and use of rainwater runoff for both domestic and agricultural use. This definition implies that the catchment area from which the water is drawn is larger than the command area, where it is collected and used[2]. The ratio of catchment to command is inversely related to the amount and intensity of rainfall, the impermeability of soil and the slope of the land on which it falls. Rainfall intensity is particularly important, since intense storms generate the most runoff.

A watershed is an area that drains to a common point. It may be managed for various objectives, depending on local needs, including capturing runoff, minimizing erosion, and reducing pollution. In management of small watersheds, capturing runoff for local use is conceptually equivalent to harvesting water. Water-harvesting systems either concentrate water into a storage reservoir or apply water directly to the soil in the cropped area. Both types of systems can vary in scale from a few square meters benefiting a single household to a few square kilometers serving a larger group of people.

Reservoir Systems

Systems that concentrate water into storage reservoirs can be used for a variety of purposes, including household, irrigation, or livestock consumption. Rooftop water-catchment systems provide domestic water in many places, especially dry areas with inadequate municipal supplies. A rooftop that is 50 square meters can supply an average of 50 litres of water per day with 500 millimeters of annual rainfall. This supply can help reduce competition between agricultural and household demands and can free women from the chore of collecting water.

Harvesting water to collect it in village ponds is an ancient system in many places in India. The principle behind this approach is to concentrate water spatially and temporally. In arid and semi-arid areas where rainfall is low and variable, harvesting rainwater allows users to conserve it until enough water is collected to reliably support a crop. The volume of water stored at the end of the rainy season determines how much area can be cultivated.

Traditional water-harvesting systems store water in ponds and reservoirs and deliver it by gravity. With the spread of motorized pumps in recent decades, water can now be captured in some places using the same techniques but allowing collected water to percolate into groundwater aquifers. The systems are small in scale, drawing from gullies and micro catchments to recharge groundwater that supplies a number of local wells.

Soil Moisture Storage Systems

In many traditional water-harvesting systems, runoff water is channeled directly to the cropped area during rainfall and stored in the soil. Where rainfall is unevenly distributed and soils have high water-holding capacity, this system may store water until the end of the rainy season, when a crop is grown under gradually receding moisture. Where soils are sandier and do not retain moisture for a long time, moisture may be channeled spatially to the location where crops or trees can take advantage of it. These systems can remain productive several months into the dry season when the surrounding land is barren.

Water-harvesting systems vary a great deal in the ratio of catchment to cultivated area. In dry areas with less available water, this ratio can easily reach 20:1 In such water-harvesting systems, cultivation systems must be extensive and located in sparsely populated areas so that command and catchment areas do not interfere with each other. Where rainfall is higher, the ratio declines sometimes to the point that the entire catchment area lies within a single plot.

Other Different Ways

There are numerous other ways in which rainwater can be harvested, which would also enrich the ground water. These are as follows:

(a) For a village or small locality, channelising rainwater into a pond with a well or deep bore.

(b) If a medium size plot is available with the homestead, then a small tank (5 to 10 cu.m) can be used, preferably with a deep bore in the tank.

(c) Rooftop harvesting leading channels to a small tank, preferably with a bore well, or directly to a deep bore.

Colony or Village Ponds: This is a traditional method of water harvesting in India. The lowest area near the village or colony is chosen to construct a tank, at the center of which is a well or a bore-well. All rainwater is lead to the tank bearing in mind that household drains are away from the tank. Fruit trees and those for firewood purposes can be planted in the land around the tank.

Rain Water Harvesting in A Home-stead or Building with Large Compound (one acre and above): Rain water from buildings in a large compound, which begins to flow due to excess rains, should be canalized to a small pond with a bore well. Ideally, such a tank should be 5 to 10 cu.m in size and be constructed around a disused bore well. A traditionally open well could also be used for this purpose, if one exists on the plot. In the case of the disused bore-well, the top portion of the metal pipe should be removed or large holes should be bored in the pipe.

Roof Top Harvesting in Houses with Small Plots: Rainwater harvesting is achieved in houses with small holdings by collecting water from most of the roof area, in channels through one or two pipes coming down from the roof. This is done using half open metal pipes at the edge of the roofs to lead the water to the appropriate drains. On the ground the drains are channelised towards small ponds of 1 to 2 cu.m. size, with an injection bore. A disused tubewell boring or a fresh boring may be used for the purpose. Government assistance could be provided for drilling bore wells, or a rebate on piped water supplied should be given to households who carry out roof top harvesting without such assistance.

Delhites can mitigate their water shortage problems only if they show more concern and adopt simple techniques to harvest rain water at their homes. Experts at the public hearing conducted by the Center for Science and Environment (CSE) said that despite Delhi receiving a large quantity of rainfall, the city always faced a water shortage during summers. A large segment of the population depends upon groundwater which also has been found to be contaminated with chemicals.

The data for Delhi reveals that the groundwater has gone down by 25 feet in the city and almost 50 per cent of this is polluted. Experts at public hearings suggested that the collected rain-water could either be directly recharged into the ground, or could also be used straight way. There is an example of successful rainwater harvesting. CSE office building in Tughlaqabad institutional area annually harvests 3,66,000 litres of water. And even the short spell of rain on May 12, 2002 which was only 8.5 mm. was enough to provide six days of drinking water for the 110 CSE staffers.

Target and quick results that bureaucrats expect will not make the water harvesting projects successful. The first step in water harvesting is social mobilization and before undertaking any such project, community involvement is necessary. Communities have been harvesting water in the country for the last 3,000 years-right from Ladakh, to Tamil Nadu, and they have done it by collecting and storing rain water, run-off water and even water from the flooded rivers.

Technical Feasibility of Rainwater Harvesting

The key technical question for agricultural water-harvesting systems is whether capturing and storing enough water to raise crop yields is feasible. Much research has gone into measuring the water available to plants under different rainfall, soil, and slope conditions and into developing ways to increase runoff, reduce erosion, or improve storage. In field conditions, technical feasibility depends in large part on local agroclimatic conditions. For example, in India the Into-German Watershed Development Project works only in villages where the terrain is conducive to water harvesting—that is, there is a sloped, uncultivated catchment area and a good opportunity to build storage ponds.

India's Rainfall Pattern and Status of Rainwater Harvesting

The British called Indians a "hydraulic" society. The simple reason is that our rainfall is variable and rainy days are few: 50 per cent of the rainfall in about 15 days and less than 100 hours out of a total 8,760 hours in a year. The total number of rainy days can range from a low of five days in a year in the desert regions of Gujarat and Rajasthan—though on some of these days there can be high intensity rainstorms—to 150 days in the Northeast.

In India rainfall distribution over time and space is not thus uniform and most of the rainfall is lost as run off or as evapo-transpiration. Further, most of the rainwater, which infiltrates during the rainy period, is entrapped in the soil without reaching the water level and ultimately lost in evapo-transpiration. Moreover, quantum of rainfall also varies very widely from year to year and drought in one part of the country or the other is a common feature, severely affecting agricultural productivity. Despite this peculiar pattern of monsoon rains, India is considered as one of the wettest countries in the world and yet a country that is facing ever-growing water scarcity—receives 400th million hectare meters (mhm) of precipitation and we use only a small part of it. The average flow in the river systems of the country has been estimated at 188 mhm and the actual quantum of water utilizable through surface structures is only 69 mhm.

Therefore, it is very important to capture this rainwater, which just comes and goes in a few hours. If water is harvested on just one hectare of land in, say, Jaisalmer, with rainfall level as little as 100 mm in a year, this means that as much as one million litres of water is captured, will be enough to meet drinking and cooking water needs of 182 people at 15 litres per day.

Cherrapunji (Meghalaya) which receives about 15,000 mm rainfall annually suffers from acute shortage of drinking water. This is because the rainwater is not conserved and allowed to drain away. It has therefore, now been increasingly realized that to meet our increasing demand of water in urban areas, to augment the ground water storage and to improve the quality of ground water, rain water harvesting is required. As a matter of fact, the right solution to overcome water scarcity and be self-sufficient in water needs is to save the precious drops falling from the sky. It is necessary to give much more importance in harnessing the rainfall to meet the increasing needs for water, thereby achieving a better balance between surface and ground water. In fact, in some areas, the primary source of water is rainfall and as such there is no other way but to exploit fully the available rainwater potential if we are required to attend to the escalating water needs, more particularly in the urban areas.

In fact, water harvesting is the activity of direct collection of rainwater. The rainwater then collected can be stored for direct use or can be recharged into the groundwater. It may be pointed out that rain is the first form of water that we know in the hydrological cycle, hence is a primary source of water for us. Rivers, lakes and groundwater are all secondary source of water. Presently, however, we depend entirely on such secondary sources of water and primary source is being neglected to a great extent.

As mentioned earlier, in India, rain water harvesting is not a new concept. The method of 'collecting and conserving water' is ancient. But slowly with the advent of tap water, rain water harvesting has lost its importance. Now our country is in a situation where efficient management of water resources has been a necessity, rainwater harvesting has come to the limelight again. Water harvesting as a strategy for meeting human needs essentially means re-establishing the relationships between the people and their environment and considering water as an element of nature to be used judiciously. Today, what India needs is a revitalization of all its ancient traditions of conserving and harvesting water.

Frequent deficient monsoons serve as a sharp reminder of the rain dependence of agriculture in India, and ultimately, the rest of the economy. Despite the fact that there has been remarkable expansion of irrigated area over the past 30 years and more, the moot point is that whether the expansion is efficient and sustainable. Further, in a situation where ground water is not adequately replenished, the practice of irrigation by pumping groundwater is clearly not sustainable in the long run. As it is, the increasing use of chemical fertilizers adds to the salinity of groundwater. The situation has become almost irreversible. This brings us to the question of how efficiently do we use water received through precipitation. According to the Central Water Commission, India received an annual precipitation of 4,000 billion cubic meters (bcm). However, the usable water resources from this precipitation are only about 1,100 bcm., *i.e.* roughly 28 per cent. Even, if this proportion can be raised by 5 per cent, it would mean an additional 200 bcm. of usable water. This is 70 bcm more than the total reservoir capacity in the country (130 bcm), taking the total of 17 major states. The method is not to use large scale expensive techniques, but if enough water harvesting infrastructure is built at the local level to satisfy local needs (say, enough for one month of requirements to counter dry spells), a higher proportion can be put to use than at present.

In most of the cities, water is becoming scarce commodity day by day and will be difficult to cope with the demand of the burgeoning population. In fact, many cities in the country face water famine during summer months. It is necessary that every individual should work for himself/herself to harness rain water and conserve it for its own use.

Rain water may be captured, preserved and stored on the land below the ground. Water harvesting may improve moisture and recharge in the downstream areas. When rain water is harvested it percolates into the ground, raises the level of the water table, making it available for agricultural and domestic use.

Research suggested that rainwater harvesting could be the best way out for mitigating the water scarcity. Various studies show a potential of making available an additional 68,000 million gallons per year through this technique. It may be reiterated that rainwater harvesting-simple, economical and eco-friendly method of water conservation is an ideal solution to recharge the groundwater. The rainwater that falls on the surface/roof top is guided to borewells or pits or new/old/abandoned wells through small diameter pipes to recharge the underground water which can be harvested to the extent of 55,000 litres per 100 sq. metres area per year.

Benefits of Rain Water Harvesting

Benefits of rainwater harvesting may be summarized as below:

- Quality of ground water improves
- Raises the water levels in wells and bore wells that are drying up.
- Mitigates the effects of drought and achieves drought proofing.
- An ideal solution to water problem in areas having inadequate water resources.
- Reduces the soil erosion as the surface runoff is reduced.
- Choking of storm water drains and flooding of roads decreases.
- Saving of energy. To lift ground water, one metre rise in water level saves about 0.40 kilo watt hour of electricity.

POLICY ISSUES

Over the last one hundred years or so, we have seen two major paradigmatic shifts in water management. One is that individuals and communities have steadily given over their role almost completely to the state even though more than 150 years ago no government anywhere in the world provided water. The second is that the simple technology of using rainwater has declined and in its place exploitation of rivers and groundwater through dams and tubewells has become the key source of water. As water in rivers and aquifers is only a small portion of the total rainwater availability, there is a growing, and in many cases, unbearable stress on these sources.

In view of vanishing water resources and decelerating water table, it was necessary to harness water run-off for water conservation and ground water recharge so that water requirements of men, livestock and plants were adequately met. This would call for a nation-wide people's movement and active participation of each and every citizen in addition to the governments, Panchayati Raj institutions and NGOs, business people and housing cooperatives-to tackle water scarcity in many parts of the country. No single initiative is enough to solve the problem of drinking water. Diverse routes and a plurality of programmes are needed to achieve the objective. It is in this context that Rotary in District 3,010 has taken a lead in this project of national importance.

Scheme to Harvest Rain Water

The Union Government has decided to launch Rs. 1,000 crore country-wide programme to harness rainwater for irrigation, fisheries and tree plantation at the village level. The resources generated from these ventures will enable panchayats to undertake developmental activities. The project is being launched with the dual objective of tackling the problem of water scarcity and augmenting the resources of pachayats in an attempt to strengthen them. Under the scheme, panchayats will undertake rainwater conservation measures with technical assistance from state administration and use the impounded water for irrigation, fisheries and tree plantation, including horticulture. The income from water-user charges and the sale proceeds of fisheries will add to panchayat's resources.

Need to Catch Big Fish Participation

If the Ministry of Defence and the Northern Railway, occupiers of the largest segment of the National Capital Region of Delhi, adopt water harvesting wholeheartedly, there is a hope that the water table will stop falling some decades hence. Today, they are high on extraction and extremely low on conservation of this fast-depleting natural resource. It is no accident that both these institutions own huge tracts of land and have very viable operational units in Tughlakabad in South district where the water table has fallen to between 35 and 40 metres due to over—extraction. The Indian Air Force has residential areas and airbase in this segment and the northern Railway has one of the largest marshalling yards in northern India here.

The IAF has woken up to the looming disaster in and around Tughlakabad and has installed four trenches and recharge wells to be able to catch the runoff during the rainy season, the potential of which is estimated to be nearly 18,200 cubic meters. The Northern Railway is as yet dormant. The IAF has also decided to set up recharge systems at Rajokri and Faridabad within the NCR and Subroto Park on the outskirts of the Delhi Cantonment on the way to the airport. But, inexplicably, the Directorate-General of Defence Estates has not gone beyond contemplating water harvesting in the Nicholson Ranges and the Shankar Vihar Complex although the Contonment Board has jurisdiction over the second largest slice of the NCT (42.79 sq km) after the 1.397 sq. km under the Muncipal Corporation of Delhi and marginally larger than the NDMC area. Within the Delhi cantonment, the daily consumption of water is 436 litres per capita from groundwater sources. If the Dhaula Kuan flyover project does establish the feasibility of introducing water harvesting within its catchment, which lies on the periphery of the Delhi Cantonment, there is no reason why similar experiment cannot be undertaken within the military perimeter as well.

The Northern Railway has substantial real estate in about a dozen locations within the NCT, the largest being the residential colonies at Kishanganj and Shakurbasti. The Delhi Government has made it mandatory that all new structures within the NCR must incorporate water-harvesting schemes in their plan layout. However, the full utilization of approximately 475 million cubic meters runoff during the monsoon will not happen unless existing establishments also introduce water harvesting methods in the areas within their control. Given that there is only very tardy progress in that direction especially among the larger landholders within the NCR, the government ought to nudge them to conform to the need of the times.

SOCIAL ORGANIZATION/PEOPLE'S PARTICIPATION

Water harvesting requires collective action in densely populated areas where more than one person uses catchment and command areas for multiple purposes. The key question in such cases is whether the benefits of water harvesting exceed the costs of coordination among users. Often this can be the most daunting constraint of all, especially where some resource uses are mutually incompatible and any intervention will impair at least one potential use. Successful water harvesting requires protecting upper reaches of small catchment against erosion that would reduce water-storage capacity in the lower reaches. Typically, upper catchments are denuded; so, protecting them requires limiting their use for grazing and fuel collection. This restriction imposes the greatest costs on landless people who depend most on these areas and who do not stand to gain directly from water harvesting. Long-term success requires devising institutional approaches to ensure that landless people gain from the water-harvesting intervention.

Various State Governments have promulgated laws and policies of rain water harvesting to replenish the fast depleting ground water resources and facilitate its movement into the ground water. Some of the government offices/buildings in various states have been identified as demonstrative schemes for preparation of concept plan and designs for roof top rain water harvesting for artificial recharge. The people's responsibility is to help revive, propagate and modernize the rain water harvesting system which can be an important key to our water future.

The system offers the ability to operate independently and is useful in remote areas and difficult terrain. Keeping in view the complexity of the problem of depleting ground water resources, municipal bodies, Ground Water Board etc. should concentrate its activities for taking the programmes/ projects on artificial recharge as a management strategy in order to create mass awareness among the public and other government organizations and to make it a mass movement. People can follow roof top harvesting techniques to meet the growing needs of a city particularly during drought.

The government has to work with the people and modify policies to facilitate community participation in developing and maintaining rain water harvesting systems. The government should make it compulsory by framing law to prepare area-wise water harvesting structures so that not a drop of water is wasted. Citizens can play a significant role in mitigating the impending water crisis. Unless the citizens are socially mobilized and participate in the rain water harvesting efforts, success could not be ensured.

> According to the WHO, water management in India should be privatized before an alarming situation arises. People should be given monetary help for building underground tanks for harvesting rain water. They can also be educated to treat this water in such a way as it can be used during summers for bathing, washing, or watering kitchen gardens

With less than 100 hours of rainfall in a year in the country, the key challenge is to store this precious water for dry season. Community based rainwater harvesting—the paradigm of the past—has much strength today. Rainwater harvesting through tanks, ponds, check dams, percolation dams, and many other structures designed to capture and store rain water, however, necessitates an integrated land-water policy.

Rainwater Harvesting—Various Options

The rainwater that falls on the rooftops can be harvested for groundwater recharge using appropriate techniques. This method may be applied for the roofs of existing houses/buildings as well.

Lakes and Depressions

Lakes have often been misused for dumping of solid waste and sewage disposal. Water quality can be improved by replicating the principles of wetlands by the introduction of aquatic plants and fish in a systematic manner.

Floodplain Reservoirs

It may be possible to create recharge basin and reservoir on the flood plains by scooping out earth from the flood plains at appropriate locations and letting these reservoirs to be filled up by the expanded monsoon season flow of the river.

Ecoparks

This involves the creation of artificial wetlands using aquatic plant root zone system to treat sewage waters up to tertiary levels for recycling to the irrigation and horticultural sector and possibly some industrial estates.

Rainwater Harvesting in Urban Areas

Growing water scarcity in the urban areas in India is forcing the search for simple, economical and self-reliant solution. While governments are looking for mega-solutions, individual campuses can harvest the free and pure rainwater falling in their premises, thus augmenting their water supply in a self-reliant way. Groundwater extraction in excess of natural recharge is depleting the water table and it is incumbent upon people to aid natural storage and recharge processes. As rainfall in India is concentrated in monsoon season and unevenly distributed, storage is the key to spreading resource availability into the lean season. Hence, capturing rainfall run-off for local use is now the key concept. Surface flow that originates as a result of rainfall is called run-off. The portion of rainfall that does not seep into the ground or evaporates is available as rainwater run-off for harvesting.

Rainwater falling on hard surface, such as roofs and pavements that decrease groundwater percolation, constituting the catchment in the urban areas and generating enhanced run-off, has to be guided in storage reservoirs or recharged into groundwater table through simple filtration systems for subsequent extraction. The water thus stored can be used for potable/non-potable uses. Rainwater harvesting in urban areas is very much needed because:

1. The maintenance and running of the existing system of water supply are deteriorating and degrading for paucity of funds.
2. Growth of water usage particularly in agglomerates and urban fringe areas is outstripping the ability of Government to expand existing water treatment and distribution system.

Roof-Top Rain Water Harvesting

Most of the urban areas face water scarcity. Dependence on ground water has increased manifold, but the natural recharge to ground water has decreased due to urbanization, construction of buildings and paved area. There is a decline of ground water level and to counter this, artificial recharge has to be adopted. Nature has provided us cheapest source of water through rains. Roof top rain water can be harvested to supplement the ground water recharge. Some state governments like Government of Rajasthan has made compulsory provision for roof top harvesting. Gujarat and Maharashtra have been pioneer states in water harvesting structures.

The number of houses in towns become the basis of calculation of roof top area. Generally the exact size of individual house is not available, therefore an average roof size is adopted for calculation of roof area. The estimation of water available from roof top harvesting is worked out by multiplying the roof area with normal rainfall available for monsoon period. Computation of roof top areas needs to be done properly. Remote sensing technique can also be utilized to have a broad assessment of such areas. However, for precise assessment, data base of roof top area of each dwelling or group of dwellings and institutional buildings need to be developed.

Collection of rainfall on roofs of houses is done by using galvanized sheet iron shaped into a channel at the boundary of the roof. The water collected is stored in a small well adjacent to the building. This approach requires connecting the outlet pipe from roof top to divert the water to either existing tubewells/borewells or especially designed wells. The urban areas are feasible for roof top rain water harvesting to recharge ground water through injected wells. Injected wells are similar to tube wells but with the purpose to recharge the aquifer directly from the surface water.

Roof top water harvesting can enable households to save water for drinking purposes upto 4-5 months. People can use roof top harvesting techniques to meet the growing needs of water particularly during drought. The quantity of rain water received at roof top is not available down the roof due to various losses in the form of moistures absorption, evaporation losses and leakage etc. Therefore, 90 per cent or less is considered available for harvesting the rain water which is taken as a source for artificial recharge to ground water at feasible locations. The main problems in promoting roof top water harvesting technology is the cost of water tank, water conduit system and gutter.

Thus, advantage of rooftop water harvesting may be summarized as follows:

- Ideal for locations with inadequate ground water and surface water resources;
- Utilization of storm water run-off that goes waste;
- Reduction of flood hazards;
- Stored water can be utilized in the lean season; and
- Harvesting structures are simple, economical and eco-friendly.

Run off Rain Water Harvesting

The rainfall runoff flowing from the roads and open grounds is substantial during rains. This water often creates water logging and the drainage system is put under stress in the urban agglomerates. Most of the rain run off water flows out of the city unutilized. This water, if conserved and utilized properly for recharging the ground water reservoir, may bring much needed relief to the water scarcity areas of the city. The harvesting of run-off water not only augments the ground water resources, but also is helpful for avoiding pooling of water which causes inconvenience to the public, and frequent damage to the roads.

A city has a number of drains, the stretches of these drains which are not polluted due to municipal waste water, sewerage run-off waste water and industrial effluents, can be taken up for harvesting of water run-off accumulating during the monsoon period. In a city during rainy season pools of water accumulate at many places because of inadequate drainage system and such locations can be identified where run-off water can be utilize for artificial recharge to ground water through suitable recharge structures such as shafts, trenches, diffusion well, deep sink holes etc. Abandoned dug well and tube well can also be utilized for this purpose.

In order to tap the water available from rains and conserve into the underground reservoirs to avoid any losses, exact volume of runoff water availability is to be computed first. The design of any artificial recharge structures should be such as to take care of peak discharge rate. To prevent artificial flooding by accumulation of surplus rain water on roads and to recharge groundwater, drains on both the sides of roads can be constructed by the people themselves or by the government organizations. In these drains recharge pits at hydrogeologically suitable places can be constructed with uncemented bottom. These pits will prevent long travelling of water and will increase the seepage.

The citizens can be instrumental in conserving water by constructing small structures like anicuts, gully plugs on existing drains to tap surface run-off. These structures are successful as they provide water for immediate drinking, irrigation etc. and also recharge ground water simultaneously. The people should be restricted not to spread water freely in any direction but to construct drains in such a manner so that surplus water of their households flow towards drains constructed on roadsides.

Catching Surface Rainwater in Delhi

While rooftop rainwater harvesting has been made mandatory in Delhi, the potential of catching surface rainwater is also worth exploring. At present rainwater is drained out through storm water drains to the Yamuna. Central Ground Water Board (CGWB) scientists agree that this is a possibility. What needs to be done is to identify priority locations and quantum of surface run off rain water recharge structures for harvesting this properly. Waterlogged areas could become a blessing in disguise and the Government should wake up to the opportunity that has been presented to them. Various agencies including the Municipal Corporation of Delhi, Public Works Department, Central Public Works Department, New Delhi Municipal Council and Delhi Development Authority should work together to make use of fresh water to recharge ground water. To collect and store the surface water during the rains, stretches of road on both sides can be selected where water is generally pooled because of slope on the roads. From these pools it can be diverted to rechargeable trenches.

Law on Water Harvesting

According to Central Ground Water Authority (CGWA) and the Union Ministry of Urban Development, rain water harvesting structures should be installed in buildings built in a plot area of 100 square meters or above where the ground water is below eight meters. CGWA has made harvesting mandatory in south and southwest districts of Delhi. The ground water extraction systems of those who do not do so in these areas can be sealed. Due to rapidly falling ground water levels, CGWA has banned the construction of any new tubewell in south and southwest Delhi, Gurgaon town and adjoining industrial areas, municipal corporations of Faridabad, Ballabgarh and Ghaziabad and the Yamuna flood plain area. Water harvesting is mandatory in these places also.

Annual Rainwater Harvesting Potential (ARHP)

The amount of water that shall be available from the campus/house/ industrial complex for harvesting annually is termed as the Annual Rainwater Harvesting Potential. To arrive at ARHP, however, a few basic data are required[3]. These are:

(i) Average Annual Rainfall (R)
(ii) Catchment Area of the Campus (At): with details of Kuchha, Paved Areas and Built-up roof (Ak…Ap…Ar.):
(iii) Type of roofs: (Kuchha, Pucca, Sloping)
(iv) Site Plan, Stormwater Drainage Plan and Location of Tubewells and/or dugwells, if any
(v) Lithological Logs of existing tubewells or details of soil characteristics and/or depth to bedrock & nature of bedrock/ soil strata
(vi) Location of rainwater down pipes marked on the site plan

Calculating Annual Rainwater Harvesting Potential (ARHP) = R × Rf × A

Where
ARHP— Annual Rainwater Harvesting Potential (in cubic meters)
R — Annual rainfall (in mts.)
Rf — Run-off Co-efficient (per centage)
A — Area (in sq. mts.)

The Run-off Co-efficient (Rf) is taken as

for Roof Catchments Tiles: 0.8-0.9
Corrugated Metal Sheet: 0.7-0.9

for Ground Surface Covering
Concrete: 0.6-0.8
Plastic Sheeting (Gravel Covered): 0.7-0.8
Butyl rubber: 0.9-0.9
Brick pavement: 0.5-0.6

for Treated Ground Catchments
Compacted and smoothened soil: 0.3-0.5
Clay/cow-dung threshing floors: 0.5-0.6
Silicone treated soil: 0.5-0.8

for Untreated Ground Catchments
Soil on slope less than 10% : 0.0-0.3
Rocky: 0.2-0.5

(The above averages are taken for the sake of simplicity. An intense rainfall generates more run-off than a light but prolonged rain. Also the frequency slope surface harness and lack of vegetation cover of the catchment contribute to increased run-off. Deliberate increase in slope and hardness of the catchment area will yield greater run-off).

Example—1 (house)

Calculating Annual Rainwater Harvesting Potential

Total Catchment Area = 500 sq.m
Roof Area : 150 sq.m
Paved Area : 150 sq.m
Kuchha Area : 200 sq.m

Average Annual Rainfall = 60cm/0.60m

Annual Rainwater Harvesting Potential (ARHP) =

$$(150 \times 0.8 \times 0.60) + (150 \times 0.70 \times 0.60) + (200 \times 0.1 \times 0.60)$$
$$= 72 + 63 + 12$$
$$= 147 \text{ cubic metres/ } 1{,}47{,}000 \text{ lts.}$$

Example—2 (Campus)

Calculating Annual Rainwater Harvesting Potential

Total Catchment Area = 5,000 sq.m
Roof Area : 500 sq.m
Paved Area : 500 sq.m
Kuchha Area : 4,000 sq.m
Average Annual Rainfall = 60cm/0.60m
Annual Rainwater Harvesting Potential (ARHP) =

$$(500 \times 0.8 \times 0.60) + (500 \times 0.70 \times 0.60) + (4{,}000 \times 0.1 \times 0.60)$$
$$= 240 + 210 + 240$$
$$= 690 \text{ cubic metres/ } 6{,}90{,}000 \text{ lts.}$$

Conveyance System

The rainwater needs to be directed from the rooftop and rest of the catchment area via the rainwater downpipes and drainage channels through the simple filtration system into the storage tanks or recharge structures. In order to maximize the entrapment of the runoff the roof should be:

(i) Gently sloped to drain towards the downpipes. Floor trap can be placed at the time of casting roof just near the inlets of downwater pipes.
(ii) Sloping roofs should be provided with a gutter at the edges connected to the downpipes.
(iii) Roofs should be uncluttered and swept clean before monsoons.

For downpipes, 90mm PVC pipes resistant to UV rays are a suitable option. Similar PVC pipes can be utilized to connect all the downpipes into the storage/recharge structure. The initial run-off from the very first rain of a season may not be utilized for storage or recharge and may be diverted to the overflow drain directly as it would contain a fair amount of suspended and dissolved matter or other deposited impurities.

The run-off from the open areas, hard & soft surfaces is to be channelised into the filtration pit through a system of open/covered drainage channels.

The Recharge/storage structures may be located somewhat away from the building foundation. They can be located at the end of downpipes, low-lying areas or across storm water drainage lines.

Filtration

Water from the total catchment is to be directed into a filtration mechanism before storage/recharge. A simple filtration pit can be made for removal of suspended particles and micro-organisms from the rain water run-off. The filters can be vertical or horizontal and can be easily constructed as shown in the figure. The sandy interface with the water needs to be cleaned periodically to prevent clogging of the pores.

Rainwater Harvesting Options

1. Storage of Run-off for Direct use
2. Groundwater Recharge

Storage for Direct Use

The rainwater run-off from the roof-top can be harvested in surface/ sub-surface storage tanks for subsequent use. This kind of harvesting is suitable for individual households in areas where rain falls almost throughout the year. The size of the storage tank shall depend on the water demand of the household.

The rainwater can be stored in any commonly used storage tanks of plastic, RCC or masonry. The tank can be above or below ground depending on space availability and cost considerations. Cleaning and disinfecting at regular intervals is a must.

In a campus, where space is available, the water can be stored in an open excavated pond which also acts as a settling pond for impurities.

Storage Tank Below Ground

Recharge Options

Where storage options are difficult or costly recharge options can be practiced. The groundwater table is the cheapest form of storage as well as immune to evaporation losses.

1. Abandoned/Running Handpumps
   (i) Suitable for small buildings with roof area upto 200 sq.m.
   (ii) Water from the catchment area is to be diverted in the handpumps to 50-100 mm dia pipe.
   (iii) A regulating valve needs to be fitted near the handpump to avoid entry of air in the suction pipe.
   (iv) It is advisable not to use the handpump during the recharging period. In case of no other alternative the water may be utilized after chlorination.
2. Recharge Pit
   (i) Suitable in areas with shallow groundwater table/aquifer for smaller buildings with roof area 100-150 sq.m.
   (ii) A pit of size 1-2 m width and 2-3 m deep is to be excavated at suitable location in the plot.
   (iii) The run-off water from the catchment area is to be diverted into the recharge pit.
   (iv) The pit is to be filled with crushed gravel and coarse sand and the top layer of sand needs to be cleaned annually periodically.
3. Recharge Trench
   (i) Suitable for shallow groundwater table areas and with buildings upto 350 sq.m of plot area.

(ii) The trench may be excavated along the boundary of the plot or along the lowest side so as to capture maximum runoff.

(iii) The trench may be made 0.5-1.5 m wide and 1-2 m deep

(iv) The trench is to be filled with sand, gravel and brickbats and the top layer needs to be cleaned periodically

4. Recharge Shaft

(i) Suitable for areas with low groundwater tables.

(ii) The shaft is to be dug manually or drilled upto or near groundwater level or above the highest impervious strata.

(iii) The diameter of the shaft may vary from 0.2-0.3 m depending upon the porosity of the receiving strata.

(iv) Slotted pipes need to be put in the shift to enable permeability in the groundwater table. The slots should be located to coincide with favourable strata.

(v) Recharge shaft should be constructed 10-15 m away from built area.

(vi) Shaft should be driven upto the porous strata or rock fracture for effective recharge

5. Recharge Shaft with Pressure Filter

(i) Suitable for areas with low groundwater tables.

(ii) The shaft is to be drilled upto groundwater level or just above the normal tube-well withdrawal level.

(iii) The diameter of the shaft may vary from 0.15-0.3m depending upon the porosity of the receiving strata.

(iv) Slotted pipes need to be put in the lower part of the shaft to enable permeability in the groundwater table. The slots should be located to coincide with favourable strata.

(v) A settlement tank of size varying from 1.5-5cum is to be constructed for collecting the rainwater run-off.

(vi) A dual media pressure filter is utilized to filter the collected water and pumpit into the ground (This is a sophisticated device. Recourse can be had to simple sand filtration)

6. Combination of Recharge and Storage Options

(i) Suitable combination of various techniques explained earlier can be utilized for site specific conditions.

(ii) The various combinations shall be dependent on the imagination of the planner.

(iii) Few typical combinations have been described below:

a) Underground storage tank with recharge shafts at the base to detain runoff as recharge is a slow process.

b) Recharge trench with recharge shafts at intervals.

c) Combination of storages, recharge pits, recharge shafts, recharge trench in an institutional campus.

d) Storage weirs in combination with recharge shafts on stormwater drains in a campus

Thus, surplus water during monsoon must be conserved and artificially recharged to ground water system keeping in view the various constraint and adverse situations like low and erratic rainfall, quality hazards, water logging and pollution of ground water. Rainfall is highly erratic in some urban areas, storage must be an integral part of any water harvesting system. Some suggestions are as:

1. Rain water harvesting should become an integral part of every home, city and municipality.
2. People should follow roof top harvesting techniques to meet the growing needs of city particularly during drought.
3. Roof top rain water harvesting should be made mandatory in urban areas and bylaws should be amended.
4. Cost effective methods of rain water harvesting as illustrated above should be used.
5. Urban local bodies—municipality, municipal corporations-should assess rainfall run-off quantity loss to flowing/evaporation.
6. Wide publicity about the rain water harvesting concept should be made through mass media.

Cost of Water Harvesting

Typically, installing a water harvesting system in a building the cost may vary between Rs. 2,000/-and 30,000/-for buildings of about 300 sq. meters. It is difficult to make an exact estimate of cost because it depends on the availability of existing structures like wells and tanks which can be modified for water harvesting use.

Precautions

Whether harvested water is used for usage or for recharging the groundwater, it is of the utmost importance that the rainwater collected is free of any pollutants that might be added to rainwater from the atmosphere or catchment. While polluted water directly used for consumption would have an immediate impact on health, polluted water recharged into the ground water would cause long-term problem of aquifer pollution. Damage done to aquifers by recharging polluted water is irreversible.

The precautions that should be taken are:

1. At the catchment level:
   - Keeping the catchment clean
   - Using gratings to trap debris at the catchment itself
   - Paving the catchment with ceramic tiles, stone tiles or other such non erosive material
2. At the conduit level:
   - Provision of first flush to drain runoff from initial spell of rain
3. Before Recharging:
   - Allowing for sedimentation of water
   - Filtering the water.

RECENT EFFORTS FOR WATER HARVESTING IN INDIA

In view of recurring droughts in several regions of the country, the Union Water Resources Ministry has prepared an ambitious Rs. 550-crore scheme for artificial recharge of groundwater under which it will finance about 10,000 water harvesting structures in various scarcity regions. Under the scheme, the Ministry will directly fund various village beneficiary groups and water users associations that want to set up water harvesting systems in their areas to recharge the depleting aquifers or ground water collection. The scheme will cover areas in 17 states that have been identified as over-exploited or deficient in terms of water resources. These include Gujarat, Rajasthan, Andhra Pradesh, Orissa, Uttar Pradesh, Bihar, Delhi and all the seven states in the North-Eastern Region.

A district-wise master plan has been formulated for the first time in Rajasthan to facilitate launching of water harvesting projects in each district. The master plan has covered each hectare of the district and every drop of rain water. Rajasthan is the only State to prepare such a master plan. Moreover, it has been formulated by the irrigation department without incurring any additional expenditure. Under the master plan, no construction activity will be allowed in the catchment areas where watershed projects are taken up. The district administration would be held responsible for any unauthorized works. Copies of master plan for each district have been handed over to the respective District Collectors.

Ruchi Experiment

Rural Centre for Human Interests (RUCHI), a Haryana-based NGO is one of the pioneers in advocating roof water harvesting in the State. In addition to implementing the system in their project areas of Solan and Sirmour districts, RUCHI has constructed about 500 ferro-cement tanks of various sizes for others for storing rain water.

The 14-year-old NGO's interest in rain harvesting stemmed from facing acute water shortage. In Shalana village where their office is located there was no source of water in the vicinity. When they began the construction of their office in 1988, they had to get truckloads of water from 16 km away. Initially, it cost them Rs. 150 for a load of 4,000 litres. "Realising our predicament", recalls Dharamvir Singh, RUCHI director, "the suppliers started blackmailing us. The price was soon hiked to Rs. 350; later, when we ran out of water; we had to cough up even Rs. 500". This prompted Mr. Singh to learn the techniques of collecting roof water. The main hitch was storage. This required a considerable amount. It was then that he learnt about the ferro-cement technology.

Structural Engineering Research Centre (SERC) of Ghaziabad, Uttar Pradesh, has been propagating this method. In 1991, RUCHI constructed a number of ferro-cement tanks. A bigger tank would have been cheaper; but lack of space compelled them to go in for several smaller tanks of 6,000 litres capacity. A filter was fitted on top of each collection tank. The slow sand filter designed by them uses gravel, sand and coconut husk or pine needles.

RUCHI conducts a series of training classes. Before the actual training starts, the trainees are given an unusual task, that of making do with 40 litres of water a day. Is this quantity adequate? With proper education, it works. In the toilet, there is no flush tank. A seven litre bucket is provided in its place. The break-up of 40 litres goes like this: 15 litres for bathing, 14 litres for the toilet, six litres for drinking and cooking and five litres for washing clothes.

On Track

- In rural areas, continued investments in infrastructure hold the key.
- Tapping groundwater reserve with no control on the same would prove fatal in the long run.
- Industry and agriculture need to exercise control on water usage.
- The shift to rain water harvesting needs to be encouraged

This region gets an average rainfall of 1,300 mm. Luckily for residents, it is spread over nine months. RUCHI has a roof catchment that can collect 4,20,000 litres of water in a year. The dry period stretches for about 80-90 days. Earlier, this period was even shorter. For most part of the year, the intermittent rains fill the tanks. In all, they harvest 3,50,000 litres-about 77 per cent of their requirement-with a storage capacity of 86,000 litres.

RUCHI also conducts six day RWH courses on request from other NGO's. Now people are receptive and have accepted this technology. Some of the families have borne the full cost of such tanks. It costs Rs. 3 per litre now. There is tremendous scope for roof water harvesting especially in high rainfall areas. Technical know-how in remote areas and funding are two constraints. Initial funds need to be given to users, which can be slowly realized. Individual tanks are a good idea. Community effort do not succeed because joint-ownership does not work in our situation.

RAJIV GANDHI FOUNDATION

The Rajiv Gandhi Foundation has launched an ambitious new project to strengthen the water harvesting movement in the country. It is building up a cadre of rural volunteers in different states, collectively called the 'Green Corps', dedicated to helping the rural communities undertake small projects for rainwater harvesting on a sustainable basis. Environmental regeneration will help to alleviate poverty and reverse rural-urban migration. City dwellers know what water shortage does to their lifestyle.

The Foundation has already recruited 55 volunteers who are currently working in different parts of Rajasthan. They constitute the first batch of the Green Corps. All have been given rigorous training on the technical aspects of water harvesting and watershed development. Even more important, they are being taught how to build community organizations, promote community initiatives and secure long-term support for the future. They will also be trained in other aspects of rural development, to become long-term agents of change. The Foundation hopes to add 50 new volunteers every year in different states of the country over the next five years. The Green Corps is therefore not just a project, but a movement. The initial response has been excellent. Through the volunteers, Jal Sabhas and Jal Samitis have been formed in 63 villages. In many of the Jal Samitis women constitute 30 per cent of the total members.

Progress in Delhi State

In many states, the practice of rainwater harvesting has been initiated, though it is still very insignificant. However, in Delhi, where water scarcity has become very acute, rainwater harvesting is now being considered as the only way in meeting the increasing demand for potable water. The present population of Delhi is around 140 lakh and based on daily requirement of 60 gallons per capita per day, works out of 800 MGD. With the present supply of 650 MGD, there is a shortfall of 150 MGD. Main reason for the shortfall is non-availability of raw water. With surface water contributing to over 70 per cent of Delhi's total water supply and the groundwater receding rapidly, the need for saving water through harvesting has been widely felt and the Delhi Development Authority (DDA) has undertaken a major rainwater harvesting project for recharging 1 million cubic meters of rainwater annually by covering an area of 190 hectares. Under the plan DDA has executed two projects of rain water harvesting at Sanjay Van and Vasant Kunj, Phase II, both in South Delhi. In Vasant Kunj project, an institutional mechanism has been adopted by the DDA where technical and financial assistance amounting to 50 per cent of the project cost is provided to the Resident Welfare Associations (RWAs) in the form of a one-time subsidy.

Simultaneously, this civic agency has also drafted a scheme for rainwater harvesting in the under-construction houses in a new pocket of the same Vasant Kunj locality. These schemes are expected to benefit the residents of water-starved Vasant Kunj. Further, DDA has also initiated a study on rain water harvesting in Dwarka through INTACH. It has also made several alterations in its construction plans for optimum use of available water supply and its regeneration. As such, the agency has decided that in future all its houses will have a provision of double piping system which will reutilize the disposed water from kitchen and bathrooms.

An Exemplary Success Story

One of the most astounding success story of rainwater harvesting is that of the Arvari river in Rajasthan's Alwar district which was brought back to life a few years ago by the joint efforts of the local villagers and the Magsaysay Award winner, Rajendra Singh. He has bagged the International River Prize. The river in Thana Gazi tehsil, which is managed by a "Parliament" of 72 villages, was among the five rivers chosen for the River Prize at the recently concluded International River Festival in the river City of Brisbane in Australia. This is not the first time that the tiny Arvari, which had remained dry for nearly two decades before it started flowing once again in the late 1990s, has brought accolades to the village community sustaining it. In the year 2000 the then President of India, K.R. Narayanan himself visited the area to felicitate the inhabitants of Bhawta-Kolyala villages in the upstream of the river with the Joseph John Award of Rs. 1 lakh.

The River Prize cites the sustainability of the Arvari river in various aspects of water management while keeping intact and at time reviving—the old traditions and historical practices in the community's interest. The Arvari Parliament, with representatives from all 72 villages of the river basin has been taking the responsibility of planning and enforcing the sustainable use of water from the river in agricultural practices. The 15-day long river festival in Brisbane also honoured the Alexandra river of the middle east, the Danube of Europe, the Hung Ho of China and the Nile of Africa. Each of them found favour with the selection committee for one or the other aspect of conservation and community participation. The Danube won the prize for generation of sustainable wealth while the Hung Ho, for flood control. The Colorado river was picked up for a case study presentation along with the Singapore river and the Sawan river of Australia at the festival in which about 7,000 persons participated.

While the rest of the rivers are big and mighty, the small Arvari, which is only 47 kms in length, caught the fancy of the experts at the festival—attended by representatives from as many as 250 rivers the world over—for its community participation. The award committee watched the river for a few more years before the prize money—2 lakh US dollars-was finally handed over. The River Prize came as a boost to the morale of Mr. Singh, who had been under attack from his detractors for the breach of the controversial dam at Lava Ka Baas in Thana Gazi tehsil of Alwar.

Madhya Pradesh Experiment

The Deendayal Research Institute, in collaboration with the District Rural Development Agency, Majhgawan (Satna), took up an area of 12,536 hectares and formed 17 micro-watershed schemes to tackle low productivity, lack of irrigation, a receding water table, shortage of drinking water, high soil erosion etc. The 'watershed area'—the area over which rainwater collects—was taken as a unit. Contour trenching and bunding check the flow of water. Trenches are excavated and bunds erected (with the soil taken out) to check the water's free flow. This prevents soil erosion while plenty of percolated water raises its sub-soil level.

The vision of water management programmes, thus, is as simple as they can be. Instead of tanks or dams, water is kept where nature keeps it-underground, Thus, there is no loss due to evaporation. The old philosophy of 'water should not be allowed to run, it should walk' has been replaced by the new one which states that 'water should creep and, ultimately, come to a halt'. The whole point is to ensure that not a single drop is allowed to flow more than 10 metres above the surface. The paradoxical scenario will be of dry streams and unfilled tanks in the rainy season. Instead, the water that seeps in, reappears in wells, tubewells, ponds and used later when and where wanted.

This strategy is the result of a calculated comparison of costs. An assessment was made of the amount of water saved and the expense incurred in each of the approaches available in the village of Dantha in Khandwa district. The cost of water secured through contour trenches, earthen checks etc. worked out to just 4 paise per litre. The system is so water-tight that even if it rains up to 10 centimetres a day-double the average daily downpour at Cherrapunji, one of the heaviest rain receivers in the world—not a drop of water will flow down the drain. The result is a water surplus even in conditions of low rainfall.

It is thrilling to visualize what this means. "Before 1996, water was a rare commodity especially in summer," says an official publication, "despite low use in agriculture, humans and cattle lived in distress... villagers used to leave the cattle in the jungle to fend for themselves... whereas there was not a single tube-well in the village, now it has 12. Twentytwo, out of 56 dug wells, never had water even in a year of normal rain. Now they not only have water in a year of drought—all the 102 dug wells have bountiful water—but farmers can (even) raise rabi, a second crop."

Something unimaginable a few years back is happening. One finds very small farmers creating bank deposits of Rs. 8,000-10,000 within a year.

Where water flow is swift, a gabion structure—a cylindrical container filled with earth, stones, or any other materials—is built with boulders tightly packed in wire mesh cages. People's participation is paramount. Majhgawn's watershed committees organized 78 self-help groups, imparted vocational training to them and employed 287 rural youths-each earning between Rs. 1,000 and Rs. 2,500 per month. Earthen bunds now irrigate 1,504 hectares, and people have enough drinking water round the year. Productivity, too, has increased by 52 to 60 per cent in paddy, mustard and gram cultivation and 40 per cent in wheat cultivation.

Kuthambakkam Village in Chennai—A Success Story

President APJ Abdul Kalam has described Kuthambakkam village as the signpost for modern India. About 30 km from Chennai in Tiruvallur district, the predominantly Dalit village of Kuthambakkam is a text-book example of what Gram Swaraj can do to make sustainable development meaningful with the help of technology. Kuthambakkam's success flows from its homegrown wisdom—optimize the use of locally available resources through technology, make grassroots empowerment the key to community development, respect the environment you live in and everything else will follow. It's one of the very few villages that has a website www.modelvillageindia.org-that the rural community uses to both improve its resources and sell its produce.

But its in optimizing environmental resources that Kuthambakkam stands out. Through an intricate series of check-dams and rainwater harvesting structures, it has managed to dramatically increase the groundwater level—potable water is available at a depth of less than 10 feet. Seven water tanks and pumping stations have been built and all nine village ponds desilted. The result—2000 acres of the village's wetland now sport a lush green look and the village is moving towards optimizing dryland farming. The villagers aim is to ensure food security for the village and cent per cent access to food for all people through the Gandhian model of rural economy. Organic faming is all set to receive a big push and the pioneering eco-friendly 'hutless village' concept is catching up. To satisfy rural housing needs, they came up with the idea of using locally manufactured compressed mud and cement blocks for building houses.

In July 2003, a fully equipped rural technology resource center—complete with a soap moulding machine, oilseed press, mini oil-mill, milk processing unit and a bakery began functioning. The produce will be sourced and sold locally. This will be the prototype for a networked village economy, based on the concept that villages themselves are considerably big marketplaces, and by interlinking them and sourcing and selling products locally, they will be able to maintain demand and supply. The internet is also playing a huge role in village self governance.

Jain Experiment at Jalgaon

The experiments by transferring water from outside commenced in 1988. There was one open well and one bore well in the area till 1992. The well was abandoned; whereas maximum yield from bore was 70 thousand l/day. This was not sufficient to irrigate 15.5 acre of fruit trees and 9 acres of agro-forestry. It was required to bring water from outside to cater the demand of the crops. In summer this was also not sufficient. There was one small earthen bund which was insufficient. Under these conditions, the construction work of Jain Sagar percolation tank was taken up. Cut Off Trenches (COT) of 20 ft. depth was excavated and filled with black soil. It took 6-7 months round the clock work to complete the tank. Water was stored in tank for the first time in 1993. The yield of the bore well increased from 70 thousand l/day to minimum 2.5 lakh l/day. One big well was excavated downstream of the tank. New five bore wells were excavated. Irrigated area rose to 38 acres from 24 acres on account of availability of water. The benefits of watershed development could be realized after the construction of percolation tank (Jain Sagar) and down stream wells. The success through this Percolation Tank gave inspiration for continuing further watershed development. Today's requirement of water in the area for irrigation is 600 to 700 million litres (0.6 to 0.7 mcum) per year which is conserved here itself. Not a single litre of water is required to be brought from outside. In addition of supplying water for irrigating over 500 acres, water is also used in the training centre, residential quarters and guest house etc.

Initiatives by Delhi Jal Board

Delhi Jal Board (DJB) is committed for augmentation and conservation of water supply in Delhi and has taken number of steps in this direction. One of these steps is Roof Top Rain Water Harvesting for augmenting ground water storage in the National Capital Teritory (NCT) of Delhi. A special cell has been created for the purpose, which is manned by qualified engineers. The cell is disseminating information about rainwater harvesting in NCT of Delhi and nearby states. It is providing technical assistance to schools, institutions, RWAs and individuals. Since 2000, the Board has provided Rain Water Harvesting structures in its 163 buildings covering an area of about 8 lakh sq. metres, which will harvest about 4.5 lakh cubic metres of rainwater annually for recharging ground water. Work is also in progress at another 27 sites of DJB (as on 30-11-2002)[4]. The complexes where rainwater harvesting efforts have met with success are Rashtrapati Bhavan and Shramshakti Bhavan. Rain water harvesting at the Indian Institute of Technology and Jawahar Lal Nehru University has led to a rise of 0.51 to 1 metre in groundwater level[5].

Mumbai Takes to Rainwater Harvesting

Mumbai has become the latest advocate of the concept of rainwater harvesting following the examples of Delhi and Chennai. Since October 1, 2002, any new building coming up on a plot of more than 1,000 square metres should have, by order, a rainwater harvesting facility and authorized consultants prepare the design of the system. The Municipal Corporation of Greater Mumbai is also planning to rope the owners of the existing building as the new buildings account for only 4-5 per cent of the total buildings in the city. The emphasis of the

Municipal Corporation is more on collection and storage of rainwater rather than on percolation. Further, installation of waste water treatment plants has been made mandatory in Mumbai for new buildings having a plot area of more than 2,000 sq. metres.

SOME MODELS

Water harvesting, which can easily be done, has the potential of solving Delhi's water crisis to a great extent. The average annual rainfall in the city is about 600 mm. The Central Ground Water Authority (CGWA) estimates the ground water recharge from a house with a rooftop of 100 sq metres will be 55,000 litres per year. This amount can meet the demands of a family of five members for four months. For the entire city, the recharging potential is about 7,65,000 million litres per annum. The demand for water in the Capital is about 800 million gallons a day, while the Delhi Jal Board (DJB) manages to supply about 650 (MGD). The gap between demand and supply is met by ground water. DJB warns that the supply from all available sources can be raised to 950 MGD by 2007 and after that any further augmentation will be very tough. With the city's population increasing by almost five lakh every year, the dependence on ground water will only grow. But ground water is not an endless resource and its levels are falling at an alarming pace. While in 1960 over half of Delhi had ground water at a depth of between two and five meters, the levels have today fallen up to 20 meters in those places. Ground water is like money in bank. Maintenance of the level is possible if deposits are made regularly.

Falling water table means that one has to run the tubewells for longer durations and use more electricity. If ground water level falls by a meter, an additional 0.4 kilo watt of energy is required to pump out the same amount of water as was being done before the level fell, CGWB says.

Model 1

Here's a simple water harvesting model for those who live in a single-unit house. A network of pipelines made of PVC asbestos or galvanized iron will be laid on the roof. These pipes will carry rainwater down to a pit. The depth of the pit will depend on the water requirement of the house and how much rain the area receives. Typically, it would be three meters deep. The pit would have layers of sand, gravel and boulders. The rain water would get filtered before it percolates into the ground and recharges the aquifer. While the cost will depend on the height of the building and the extent to which the water table has fallen, one can expect to spend at least Rs. 5,000.

Model 2

This one is for big buildings that may house apartments, offices or industries: Since all buildings have pipelines that carry water from the roof to drains usually, a fresh network will not be required. At the space where the pipes discharge the water on the ground, a chamber will be built. A tubewell, deep enough to reach the depth of the ground water, is also dug. Water from the chamber will fall into this tubewell, which is usually 10 to 20 meters deep. If the water has silt in it, it will be made to pass through a filter consisting of layers of soil and boulders before it enters the storage. The harvesting structure's cost for a building spread over four to five acres that has 80 to 90 apartments will be about Rs. 40,000. For a bigger plot which houses a bigger office complex or up to 150 apartments the expense could be about Rs. 80,000.

Model 3

Unity is good for water also. Entire colony can also undertake harvesting by following these simple steps. As harvesting structures for a colony have to be made on public land and the expenses involved are in lakhs of rupees, it is better to involve the local agency in a partnership with the community water harvesting structures. The rain water of a planned colony flows away in storm water drains. Barriers will be made there to make that water flow into a specially constructed tank. The size of the tank will depend on the monsoon run-off generated in the colony. If the drain is very dirty or choked, tanks will be made to trap the rain water at a place just before the drain.

A long recharge pit will be made. This will have two to three tubewells, usually 10 to 20 meters deep. The tubewells will be covered with a filter chamber comprising layers of boulders, coarse sand and gravel. The pit will be covered with detachable slabs. The water from the tank will flow into the pit. After getting filtered, it will enter the tubewells and recharge the ground water.

Model 4

If one lives in an area that gets water logged even after a short spell of rain, then he can solve his water problem very simple. A tank will be dug at a place from where water can flow into it. The dimensions of the tank will depend on the monsoon run off generated in the area. A filter chamber consisting of layers of sand, gravel and boulders will be made at the bottom of the tank. If the water table is low, a tubewell will be dug in the centre of the tank. The tubewell is usually 10 to 20 meters deep. The pit will be covered with detachable slabs. After getting filtered, the water will flow into the aquifers. The expenses incurred will depend on the volume of water getting collected during the rainy season at the place and the cost would be anywhere between Rs. 10,000 and Rs. 30,000.

Conclusion

Water harvesting is certain to grow in importance in coming years as policymakers and planners seek cost-effective solutions to water supplies. Many simple systems can easily be put in place to overcome water scarcity, such as, rooftop collection for domestic use. As water becomes increasingly scarce, treatment of storage ponds to reduce leakage may become more financially viable. On the other hand, systems that require cooperative management and strict limits on land use will probably not become any simpler as local economics and land use systems become increasingly complex.

Water harvesting does not make large dams and ground water extraction obsolete. Rather, each system has its own advantages and disadvantages, and all systems can complement each other. The major advantage of water harvesting is that it begins to address problems locally, relieving the pressure on large-scale, centralized systems. Integrating water harvesting with other systems, as well as improving the management of demand, will lead to reliable, cost effective water systems.

There are some benefits that can be derived from rain water harvesting in urban areas. First benefit is that rain water collection is simple in concept, and necessary structures can be built and maintained, the second benefit is the high quality water which can be collected and stored for lean period. Public participation coupled with suitable modifications in building by-laws can help in enhancing ground water recharge which has substantially been reduced due to urbanization. Roof top harvesting in urban areas can offer a good source of water for artificial recharge of aquifers. In urban areas the rain water run-off usually goes waste instead of recharging the ground water in the area. Construction of rain water harvesting structures for tapping roof top run-off should be made mandatory. Given growing water scarcity and the high financial, social, and environmental costs of alternatives, governments may find subsidizing water-harvesting systems attractive, especially those for domestic supply. DDA already subsidizes rooftop collection systems.

REFERENCES

1. Population Reports, February, 1999, Population Information Programme, The Johns Hopkins School of Public Health,
2. Kerr John and Pangre Ganesh: Water Harvesting and Watershed Management; Overcoming Water Scarcity and Quality constraints; vision 2020, Focus 9, IFRRI,
3. Resource, Earth Management Ple: Rainwater harvesting—A Basic Manual for Urban Built-up Areas (Houses, Institutions, Industries, Campus, Complexes).
4. Delhi Jal Board Souvenir on the occasion of seminar on Rain Water harvesting held at India Habital Center, New Delhi on 30th December, 2002.
5. *Hindustan Times*, New Delhi, April 21, 2002.

13

Integrated Watershed Development

"Gracious be divine Waters for our protection, be they for our drink, and stream on us bliss and happiness"

—Rig Veda

Introduction

Watershed management refers to managing an entire land area served by all the rivers and aquifers that drain into a particular body of water (such as a semi-enclosed bay). River basin management is essentially the same concept applied to one river system, although the two terms are used interchangeably.

The US defines a watershed as the entire area drained by a river system or one of its major tributaries. The UK defines a watershed as the divide between river basins, a potentially larger area. International Food Policy Research Institute (IFPRI), Washington, on the other hand, enunciates watershed (or catchment) as a geographic area that drains to a common point, which makes it an attractive unit for technical efforts to conserve soil and maximize the utilization of surface and subsurface water for crop production. A watershed is also an area with administrative and property boundaries, lands that fall under different property regimes, and farmers whose actions may affect each other's interests. Boundaries defined by humans, however, normally do not match biophysical ones[1]. No matter how it is defined, "we need to see a river or lake, along with its entire watershed and all its physical, chemical, and biological elements, as a part of a complex, integrated system," according to Janet Abramovitz of the Worldwatch Institute[2].

Everyone has a watershed address: We all live in basins that drain rainwater into streams and rivers that eventually send the water back to the see or into inland lakes. The people living in most of these addresses have radically altered the natural drainage systems around them. Tempering with watersheds has proved ruinous for many developing countries including India, where hillsides denuded of vegetation empty tons of soil into water courses every year, causing floods during the wet seasons and suffocating aquatic life during the dry seasons. Deforestation has ruined land and altered climates, causing less rain to fall in some areas. In others, rainwater runs off so far that little can be collected for use. In sub-Saharan Africa, for example, the albedo effect—the drying of the landscape as a result of the wholesale clearance of tropical forests and poor farming practices—has resulted in below-average rainfall over the past 40 years compared with the century as a whole[3].

Rainfed agriculture in India's semi-arid tropics is characterized by low productivity, degraded natural resources, and widespread poverty. Most of the hundreds of millions of people living in the Indian semi-arid tropics depend on agriculture and natural resource management for their livelihoods, so development planners are now eager to implement productive, environmentally sustainable land and watershed management systems.

Concept

A Watershed is a geographic area that drains to a common point, which makes it an attractive unit for technical efforts to conserve soil and maximize the utilization of surface and subsurface water for crop production. Watershed Management refers to managing an entire land area served by all the rivers and aquifers that drain into a particular body of water (such as a semi-enclosed bay). River basin management is also essentially the same concept applied to only one river system, although the two terms are used interchangeably. Watershed may be managed for various objectives, depending on local needs, including capturing runoff, minimizing erosion and reducing pollutions. In management of small watersheds, capturing runoff for local use is conceptually equivalent to harvesting water.

This approach demands integrated development of non-arable land, arable land, rain water, vegetation, livestock, local materials, common property resources, human resources and programs for landless in a participatory process. Livelihood gatherings by increasing production of biomass, productive employment generation and conservation of resources are the most important concerns of any watershed development project. Emphasis on the landless, gender issues and biodiversity utilization is also enshrined in the basic principle of watershed management.

The Need

Integrated Watershed Management is a multidisciplinary approach for rational utilization of natural resources existing in the watershed for optimum and sustained production with minimum hazards, considering holistic development for all the users of watershed. It essentially relates to management of resources which means scientific management of land use, protection of land against all forms of degradation, enhancement and maintenance of soil fertility, conservation of water for watershed use, flood protection, sediment reduction, improvement of productivity under all types of land uses and sustenances of animals as well as peoples within a national physiographic unit[4].

Ideally, a comprehensive watershed management plan mobilizes communities and individuals and gains broad public acceptance at the national level. Thus, watershed management takes into account social, economical, environmental sustainability and institutional factors operating within and outside the watershed for effective management of natural resources. The essentials of Watershed Management includes identification of priority area and integrated development approach, community driven, agencies and development departments involved for their interaction with stakeholders for harmonizing with traditional knowledge; bottom-up approach, having capacity for making interim corrections; executions of programme through village institutions, people's participation and gender balance; and judicious and equitable sharing of costs and benefits of the project.

The water harvesting and the watershed management coupled with selective sowing and cropping patterns can help improve productivity and usage efficiency. These practices can also help recharge the ground aquifers and minimize the increasing gap between groundwater withdrawal and recharging rates. Thus, watershed development projects are designed to harmonize the use of water, soil, forest and pasture resources in a way that conserves these resources while raising agricultural productivity, both by conserving moisture in the ground and increasing irrigation through tank and aquifer-based water harvesting.

Prerequisites and Problems of Watershed Management

At the very outset, it may be pointed out that watershed projects cannot succeed without full participation of project beneficiaries and careful attention of social organizations. This is because the costs and benefits of watershed interventions are location specific and unevenly distributed among the people effected. Watershed, as defined earlier, is an area with administrative and property boundaries, lands that fall under different property regimes, and farmers whose action may affect each other's interests. But boundaries defined by humans, however, normally do not match biophysical area.

Secondly, economic conditions and access to infrastructure may have as great an impact as a watershed project in determining the outcomes that projects seek to achieve because such factors determine the incentives for people to manage and protect natural resources and invest in increased agricultural productivity. Socio-economic relationship among people in a watershed can complicate efforts to introduce seemingly straightforward technical improvements. This is because a watershed encompasses many decision makers who are affected unequally by development. When a watershed project is introduced, often the bulk of the work is done in the upper reaches, while the benefits accrue primarily to those in the lower reaches. For example, revegetating the upper reaches involves imposing a ban on both grazing of animals and felling of trees so that plants can become established. As a result, the people who use the upper watershed—typically relatively poor people with little or no land-bear the brunt of the costs of watershed development, which mainly benefits wealthier farmers in the lower watershed. Those who are made worse off by a watershed project can undermine its efforts if they refuse to go along with it. Herders, for example, might refuse to abide by grazing bans and trespass on the common lands if they can. In general, watershed technologies are likely to fail if they divide benefits unevenly but require near-universal cooperation to make them work. In this case, equity becomes a prerequisite to efficiency[5].

While early watershed projects failed to recognize the socio-economic dimensions of watershed development, this has changed significantly in the last decade. In recent years, there has been a growing appreciation of the need to organize communities to work collectively, to ensure that everyone takes an interest in the work that is done and benefits from the project. In the 1990s, every project defined participation differently. For government programmes, typically it meant making the effort to convince people of the soundness of an approach that was essentially predesigned without any input from those who would be affected. Some projects took people's involvement a step further: local committees were established to mobilize labourers to move earth and plant vegetation and to facilitate communication within the village to improve the management of common lands. At the extreme, many new projects operate under the assumption that local people know best how to care for their land and simply need outside assistance to help them organize and gain access to resources, including funds and social services. It may be stated in this context that managing watersheds sustainably means building institutional capacity, including the creation of cross-sectoral data collection and monitoring systems. Capacity building is a key theme of international organizations promoting change, including the World Bank, and the Global Water Partnership[6].

Implementation Hurdles-Government Vs NGO

Watershed management is not an easy to accomplish. It is a complex and contentious process that involves many stakeholders with competing views about water use. Moreover, in India, the programme implementation through contractors create a number of conflicts among the contractors and people, and the people naturally get a very wrong message about watershed programme. No effort is usually made to resolve the conflicts. There is a total adhocism in the implementation of the programme. Lack of interaction among the projected beneficiaries, improper and selected distribution of inputs, conflict among the farmers over sharing of common benefits, and low status of women in society, etc. are perceived as important socio-economic issues. The project staff have no real concern for the development of the people. They are interested only in fulfilling the target as envisaged in the programme. On the other hand, the performance of the non-government organizations (NGOs) are found to be better than the government agencies. The reasons for the success of the NGOs in this area are, *inter-alia,* small organizational structure, strict supervision, better facilities for staff especially to visit the farmers, rational division of labour, better team coordination, flexibility in policy and linkage with development departments, incentives for good work and periodical evaluation of the work. Technologically NGOs may not be very sound, but they have approach for technology dissemination. At village level formation of informal groups, regular night meeting of farmers and cultural programmes, good training supported with better audio-visual aids, use of special exercise for convincing the farmers are some other specific activities of NGOs.However, NGO programmes are by no means uniform, but they all place strong emphasis on social organization. Their guiding principle is that without proper social organization, efforts to introduce watershed technology will be fruitless.

The two features that most distinguish NGO watershed programmes from government programmes are their scale of operations and their staffing structure. While government programmes have huge budgets and work in hundreds of villages, most NGOs work in only a handful of villages. They devote more staff time per village, and they often work on a variety of activities in addition to watershed management. While government employees involved in watershed management are almost exclusively trained in agricultural sciences and engineering, NGO staff members include many more nontechnical staff trained in community organization. They believe that social organization contributes as much to successful watershed development as technical input. Some NGOs collaborate with government agencies that provide technical expertise, but others do not. Some NGOs also aim to help villagers gain access to variety of government services such as health care and education.It is important to note that NGOs vary a great deal. Some are large and well established, with access to substantial funding, whereas others are small, less experienced, and underfunded.

Activities Under Watershed Management

Aims and objectives of Integrated Watershed Management have already been mentioned earlier. In watershed management projects, mechanical or vegetative structures are installed across gullies and rills and along contour lines, and areas are earmarked for particular land use based on their land capability classification.

Cultivable areas are put under crops according to strict principles of contour-based cultivation. Erosion-prone, less favourable lands are put under perennial vegetation. This approach aims to optimize moisture retention and reduce soil erosion, thus maximizing productivity and minimizing land degradation. Improved moisture management increases the productivity of improved seeds and fertilizer, so conservation and productivity-enhancing measures are complementary.

Excess surface runoff water is harvested in irrigation or percolation tanks while subsurface drainage recharges groundwater aquifers. So conservation measures in the upper watershed have a positive impact on productivity in the lower watershed. Reducing erosion in the upper reaches also helps to reduce sedimentation of irrigation tanks (ponds) in the lower reaches. The watershed approach enables planners to harmonize the use of soil, water, and vegetation in a way that conserves these resources and maximizes their productivity. This system-based approach is what distinguishes watershed management from earlier plot-based approaches to soil and water management.

Watershed Components

- Land development including in situ soil and moisture conservation measures like contour and graded bunds fortified by plantation, bench terracing in hilly terrain; and nurseries for fodder, timber, fuel wood, horticulture and non-timber-forest produce.
- Afforestation including block plantations, agro-forestry and horticultural development. Shelter-belt plantations, sand dune stabilization etc.
- Drainage line treatment with a combination of vegetative and engineering structures.
- Development of small water harvesting structures such as low-cost farm ponds, nalla bunds, checkdams and percolation tanks and ground water recharge measures.
- Renovation and augmentation of water resources, desiltation of tanks for drinking water and irrigation.
- Pasture development either by itself or in conjunction with plantations.
- Repair, restoration and upgrading of existing common property assets and structures in the watershed to obtain optimum and sustained benefits from previous public investments.
- Crop demonstrations for popularizing new crops and crop varieties or innovative crop management practices.
- Promotion and propagation of non-conventional energy saving devices and energy conservation measures.

Source: Guidelines for Watershed Development (Revised-2001) Department of Land Resources, Minsitry of Rural Development.

Operational Guidelines

All development activities within a watershed should be pooled into a unified group action. Independent and uncoordinated sector's development does not fit into this concept. Integrated development of watershed through participatory approach has been conceptualized as a way of real development of rural sector by its policy and decisions. During eighth five-year plan emphasis was given that planning for development should be undertaken on watershed basis and in view of that new guidelines were delineated to achieve success in improvement of socio-economic condition of the people. Since then government and non-government sectors have given their serious attention for development through watersheds and now it has been realized that watershed is the real approach to deliver the desired resources to the farmers. The operational guidelines have already been decided and they are:

1. Emphasis on people's participation. This includes
   - (a) Identifying problems and planning of activities by people,
   - (b) Contribution from people living in the watershed,
   - (c) Implementation of program by people,
   - (d) Operation and maintenance by people; and
   - (e) Transparency in utilization of funds and maintenance of accounts.
2. Involvement of different agencies in development process.
3. Focus on group approach.
4. Convergence of development and welfare services of watershed area.
5. Project Implementation Agency should be made responsible for planning and implementation of watershed development projects.
6. Provision for entry point activities to build credibility.
7. Full time watershed development team (WDT) exclusively for watershed development.
8. Provision for human resource development, training, gender issues, and income generating activities.
9. Focus on local institutional building through participatory approach in the watershed.

History of Watershed Management

The water harvesting and afforestation approach to watershed management is gathering momentum in various states. However, Drought Prone Area Programme (DPAP) is the earliest area development programme launched by the Government of India in 1973-74 to tackle the special problems faced by those fragile areas which are constantly affected by severe drought condition and the objectives of the programme are being addressed in general by taking up development works through watershed for land development, water resource development and afforestation/pasture development. These approaches were led by on-station research undertaken by institutes under the Indian Council of Agricultural Research (ICAR) such as the Central Research Institute for Dryland Agriculture (CRIDA), as well as the International Crops Research Institute for the Semi-Arid Tropics (ICRISAT) which is located at Hyderabad (Andhra Pradesh).

ICAR Model Watersheds

In the mid 1980s, ICAR decided to implement the findings of its dryland agricultural research in 47 model watersheds around the country. Many of these pilot sites were treated as research watersheds, where the work undertaken was closely monitored and changes in land and water conditions were analyzed. Physical costs of the watershed works were relatively low, but supervision was intensive, with persistent efforts to introduce new varieties and other improved technologies and management practices.

CRIDA Projects

CRIDA took up three such watersheds in Andhra Pradesh and one model watershed was launched in Maharashtra.

World Bank Pilot Project for Watershed Development in Rainfed Areas.

This project was initiated in 1984 in Karnataka, Madhya Pradesh, Maharashtra and Andhra Pradesh. Like the ICAR model watersheds, it sought to introduce improved rainfed agricultural technology. The project's guiding philosophy was that low-cost soil and water conservation measures, including improved agronomic practices like contour cultivation and vegetative rather than mechanical bunds, could make a strong contribution to rainfed agricultural development at a relatively low cost[7]. While the work was undertaken on a watershed basis, the project also emphasized proper treatment; within each plot, as the project's design team felt this was missing from the watershed approach pioneered in western Maharashtra. A major thrust of this project would be to promote contour-based cultivation, which would conserve soil and concentrate moisture at a very little monetary cost. The improved soil moisture regime in turn would make improved seeds, fertilizers, and other inputs more productive.

The project also applied lessons learned from earlier projects regarding institutional approaches. Efforts were made to streamline state government operations to support the project; special offices were established at both the central and state levels in order to coordinate the administrative needs of the project.

Like other early government projects, the pilot project aimed for universal implementation of a single, centrally developed plan, with efforts made to convince local people of its merits. The project document stressed the need to adapt proven technologies to local conditions, but in practice there was little flexibility. Techniques not pre-approved under the project design were not supported. Recognizing that pasture development and afforestation work would have no lasting impact unless people supported it, the project mandated that no pasture development work would be undertaken without local people's consent. But that was the extent of efforts to gain local participation.

IWDP

When the pilot project ended, a second phase of the World Bank project was introduced in 1989-90. This project was called the Integrated Wasteland Development Project (IWDP), with separate components in the hills and plains. The plains portion of the project was undertaken in Rajasthan, Orissa, and Gujarat. The IWDP took essentially the same approach to developing rainfed crop production as the earlier pilot project. Its main difference was that it focused greater attention to developing and strengthening local organizations as the means of garnering people's participation and approval of collective action for protecting common pasture areas. The project's administrative approach was also restructured.

Project evaluations suggest mixed performance. In Rajasthan, the IWDP suffered from the same problems of poor participation and inflexible technology choice as the pilot project[8,9]. In Orissa, on the other hand, the project is said to have performed well. A second phase of the IWDP was commissioned for the Himalayan foothills in 1999; increased transfer of control to local community groups is an important stated goal[10].

National Watershed Development Project for Rainfed Areas (NWDPRA)

The NWDPRA is one of the most important programmes as it seeks to usher in agricultural development in the resource poor vast tracts of rainfed and dry lands. This programme has been initiated as the Ministry of Agriculture's counterpart to the World Bank funded pilot project and IWDP. Similar in approach to the World Bank projects the NWDPRA promotes the same low-cost vegetative bunding techniques and contour based cultivation. Vegetative and other low costs measures are also used on nonarable lands. The NWDPRA works on a watershed basis, where watersheds do not necessarily correspond to village boundaries. The NWDPRA watersheds are smaller, ranging from 500 to 5,000 hectares, or about 5 to 20 per cent of the area of the World Bank watersheds. As a result, the NWDPRA watersheds typically cover one village entirely or nearly so, plus parts of one or two neighboring villages. This approach is considered to make the most sense from a land and water management perspective, but it raises administrative complications because project staff has to deal with multiple village administrations in one relatively smaller area. Also, organizing local institutional arrangements for managing nonarable common lands is complicated when working across village boundaries. The World bank pilot project and IWDP shared this problem and had difficulty making it work[11].

The NWDPRA is centrally funded and operates through state-level agriculture or watershed development departments. The Seventh Plan centrally sponsored scheme of NWDPRA was modified on the basis of findings of evaluation studies for implementation in the Eighth Five Year Plan and during the plan period, the project was launched in 25 states and 2 Union Territories. The project was continued during the Ninth Five Year Plan in 28 states (including 3 newly formed states of Uttaranchal, Jharkhand and Chattisgarh) and 2 Union Territories with the purpose of increasing agricultural productivity and production in rainfed areas through sustainable use of natural resources.

The broad objectives of NWDPRA are as follows:

- Conservation, development and sustainable management of natural resources including their use.
- Enhancement of agricultural productivity and production in a sustainable manner.
- Restoration of ecological balance in the degraded and fragile rainfed eco-system by greening these areas through appropriate mix of trees, shrubs and grasses.
- Reduction in regional disparity between irrigated and rainfed areas.
- Creation of sustained employment opportunities for the rural community including the landless.

Operation of NWDPRA Project

The guidelines of NWDPRA reflect that from each village of the watershed, three Mitra Kisans (contact farmers) will be selected to form the core group of farmers for acquiring skills and knowledge and to disseminate the same to other farmers. One of the three selected farmers should be an educated landless agricultural labour to be called Gopal. New Mitra Kisans will be selected every year. Mitra Kisans will be looking after private property resources management and Gopal will be for common property resource management (common grazing land). Institutional contact of Mitra Kisan Mandal (MKM) should be through training at nearby centres. The project will operate either through the Panchayat or an informal group to participate in project survey, planning implementation and post project maintenance of community assets. Community prices would be given for active participation in maintaining common property resources, besides provision of reserve fund for promoting participation through prizes. Voluntary agencies, if available, would be involved in awareness creation, group formation training and preparation of training materials.

Project Performance

During Eighth Five Year Plan, against a target of 28.00 lakh ha., an area of 43 lakh ha. was developed through implementation of the project spread over in 2,554 watersheds with an expenditure of Rs. 971,52,00 thousand.

During Ninth Five Year Plan, NWDPRA had been considerably restructured with greater decentralization and community participation, higher degree of flexibility in choice of technology and suitable institutional arrangements for ensuring long-term sustainability. During Ninth Five Year Plan, it was targeted to treat 22.5 lakh ha. of rainfed areas at a cost of Rs. 1,02,00,000 thousand. An area of 27.66 lakh ha. has been treated against a target of 22.5 lakh ha. with an expenditure of Rs. 91,08,063 thousand during IXth Five Year Plan period. The scheme of NWDP has been subsumed with Scheme for Macro-management of Agriculture-Supplementation/Complementation of the States efforts through work plans from 2000-01.[12]

The NWDPRA project guidelines mentioned the issues of people's participation and institution building, but it presented no clear strategy and only a small budget for addressing them. It appeared that the project's intentions were in line with modern views about the benefits of participation but that the mechanisms for ensuring them were not fully developed.

Restructuring of NWDPRA

The NWDPRA has been thoroughly restructured by retaining the technical strength of the older programme and incorporating the lessons learnt from successful projects, especially on community participation. The watershed development programme is now planned, implemented, monitored and maintained by the watershed communities. To bring about uniformity in programmes being implemented by various agencies, the WARASA-Jan Sahbhagita guidelines have been brought out in conformity with the "Common Approach/Principles for Watershed Development" agreed upon by the Ministries of Agriculture and Rural Development. The salient features of the restructured project are as follows:

- Revision of cost norms from Rs. 3,500 to Rs. 4,500/ha (<8 per cent slope) and from Rs. 5,000 to 6,000/ha. (>8% slope)
- Introduction of flexibility and revision of component-wise allocation of resources
- Development of Sustainable Watershed with different Project Implementating Agencies (PIAs)

- Broadbasing of Watershed Development Team (WDT) for better community mobilization
- Innovative mechanisms for utilization of research funds earmarked for watershed technology
- Thrust on Transfer of Technology
- Provision of Farm Pond Technology
- Cluster Plantation under Horticulture/Agro-forestry.
- Review of Livestock Components.
- Change in Mode of Appointment of Apprentice.
- Enlarging role of NGOs and Panchayats.
- Management of Common Property Resources.
- Project benefit and cost sharing.
- Improvement in Monitoring and Evaluation (M&E) through concurrent evaluation.
- Impact assessment through development of realistic quantifiable indicators.
- Capacity building through training and orientation.

Watershed Development Fund (WDF)

As a follow up action to the Union Finance Minister's budget (1999-2000) speech, a Watershed Development Fund (WDF) has since been established at NABARD with the objective of integrated watershed development in 100 priority districts in 14 States through participatory approach. The total corpus of the WDF is Rs. 200,00,00 thousand which includes Rs. 100,00,00 thousand by NABARD and a matching contribution of Rs. 100,00,00 thousand by Department of Agriculture & Cooperation, Government of India. The fund will be utilized to create the necessary framework conditions to replicate and consolidate the isolated successful initiatives under different programmes in the government, semi-government and NGO sectors. The WDF is being operationalised in close coordination with the Central and State Ministries and Departments as a continuum of their efforts but with a distinct identity, and contribution by NABARD and the Government of India financially and managerially. Ministry of Agriculture is the nodal ministry for this purpose.

The activities under the WDF are being taken up under the guidance of a High Powered Steering Committee constituted with representatives from the Ministry of Agriculture, Ministry of Rural Development,

Ministry of Finance (Banking Division), representatives of State Governments, NABARD and a selected NGOs from different States. The secretarial support is provided by NABARD.

Utilizing WDF mode of assistance by NABARD, it is contemplated to have a spread of the programme in 14 States, namely, Rajasthan, Haryana, Gujarat, Madhya Pradesh, Orissa, Maharashtra, Andhra Pradesh, Karnataka, Tamil Nadu, Bihar, Uttar Pradesh, West Bangal, Himachal Pradesh and Jammu & Kashmir. In the first phase the WDF is accessible to 6 States, namely, Andhra Pradesh, Maharashtra, Gujarat, Madhya Pradesh, Orissa and Uttar Pradesh. In all, about 100 districts are to be covered with WDF mode of assistance.

So far, Memorandum of Understanding has been signed by 9 States. They include Andhra Pradesh, Gujarat, Karnataka, Maharashtra, Rajasthan, Tamil Nadu, Uttar Pradesh, West Bengal and Jharkhand. State Steering Committees have been constituted in all the 8 States except Jharkhand.

The progress indicates that 300 watershed have been identified in 8 States for WDF facility out of which 183 watersheds have been sanctioned and preliminary work relating to social mobilization, exposure and orientation of the community and preparation of capacity building phase proposals are under progress in the remaining watersheds. The total grant involved in the 183 projects so far is of the order of Rs. 8,41,000 thousand. This comprises 35 grant based and 148 loan based projects. Loan based projects involve a commitment of Rs. 84,60,000 thousand.

Externally Aided Projects

A number of externally aided projects on the theme of Integrated Watershed Development are also being implemented during the Tenth Five Year Plan. Ministry of Agriculture is involved in the supervision, coordination and monitoring of these Projects. The funding for these Projects is directly given to the State Governments and is not routed through the Ministry.

Integrated Watershed Development Project (IWDP-Hills-Phase-II)

On the completion of the IWDP-Hills Phase-I on 31.3.99 utilizing IDA Credit of US $ 67.55 million accomplishing treatment of an area of 1.96 lakh ha, the IWDP-Hills-Phase-II had been launched during 1999 at an estimated cost of US $ 193.00 million equivalent to Rs. 954,60,00 thousand (The IDA Credit/Loan will be US $ 135.00 million comprising IDA Credit of US $50 million and a loan of US$ 85 million) with the objective to restore the productive potential of the Shivalik Hills in the States of Haryana, H.P., J & K, Punjab and Uttaranchal using cost effective watershed treatment technologies and community participation approach. The project approach and design of the IWDP (Hills-II) Project differed from the approach of previous project in placing greater emphasis on:

- Participatory process following 'bottom up' approach ensuring greater participation by stakeholders and effective project implementation and sustainability. Promotion of effective community involvement in selecting treatments, participating in the sustainable application of such treatments and long-term management of assets created through the project.
- Emphasis on social mobilization to ensure that benefits reach more marginalized and vulnerable elements of the rural population.
- Developing rural infrastructure including roads and market facilities etc.

The overall programme is to treat an area of 2.82 lakh ha. with an expenditure of Rs. 9,54,60,00 thousand. Till the end of October, 2002, an area of 1,33,944 ha. had been treated incurring an expenditure of Rs. 385,82,00 thousand. An amount of SDR 36.900 m. (constituting 100 per cent of total IDA credit of SDR 36.900 m) and US$ 2,358 m (constituting around 2.77 per cent of the IBRD loan of US$85,00m) had been reimbursed till September, 2002. The programme for 2002-03 was to incur Rs. 180,61,00 thousands to treat 47,084 ha.[13].

Scheme of Watershed Development Project in Shifting Cultivation Areas (WDPSCA)

The scheme of Watershed Development Project in Shifting Cultivation Areas (WDPSCA) was launched in seven North Eastern States during Eighth Plan from 1994-95 with 100 per cent as Special Central assistance to the State plan. The scheme aims at overall development of jhum areas on watershed basis. During VIII Five Year Plan, an area of 0.70 lakh ha. had been treated with an expenditure of Rs. 3,15,100 thousand.

During IX Plan, an amount of Rs. 7,50,000 thousand was released to treat 212 micro watersheds and an area of 1.59 lakh ha. have been treated at a cost of Rs. 8,20,000 thousand. During the year 2002-03, it was proposed to treat an area of 0.20 lakh ha. at a total cost of Rs. 2,00,000 thousand. During this Plan, the programme was implemented using the revised guidelines based on specific terrain, hydrologic and socio-economic conditions and common approaches to watershed development which came into effect from November, 2000. Implementation of this programme lays emphasis on adoption of low cost technology with the use of local material for soil and water conservation such as vegetative measures, improved production systems and construction of small water harvesting structures. It also includes components of live-stock development and house-hold production with due emphasis on beneficiary participation and NGOs involvement[14].

Impact of Watershed Management

It is almost more than two decades that the integrated watershed management projects are in operation in the different states in India. In many of the project areas, the integrated watershed approach proved effective for recharging aquifers, mitigating drought, moderating flood, reducing runoff and soil loss and enhancing social and environmental protection, It increased productivity of arable and non-arable lands, and rehabilitated degraded ecosystems on a sustainable basis. It was an effective tool for community participation and empowerment of village institutions to manage common property resources with benefits to all sections of society. Watershed Management programmes generated employment opportunities with positive impact on women. It proved sound economic viability with overall development. This approach can meet the challenges of food security for ever-increasing population. The positive results in different areas are depicted below.

(a) Aquifers Recharging

Watershed Management programme brought out an increase in groundwater table by recharging aquifers in different regions of the country. Rise in water table ranged from 0.8 to 2.0 m and was spectacular in low rainfall regions (Table 13.1).

Table 13.1: Impact of Watershed Management on Aquifers Recharge in different regions of India

Watershed	*Surface Storage Capacity Created (ha-m)*	*Rise in Groundwater Table Over Pre-project (m)*
Haryana		
Bazar-Ganiyar	79.0	2.0
Bunga	60.0	1.8
Himachal Pradesh		
Behdala	18.0	1.0
Rajasthan		
Chhajawa	20.0	2.0
Andhra Pradesh		
Chinnatekur	5.6	0.8

(b) Drought and flood Moderation

The Watershed Management programme proved its potential for drought mitigation in rainfed agriculture resulting in conservation of natural resources. A comparison during the severe drought of 1987 showed that the percentage loss in net area sown ranged from 0 to 35 per cent inside the watershed compared with 40 to 85 per cent outside the watershed. The reduction in yield ranged from 0 to 40 per cent within watershed and 40 to 75 per cent outside the watershed.

Integrated Watershed Management reduced the runoff and soil loss, and thus indirectly moderated flooding of downstream area. There was reduction of 7-60 per cent runoff and 17-81 per cent soil loss, which ultimately conserved natural resources (Table 13.2).

Table 13.2: Impact of Watershed Management on Runoff and Soil Loss (Flood Moderation)

Watershed	*Runoff (% of rainfall)*		*Soil loss (tonne/ha.)*	
	Pre-project	*Post-project*	*Pre-project*	*Post-project*
Uttaranchal				
Fakot	42.0	14.2	11.9	2.5
Karnataka				
G.R. Halli	14.0	1.3	3.5	1.0
Himachal Pradesh				
Behdala	30.0	15.0	12.0	8.0
Una	30.0	20.0	12.0	10.0
Karnataka				
Joladarasi	20.0	7.0	12.0	2.3

Source: J.S. Samra, Integrated Watershed Management.

(c) Environmental and Ecological Impact

Management of watershed with soil and water conservation measures, afforestation and silvipastoral system influenced the hydrological regimes, lean flow and ultimately resulted in improvement of environment. Conversion of natural Shola forest to an Eucalyptus plantation in the Nilgiri hills reduced total runoff by 16 per cent and 28 per cent during first and second rotations, respectively. Management of 4,207 ha of the Sukhna lake (Chandigarh) catchment in the Shiwalik hills reduced runoff from 29 to 7 per cent and sediment load from 140 to 18 tonne/ha over 20 years.

Similarly a limestone mined watershed of 64 ha at Sahastradhara in Doon valley treated with geojute, bioengineering and mechanical measures as Integrated Watershed Management approach resulted in controlling mass soil erosion of 550 tonnes/ha to 8 tonnes/ha/year, reduced runoff from 57 to 37 per cent, increased lean flow of quality water from 60 to 240 days and improved vegetative cover from 10 to 80 per cent in about 8 years. Besides, the State Public Welfare Department is saving about Rs. 0.1 million/year which they used to spend on removing debris to clear the road.

The programme adopted for the ravine eco-system showed that degraded ravine can be stocked with lush green fuel fodder plantation following Watershed Management programme. Shallow ravines may be terraced and field bunded with safe water disposal structures.

Medium ravines can be utilized for rainfed horticulture. Deep ravines were restored vegetatively by bamboo plantation and gully plugging, generating good biomass from ravine ecosystem.

(d) Sustainability and food security

The participatory watershed based conservation and management of natural resources exhibited long-lasting, pro-people and environment-friendly impact when monitored after withdrawal of active support or interventions. In the ORP watershed located at Fakot in the Uttaranchal, the production and environmental positive indicators continued to increase several folds especially in those cases where active community participation was involved from the beginning. Continued increase in the production of food, fruit as well as milk, decline in runoff and soil loss and dependency on forest fodder even after withdrawal of active support indicated food security and sustainability of the programme. Watershed community diversified from traditional low yielding but less risky crops of Mandua and Jhingora to other fruits, vegetables and floriculture.

(e) Community Participation

The participation of community and empowerment under Relmajra (Punjab) watershed showed that people continued to use their own resources for land development activities even after project period. Similarly, water harvested from 175 ha. catchment and irrigation in 243 ha. of the command area in Bunga (Haryana) watershed managed by registered society of the village community developed a sense of ownership and social fencing among the people. The society generated its own total income of Rs. 0.8,50,7661 million from various sources and invested Rs. 0.7,42,425 million on the common activities, and thus resulted in net income of Rs. 0.1,08,336 million over 15 years.

The incentive-based participation of people and empowerment for the management of forests was introduced in the Sukhomajri (Haryana) watershed. The productivity of the community forest land was improved by social fencing encouraged by rainwater harvesting. In situ moisture conservation and equitable sharing of benefits from the net income of Rs. 0.2,56,2666 million was realized over 14 years.

(f) Gender Issues

Watershed approach increased total working hours for the women. A case study of Bunga watershed showed that investment of their time on drudgery items of collecting fuelwood, drinking water, fodder and grazing animals decreased by 132 per cent and on child care by 32 per cent. Accordingly, the time spent or economically beneficial activities increased by 194 per cent in animal husbandry and 148 per cent in agriculture. The total time spent per day increased from 12.5 to 13.7 hr.

Critical Issues

There are no two opinion about the potential benefit of the programme provided it could be implemented as per the guidelines. But inspite of the elaborate planning and allocation of big funds, the programme is not achieving the stipulated success. In most of the locations, the programme is running on adhoc basis. After the government agencies remove their support, the programme collapses. The innovative idea of ensuring people's participation remains unsolved. The soil and water conservation goals of watershed projects are of long terms. For the sustainable development of the watershed programme, the people living in the watershed who are responsible for the situation, who are affected by the situation and who would be benefited by the programme, become very important and they should definitely be involved in the programme right from the planning to implementation and monitoring. When we critically look into on going programmes, we find that there are only a very few success stories on this count. There is a need of critical evaluation of the successful as well as not so successful-watershed development projects and identification of the underlying issues related to people's participation for sustainable development of watershed.

The Mid-Term Appraisal of the Ninth Plan highlighted some of the major concerns in the implementation of watershed programmes as below:

- Lack of people's participation.
- Field staff unfamiliar with participatory approaches.
- Insecurity about fund availability at the grass root level.
- Limited time for preparatory activities.

- Little emphasis on cohesive group formation.
- Lack of transparent criteria for selecting areas and villages.
- Limited human resource capabilities.
- Lack of involvement of senior government functionaries and line agencies.
- Week horizontal linkages among various agencies at the district level.
- No protocol exists for withdrawal after project completion.
- Plethora of watershed development programmes with different guidelines and cost norms.

Source: Mid Term Appraisal of Ninth Five Year Plan.

Participatory Experiences

We have mentioned earlier that people's participation is a prerequisite for the successful execution of watershed management programme. In fact, there is a need for critical evaluation of the successful and not-so-successful watershed development projects and identification of the underlying issues related to people's participation for sustainable development of watersheds. Participation as a process is a dynamic, non-quantifiable and essentially unpredictable element. As a process, participation will change the life of the project into a permanent dynamic movement. Participation may not be seen merely as a managerial technique, but more as a technique to 'enable' rural people to have a more direct involvement in development projects. The critical elements in this process are those of awareness creation, local capacity building, liberation and empowerment. Participation if mobilized in this context, leads to greater control of the poor over their own life situation. Through the acquisition of knowledge and local capacity building, they will be able to understand the cause of their poverty and are in a better position to mobilize and utilize resources available in order to improve their situation either through individual or collective efforts. More generally, participation as an end involves the notion of "bottom up" or the generation of influence or involvement from the grass roots level to cause change in the existing institutional arrangements.

People organize best around the problems they consider the most important in assessing needs in planning initiatives to meet those needs. They make rational economic decision in the context of their own environment and circumstances. Voluntary commitment of labour, time, material and money to a project is both an evidence of participation and necessary condition for breaking pattern of development paternalism. Local control over the amount, quality and distribution of benefits increase self sustainability.

There are many factors, which make the watershed development a successful programme. People's willing cooperation is definitely the basic one. Watershed development is essentially a group action programme involving the entire population of the watershed. In order that the programme gets going and achieves the envisaged results, it is imperative to enlist people's active participation in all the phases of planning and implementation of the programme. Therefore, guidelines of the programme need to be honestly pursued ensuring people's participation.

The people's participation in the development programmes in the past has been sought through establishing self-reliant groups, farmer to farmer exchange of ideas and through developing local action groups.

The Agha Khan Rural Support Programme (AKRSP) of Gujarat, the Society for People's Education and Economic Change (SPEECH) in Tamil Nadu, the Social Center in Maharashtra, the Youth for Action (YFA) in Andhra Pradesh and the participative integrated development of watersheds by MARYADA in Karnataka are some of the examples where the NGOs are successfully running watershed development projects. Another much talked about example of successful watershed is that of the Ralegon Sidhi in Ahmadnagar district of Maharashtra started by Shri Anna Hazare. If we critically analyze the reasons behind the success of these projects in these selected areas, then the common factor that emerges out is the people's participation, that is the involvement of the local community in the programme.

In Gujarat PRA techniques were used by the extension volunteers selected from the village and the village institutions helped in organization of the activities of the project. In Tamil Nadu, non-formal classes were used to create awareness and for organizing people in village 'sanghas' or committees to carry out the programme activities. Local songs and stories were used to mobilize people. In Maharashtra, local people themselves were involved in surveying the village resources and planning the programme activities. In Karnataka, local people were organized in subgroups who were given separate responsibilities and formed their own rules and regulations to carryout the programme. Now, all these success stories emphasize the point that people's participation in watershed management is a must for its success and sustainability. At the same time we are also aware that in majority of the programmes, we are not able to utilize local people and their involvement in the project. Therefore, it becomes important to analyze the issues, which should be dealt with to evoke people's participation in watershed development programmes. Some selected analyses are included here.

Jha (1997)[15] studied people's participatory process in a few watershed projects namely Sukhomajri and Bunga in Panchkula district of Haryana and Dasuya Langerpur and Balachaur in Hoshiarpur district of Punjab. Some of the common problems identified related to people's participation in the programme of watershed development and are listed below:

- Unequitable distribution of benefits from the project.
- Closure of forest for grazing and fuel wood.
- Marginalization of women and landless in the project.
- Lack of co-ordination among the departments.
- Lack of sustained efforts in soliciting people's participation.
- Lack of proper orientation and experience of project personnel
- Shorter period of project and larger area of operation.
- Unresolved legal issue.
- Presence of panchayat politics.
- The vested interest of some in the management of village level institution.

In 51st Annual Conference of Indian Society of Agricultural Economics held in 1991 at Hyderabad, 40 papers were submitted focusing on economic analysis and functioning of watershed projects. Some of findings relevant to the present context are:

- There is lack of adequate information about the resources and requirement of the people of the watershed area for proper planning of the project.
- People and locations are two most important factors that should be considered while designing a watershed project.
- Involvement of people in the project is possible by considering their needs and major constraints faced by them.
- The sustenance of the flow of benefits as well as impact parameters are direct functions of community participation.

- Policy modifications for improving land and water management, input supply, credit and insurance, extension education and marketing are useful for free development of the watershed programme.
- Watershed policy thrust has been more on crisis management than on developing risk aversion mechanisms.
- Weaknesses of the programme include lack of proper comprehensive plan, lack of coordination among departments, lack of involvement of local people and organizations, lack of motivated staff and flexibility in the use of funds.
- Active people's participation in watershed development depends on expected benefits and costs of participation, organization of farmers in small groups, honest and good local leadership, existence and enforcement or rules for fair and equitable benefits from collective action, involvement of non-governmental organizations in organizing training and motivating the farmers, besides willingness and ability of the government to provide financial and technical support.
- Real benefit of the programme will be known only in the long run.

Based upon the analysis of different watershed development and research studies, the important inferences that can be drawn to increase people's participation are:

- The Government at the outset must clear the legal and administrative hurdles particularly with reference to sharing of benefits from common property resources for soliciting people's participation, it should be on the mutually agreeable term.
- The project leader at the watershed level must have administrative and financial control over the various line departments involved in the project for better and timely delivery of goods to the people. Such delegation of authority will act as stimulant to have appropriate accountability in programme implementation.
- The gender issue is an important factor in people's participation. Thus, the project team should essentially include

women personnel for handling gender problems and organizing them in the watershed.

- Induction training of project staff on participatory methodology of development should be the prerequisite of the programme. Their orientation and training are of paramount importance.
- The size of the watershed project should be manageable, preferably of 2-3 villages, in view of the sociological aspects involved in the process of people's participation.
- Development of local leadership is essential for the project sustainability. Their identification and training on participatory management of watershed is important. A concerted effort is required to identify the leaders having commitment for the social cause.
- The period of the project duration should be flexible, as it requires continuous and sustained efforts linking people's interests with the project's interests. The time required may vary depending upon the local social situations.
- Adequate awareness creation should be the starting point in the project implementation. The extension methods such as frequent meetings, farmers to farmers contacts, field visits and training etc. need to be appropriately used for the same.
- There should be proper identification of felt-needs of the people. Project activities should be planned accordingly. The activities must encompass income and employment generation by utilizing common resources under agroforestry, fisheries etc.
- The incentives or benefits to the people are essential to ensure their participation. They should be quick, substantial and for long period.
- Developing a strong village level non-political institution for collective action is a must. The village level institution should have autonomy to form bylaws and also to amend them in accordance with the changing perception of the people.
- To avoid social tension and for equitable distribution of benefits, the poor and landless must be fairly treated. They may be helped to start small scale enterprises, composite nurseries etc. through organizing self-help groups of people having similar interests.

- The village level institutions must recycle the social fund for generating wealth to invest in the common cause for sustainable collective action. Certain amount of social fund must be earmarked for the development of weaker section by way of providing wage employment on regular basis to them.
- The village level institutions must be empowered to resolve conflicts and guidance by external neutral agency must be provided even after the project. The presence of a village level functionary for institutional development would be desirable.
- The continuous social research support in identifying the basis of collective action must get due place in the entire process. The result would be essential when one benefit around which the institution has been formed does not last long.

A Success Story of Watershed Management from Abroad

Restoring the Chesapeake Bay

It has been stated earlier that watershed management is complicated and can be contentious because watershed boundaries do not coincide with those of political or administrative jurisdictions. Thus management involves many levels of government and many different communities, each with its own constituencies and concerns. Balancing the multitude of interests involved is so time consuming and troublesome that watershed management rarely succeeds.

It can be done, however, as shown by the example of the Chesapeake Bay, North America's largest brackish water estuary. Managing the Chesapeake watershed presents unique and seemingly insurmountable obstacles. Although the Bay itself covers only 2,200 square miles (5,500 square kilometers) and averages just 21 feet deep (6.5 meters), its watershed is vast, sprawling over 64,000 square miles (160,000 square kilometers) through six US states. Moreover, 15 million people live in the watershed, which is one of the most heavily populated areas on the East Coast.

By the early 1980s planners were confronted with the consequences of decades of accumulated abuse of the water resources in the Chesapeake watershed. Signs of decay were everywhere:

- By 1987 the Bay was receiving about 184,000 metric tons of nitrogen and 74,000 metric tons of phosphorus a year from human and animal wastes and commercial fertilizers.

- The Bay had lost more than half of its natural tidal and nontidal wetlands and 40 per cent of its watershed forests.
- Seagrass beds, which once covered over half the Bay's bottom, amounting to several hundred thousand acres, had been reduced to no more than 34,000 acres (14,000 hectares) by 1984—10 per cent to 20 per cent of their original area.
- Chronic overharvesting, along with mounting pollution and habital degradation, contributed to the dramatic decline in the Bay's fish and shellfish populations. Commercial catches of rockfish (striped bass) dropped from 2,608 metric tons in 1970 to just 272 metric tons in 1983. The Bay's oyster fishery had plunged catastrophically. From an all-time peak of 20 million bushels in 1884, the oyster harvest declined to just 168,000 bushels in 1992.

A handful of local environmentalists launched the initiative to save the Bay in the late 1960s. At the time few people anticipated their efforts would blossom into a 6-state watershed-wide program to restore the Chesapeake. Eventually, the effort would become a multimillion dollar restoration that would involve local, state, and federal agencies and tens of thousands of people from all walks of life.

A nonprofit membership organization, the Chesapeake Bay Foundation, was one of the key groups urging state government to take a comprehensive watershed approach to the Bay's environmental problems. After a slow start in the early 1980s, the three key state governments-Maryland, Virginia, and Pennsylvania—with help from the US Environmental Protection Agency (EPA), agreed to launch an ambitious, integrated cleanup programme.

In 1987, when the EPA and the governors of the three main watershed states met at a major strategy conference, nothing on this scale had ever been attempted before in the US. The result was the Chesapeake Bay Agreement between the states and the federal government. The agreement provides solid political commitment from all watershed states for a 40 per cent reduction in nutrient loadings to the Bay by the year 2000, using 1985 as the base year. "The year 2000 goal is a permanent cap on emissions of nitrogen and phosphorus," explains William Matuszeski, Director of EPA's Chesapeake Bay

Program in Annapolis, Maryland. It is, in essence, a nondegradation policy for the entire watershed, in which each major tributary of the Chesapeake will reduce nutrient pollution by 40 per cent.

In order to meet such an ambitious target, each state crafted its own pollution reduction strategy, working within the terms of the overall agreement. The Pennsylvania legislature even passed the first nutrient management law in US history. Although neither Maryland nor Virginia took binding legislative steps, both states have encouraged farmers in the Chesapeake watershed to adopt "best management practices," which involve low-till or no-till agriculture, along with the reduced use of pesticides and a marked reduction in the application of chemical fertilizers, one of the main sources of nutrient pollution to the Bay.

The watershed states have made progress in reducing point sources of pollution-effluent from sewage treatment plants and industrial complexes. Maryland, Virginia, Pennsylvania, and the District of Columbia also have passed legislation to ban the use of phosphate detergents. These actions reduced phosphorus pollution to the Bay by 40 per cent between 1985 and 1994.

The main improvement, states Matuszeski, is that "we have halted the increase in nitrogen pollution to the Bay in spite of a growing population". This has been the result of building-3 stage sewage treatment plants around the Bay and decreasing the amount of animal manure and fertilizers seeping into the region's surface waters.

According to the EPA, many farmers in Maryland, Pennsylvania, and Virginia have cut fertilizer use dramatically, and some farmers use no commercial fertilizers at all, relying instead on animal manure and crop residues. The use of pesticides has dropped 20 per cent since 1985. The Bay is not free of problems, as evidenced by the 1997 outbreak of the toxic organism Pfiesteria piscicida, which kills marine life and degrades water quality. Still, Matuszeski thinks that progress in meeting watershed management goals will be faster "once we can get natural ecosystems to start working for us, like wetlands and seagrasses". Progress has been slower than expected, but most trends are in the right direction[16].

Panchayats and Watershed Programmes

The Prime Minister, Atal Behari Vajpayee, inaugurated Haryali, a rural programme for watershed development during the National Conference of State Ministers of Rural Development, Panchayat Raj and Public

Works (February, 2003). This initiative comes close on the heels of another scheme, 'Swajaldhara', which envisages peoples' participation in management of water supply in rural areas. The conference was held to implement the progress of ongoing Centrally-sponsored rural development programmes. Announcing this here, the Minister for Rural Development, Shanta Kumar, said 'Haryali' was being launched to strengthen Panchayati Raj institutions for implementation of watershed programmes such as the Drought-Prone Areas Programme, the Desert Development Programme and the Integrated Wastelands Development Programme.

The panchayats would contribute 10 per cent of the cost of construction of project, while 90 per cent would be funded by the Center. Through these programmes, panchayats would have to create regular income for themselves from rainwater mangement, construction of check dams, irrigation, fishery, forestry and horticulture. The Center has set aside Rs. 1,000 crores for the 'Haryali' programme, of which Rs. 885 crores will be borne by the Ministry of Rural Development and the rest by the States. Under the initiative, the gram panchayats would execute watershed projects with technical support from the block panchayat-zilla panchayat which would act as the Project Implementation Agency (PIA). In the absence of technical staff, "a suitable government line department or autonomous agency can be identified for technical assistance. However, if even this fails, then a non-government organization (NGO) of standing with adequate technical expertise and experience would be identified to act as the PIA. Representations of Self-Help Groups (SHGs) and User Groups (UGs) (especially women) in the standing committee of gram panchayats have been proposed for this initiative. Besides, the ongoing project undertaken by NGOs would continue.

Single Roof for all Watershed Development Projects

The Union cabinet has the unenviable task of deciding which ministry should take charge of watershed development. Different watershed development programmes are currently scattered among the agriculture, rural development and environment ministries. "Every department has its own views," says Shanta Kumar, rural development minister. The aim of putting all the different watershed schemes and programmes under one roof is to streamline funds and avoid duplication. Once integration has been achieved at Central level, the government would be keen to see this replicated in states where, again, departments as varied as environment, agriculture, rural development and soil conservation all have watershed development projects with them.

This integration plan was first mentioned in the President's address to Parliament in 2000. To give an example, the agriculture ministry has a decade-old national watershed development project for rainfed areas, an integrated watershed development project for hilly areas and watershed development fund in NABARD. The fund has a Rs. 200 crore corpus for integrated watershed development in 100 districts in 14 states. The rural development ministry has three projects: The desert development programme, the drought-porne areas programme and the integrated wasteland development programme, all using watershed development as the base. The environment ministry's National Afforestation and Eco-development Board focuses on improving degraded forest areas. No department wants to give up an inch of its territory.

Tenth Plan Strategy-'Watershed Plus'

During the Tenth Plan period, Watershed Development Programme would be an important component of anti-poverty strategy. The development wastelands, degraded lands and rainfed areas would be implemented in an integrated manner. It has been recognized that the watershed programmes are important for bringing land area under tree cover and the Tenth Plan has set an ambitious target for afforestation. Nearly 30 million hectares of non-forest land would have to be brought under tree cover to achieve the afforestation programme. The trend of afforestation indicates that the forest cover may fall short of target. It is necessary to create an environment that encourages people to grow trees on their farm land. It is also necessary to promote vigorously farm-forestry and agro-forestry on marginal lands.

The Planning Commission states that Watershed Programmes would require significant upscaling during the Tenth Plan to prevent further land degradation and for the restoration of carrying capacity of lands that are classified as degraded. In addition, uplands in low rainfall areas will also require water and soil conservation measures. It is estimated that 88.5 million hectares of degraded land have to be treated under watershed programmes in the Tenth and subsequent plan period. The Planning Commission has also prepared a 20 year Perspective Plan for Watershed Development.

The Perspective Plan emphasizes the fact that the watershed development programme would have to become a people's movement for its success. The 'Watershed Plus' which recognizes the need to involve the community as a necessary condition for the sustainability of a watershed programme should be the approach for implementation.

The guidelines seeking to operationalize the Watershed Plus philosophy are as follows:

- A programme-specific and focused project approach.
- Greater flexibility in implementation.
- Well-defined role for state, district and village level institutions.
- Removal of overlaps.
- A provision for keeping the watershed development projects on probation.
- An exit protocol for the project implementing agencies (PIAs)
- A twin track approach that provides for short term and long term benefits in the implementation of projects.
- A combination of government organizations/NGOs as PIA
- A greater role for women
- An effective role for the panchayati raj institutions (PRIs)
- Bringing self-help groups comprising rural poor, especially those belonging to SC/ST categories to the forefront
- Establishing a credit facility from financial institutions
- Transparency in implementation
- Effective use of remote sensing data furnished by the National Remote Sensing Agency.

Conclusion

The Green Revolution that transformed irrigated agriculture elsewhere in India had little effect in the rainfed semi-arid regions, where agricultural productivity is low, natural resources are degraded, and the people are poor. In the 1980s and 1990s, agricultural scientists and planners aimed to promote rainfed agriculture through watershed development. By capturing scarce water resources and improving the management of soil and vegetation, watershed development is assumed to have the potential to create conditions conducive to higher agricultural productivity, while conserving natural resources. Watershed development programme has become popular among development planners and agricultural scientists because it promises a win-win situation in which natural resource conservation and agricultural productivity are complementary. It is often assumed that watershed development will also have favourable implications for poverty alleviation, but it may actually impose hardships. In fact, watershed projects can actually make women and landless people worse off by restricting their access to resources that contribute to their livelihoods.

Much of the hype surrounding the country's watershed development draws, unfortunately, on a small number of extraordinary success stories that may be difficult or impossible to replicate on a large scale. Hence, it is necessary to evaluate a large number of randomly selected projects through NGOs in order to see beyond these unusual successes. Further, watershed development has evolved from purely technical, externally imposed intervention in 1980s to a more participatory exercise in which local people help design and implement management plans. The IFPRI study on Watershed Development Projects in India—An Evaluation (Research Report 127) reveals that in Maharashtra and Andhra Pradesh (where the study was undertaken), the participatory projects performed better than their technocratic, top-down counterparts. Other studies also indicate that those projects which take a more participatory approach and are managed by NGOs have performed better in conserving natural resources and raising agricultural productivity. But such success may come at the expense of the poorest people in watershed areas because improving management of a watershed usually requires restricting to the natural resources base on which they depend. Hence, the Integrated Watershed Development approach should also take into consideration this human aspect very seriously. Otherwise, those whose interests are harmed will naturally refuse to go along with the effort.

Finallly, lack of existing monitoring and evaluation data for watershed projects is hitherto an important roadblock. Needless to say, monitoring and evaluation are essential for assessing new, innovative approaches to watershed development. And this work can best be performed by suitable Non-Governmental Organizations.

REFERENCES

1. Kerr, John in collaboration with Ganesh Pangare and Vasudha Lokur Pangare: Watershed Development Projects in India—An Evaluation; Research Report 127, IFPRI, Wshington.

2. Abramovitz. J: Imperiled Waters, improverished future: The decline of freshwater ecosystems. Washington, D.C., Worldwatch Institute, March, 1996 pp. 5-66
3. Adams, W.M.: Wasting the rain: Rivers, people and planning in Africa, London, Earthscan, 1992, P. 214; Author's source: Population Report, Series M, Number 14.
4. Sarma, J.S. and Ratan Singh: Integrated Watershed Management; *Indian Farming*; December, 2002;
5. Kerr, J., N.K. Sanghi, 1992. Indigenous soil and water conservation in India's semi-arid tropics. Sustainable Agriculture Programme Gate keeper Series Paper No. 34. London: International Institute for Environment and Development.
6. Population Report, Series M, Number 14, February, 1999. The Johns Hopkins School of Public Health, Baltimore, USA.
7. World Bank, 1988: India review of rainfed agriculture and rural development, Washington, D.C.
8. ICRISAT, 1996: Economics of Pasture protection and development in Rajasthan; Interim report submitted to the Development of Watershed and Soil Conservation, Government of Rajasthan.
9. Rajasthan Agricultural University, 1999. Impact evaluation report: World Bank Integrated Watershed Development Project (Plains), Rajasthan. College of Technology and Agricultural Engineering, Rajasthan Agricultural University, Udaipur.
10. Velded, T. 2000. Participation in the Himalayan foothills: Lessons from watershed development in India. Social Development Paper No. 38. Environmentally and Socially Sustainable Development Network, South Asia, Washington.
11. ICRISAT (International Crops Research Institute for the Semi-Arid Tropics), 1996. Economics of Pasture Protection and development in Rajasthan: Interim report—submitted to the Department of Watershed and Soil Conservation, Government of Rajasthan.
12. Department of Agriculture & Cooperation, Ministry of Agriculture, Government of India, Annual Report 2002-03.
13. Ibid
14. Ibid
15. Jha, S.K. (1997): Participatory Experiences in Watershed Development Projects: Ph.D thesis in the Division of Agricultural Extension, IARI, New Delhi.
16. Population Reports, Series M, Number 14, February, 1999.

14

Interlinking of Rivers*

"The rivers... ... all discharge their waters into the sea. They lead from sea to sea, the clouds rise them to the sky as vapour and release them in the form of rain"

—Chandogya Upainshad

Introduction

Demand for freshwater is rising rapidly as population grows and becomes more urban, and as water use per capita increases. India is ranked 47th from the bottom among the 147 "Water-poorest" countries in the world in a "Water Poverty Index" prepared by the University of Keel and Institute of Hydrology at Wallingford. Some areas already face acute shortages, and more will face them in the future unless steps are taken to manage water resources better. Throughout much of the world the renewable supply of freshwater-the amount available year after year on sustainable basis—comes in the form of seasonal rains that run off too quickly for efficient use. In India, for example, except for small coastal area in the South, almost the entire rainfall occurs during three to four monsoon months. Thus, cultivation during non-monsoon months is irrigation dependent. A characteristic of the monsoon climate is variability of rainfall from year to year. India has an average of one in five below normal years.

Indian land mass receives, on an average, 4,000 billion cubic metres (BCM) of precipitation every year, 75 per cent of which is spread over monsoon months while concentrated rainfall occurs only in a period of about 100 hours duration. After providing for the losses due to immediate evaporation and ground absorption, the estimated

---

* First draft of this chapter (which has been thoroughly revised) was contributed by Mr. S.P. Malhotra.

available water resource is about 1,900 BCM. Two—thirds of this is contributed by the Ganga—Brahmaputra-Meghna (GBM) system covering one third of the country's geographical area. Consequently, the remaining two-thirds of the country has to be satisfied with the balance resource, and face the deficiencies, while the GBM basin experiences floods. Thus the main characteristic of India's water resources is its uneven distribution over space and time leading to endemic and sporadic problems of water shortages and excesses. Frequent occurrence of floods and droughts is a reflection of our failure to develop and manage the country's water resources, needless to say that water, a key element in our eco-system, is not getting enough attention except when droughts stalk the countryside and floods devastate large tracts of land habitats.

Caught between growing demand for freshwater on the one hand and limited and increasingly polluted supplies on the other, India faces a difficult choice. Whether water is used for agriculture, industry or domestic purposes, there is much room for conservation and better management. The country now needs a Blue Revolution to conserve and manage freshwater supplies in the face of growing demand from population growth, irrigated agriculture, industries and cities—just as the Green Revolution transformed India's agriculture in the late 1960's. A Blue Revolution, however, calls for coordinated responses to problems at local, national and international levels. In the existing Indian context, adopting a watershed or river-basin management perspective is considered to be a needed alternative to uncoordinated water-management policies by separate jurisdictions.

To meet increasing water demands, storage of monsoon flows is essential. Local small scale storage is useful, particularly for domestic supply, but cannot meet the full demand of high—productivity agriculture, industry or metropolitan cities. For such large-scale demand, there is no alternative to building storage reservoirs on our rivers. Many basins in the country are surplus in water resource even in the ultimate stage of development while others face serious shortages. Creation of storages and inter-basin transfer of water from surplus to deficit regions could therefore be an option for achieving more equitable distribution of our water wealth and its optimal utilization. Hence it is argued, if the resource abundant rivers like the Brahmaputra, Ganga etc., could be linked with other rivers and a National Water Grid created, the miseries of recurring droughts and floods could be contained to a great extent.

The popular appeal of interlinking rivers is based on the understanding that an enormous amount of water of our rivers flow into the sea and that if only this is prevented, and water transferred from water-abundant—rivers to water deficit areas, there will be adequate supply for everyone in every part of the country. At another level, the project is seen as promoting national integration and a fair sharing of the country's natural water wealth. The planning Commission Dy. Chairman, Shri K.C. Pant also opines that states must recognize water as a national, rather than a regional resource and there should be consensus among them on interlinking rivers as a long term strategy, in a spirit of give and take.

Whether the linking of rivers will promote integration or generate more disputes is a moot question. Besides, several obvious, but prima facie important questions about the concept, and the feasibility, desirability and viability of the proposals need to be properly addressed to before considering implementation of such a gargantuan project. The issue is now being very hotly debated. This chapter deals in details the history of the concept of interlinking of rivers, past experiences, the needs, arguments, for and against implementation of the project and other related issues.

The Need for Interlinking of Rivers

The rainfall over the country is primarily orographic, associated with tropical depressions originating in the Arabian Sea and the Bay of Bengal. The summer monsoon accounts for more than 85 per cent of the precipitation. The uncertainty of occurrence of rainfall marked by prolonged dry spells and fluctuations in seasonal and annual rainfall is a serious problem for the country. Large parts of Haryana, Maharashtra, Andhra Pradesh, Rajasthan, Gujarat, Madhya Pradesh, Karnataka and Tamil Nadu are not only deficient in rainfall but also subject to large variations, resulting in frequent droughts and causing immense hardship to the population and enormous loss to the nation. The water availability even for drinking purposes becomes critical, particularly in the summer months as the rivers dry up and the ground water recedes. Regional variations in the rainfall lead to situations when some parts of the country do not have enough water even for raising a single crop. On the other hand excess rainfall occurring in some parts of the country create havoc due to floods.

The availability of surface water in various regions in the country is uneven with the north zone (Indus, Ganga and Brahamputra) having 77 per cent of the country's water resources, the west flowing rivers with 5 per cent, the east flowing rivers with 14 per cent and the west flowing rivers in Kerala with 3 per cent only. Similar is the case with ground water, which is quite abundant in the Indo-Gangetic Plain but scarce in the western region in the south. If the ultimate irrigation potential through major, medium and minor projects/schemes in the various regions is considered in relation to the cultivable area, it comes to 116.6 per cent in the eastern region, 66.97 per cent in the north-eastern region, 84.81 per cent in the northern region, 64.37 per cent in the southern region and 58.58 per cent in the western region. As against this scenario, the total creation of irrigation potential, as per cent of respective ultimate irrigation potential these regions, as estimated at the end of the Eighth Plan works out to 53.24 per cent, 28.65 per cent, 95.32 per cent, 54.59 per cent and 39.95 per cent respectively with 64 per cent at the national level. The above data show that only the northern region is ahead of the national average. The 9th plan document indicates that against the national per capita (based on 1991 census) annual availability of 2,214 cubic metres (cu.m) of water, the average availability in Brahmaputra river basin is as high as 18,470 cu.m while it is as low as 383 cu.m in the east flowing rivers between Pennar and Kanyakumari[1]. Having 8 per cent of the country's population, Rajasthan has just 1 per cent of the country's water resource and thereby the per capita water use in Rajasthan (1991) is estimated to be only 562 cu.m., a level almost near to "absolute scarcity". On the other than hand, about 40 per cent of the utilizable surface water resources are in the Ganga-Brahmaputra-Meghana system. With 5.9 per cent of geographical area and 3.2 per cent of population of the country, Brahmaputra sub-basin alone, has 29 per cent of annual water resources. The per capita annual availability for the rest of the country excluding Brahmaputra basin, works out to about 1500 cu.m, which falls under "water scarcity" category. The Cauvery, the Pennar, the Sabarmati and the east flowing rivers are some of the basins, which fall into the category of "scarcity conditions".

Irrigation using river water and ground water has been the prime factor for raising the food grain production in our country from a mere 50 million tonnes in the 1950s to more than 200 million tonnes at present, leading us to attain self sufficiency in food. Irrigated area has increased from 22 million hectares to 95 million hectares during this period. The population of India, which is around 1,000 million at present, is expected to increase to 1,500 to 1,800 million in the year 2050 and that would require about 450 million tonnes of food grains. With the consumption level most likely to increase in the future we may have to plan for a production level of at least 500 million tonnes. For meeting this requirement, it would be necessary to increase irrigation potential to 160 million hectares for all crops by 2050. India's maximum irrigation potential that could be created through conventional sources has been assessed to be about 140 million hectares. For attaining a potential of 160 million hectares, other strategies shall have to be evolved.

Floods are a recurring feature particularly in Brahmaputra and Ganga rivers in which almost 60 per cent of the river flows of our country occur. Flood damages which were Rs. 52 crores in 1953 had gone upto Rs. 5,846 crores in 1998 with annual average being Rs. 1,343 crores, affecting the states of Assam, Bihar, West Bengal and Uttar Pradesh alongwith untold human sufferings. On the other hand large areas in the states of Rajasthan, Gujarat, Andhra Pradesh, Karnataka and Tamil Nadu face recurring droughts. As much as 85 per cent of drought prone area falls in these states.

Hence, of the most effective ways to increase the irrigation potential for increasing the food grain production, mitigate floods and droughts and reduce regional imbalance in the availability of water is considered to be the interlinking of our rivers to transfer water from the surplus rivers to deficit areas. Brahmaputra & Ganga-particularly their northern tributaries, Mahanadi, Godavari and west flowing rivers originating from the Western Ghats are found to be surplus in water resources. If we can build storage reservoirs on these rivers and connect them to other parts of the country, regional imbalances could be reduced significantly and lot of benefits by way of additional irrigation, domestic and industrial water supply, hydro-power generation, navigational facilities, etc. would accrue.

Historical Perspective

Sir Arthur Cotton who pioneered the development of water resources in Southern India from 1839 onwards, had proposed a plan for interlinking of Indian rivers for inland navigation. A small portion of the plan was implemented but was abandoned later in favour of railways. After Independence, though the idea of interlinking of rivers has repeatedly been occurring to the Indian scientists and engineers, but has always met with rejection on account of technical unfeasibility or socio-economic issues. Four decades back, in 1972, it was Dr. K.L. Rao, the then Union Minister for Irrigation and Power, who suggested a 2,640 km long Ganga-Cauvery link by carrying waters partly by gravity and partly by lift of 350 m, 1,680 cumecs of Ganga flood waters near Patna for 150 days annually; and also a Brahmaputra-Ganga link to transfer 1,800-3,000 cumecs with a lift of 12-15 m. However, detailed examination of the proposal in Central Water Commission indicated the costs to be prohibitive and power needs for pumping excessively high (5-7 million KW). Hence the proposal was not further pursued.

Then came forward the Captain Dastur Plan in 1974, popularly known as Garland Scheme. Under this scheme, a 4,200 km long Himalayan canal and a 9,300 km long Garland canal were suggested, both to be linked by pipelines near Patna and Delhi. Again the plan was found technically unfeasible and laid to rest. However, the government prepared the National Perspective plan in 1980 and in 1982 the National Water Development Agency (NWDA) was set up to carryout detailed studies on the issue. The scheme of the National Perspective for Water Development, inter-alia envisages the creation of irrigation potential for about 35 million hectares (25 million hectares from surface water and 10 million hectares from increased use of ground water) and hydropower generation of about 30,000 MW, part of which would be in Nepal and Bhutan.

The river—linking proposal is alive in the work of the NWDA for the past two decades. But the NWDA's proposals remained a non-starter. A look at the Ninth Five Year Plan document, the National Commission for Integrated Water Resources Development (NCIWRDP) Report (1999), the Prime Minister's address to the National Water Resources Council on April 1, 2002 and the Tenth Five Year Plan would show that no river—linking plan was on the anvil. In fact, the subject of "inter-basin transfers" had been specifically referred to the high level NCIWRDP set up in 1996, but in its report (September 1999) it had observed that further studies were needed on the Himalayan component; and that in the Peninsular component, massive transfers were not needed. However, the National Democratic Alliance (NDA) in its manifesto promised interlinking of rivers. The political resolution of the Bharatiya Janata Party (BJP)'s National Council meeting held at Nagpur on 27-28 August, 2000 stated: "We also urge Government to consider a time bound programme to link

Ganga and Cauvery waters." In view of the inordinate delay in implementing the proposed Plan, Dravida Peravai General Secretary, Shri N. Nandhivarman filed a writ petition in the Supreme Court of India, against the Union of India and others (Writ Petition # 496 of 2001). The petitioner stated: "Every feasibility study ordered from the days of K.L. Rao, the Minister of state for Irrigation at Union Government had only endorsed the findings after findings that favour the national dream project. The Rs. 33,600 crore project linking peninsular rivers could first be taken up. Or Himalayan component can be taken up. Or links within states and agreeable states could be taken up. Yet there is no good news from the seats of power, and farmers of this country who alone are a majority in our agrarian economy are aghast at the neglect of this national project".

In this prayer, the petitioner sought the apex court's intervention to direct authorities to initiate the implementation of this project in a phased manner and with a time frame. The petitioner also prayed that the feasibility reports prepared by the NWDA must be made public along with the reasons for long consumption of time in initiating the project. The petitioner also sought details of the total amount spent by the central and all the state governments on flood and drought relief.

The petitioner stated that another project promised in the manifesto of the NDA namely Sethu Samudram Project, which was conceived 141 years back, has seen 22 feasibility reports in these periods and now global tenders were called for to prepare fresh feasibility reports. Fearing such a fate will happen even to this interlinking of rivers project, Mr. N. Nandhivarman in his petition states that from the ongoing deliberations for decades, it is an irony that the foreigners who ruled us interlinked Indian states with railway link, whereas in independent India to interlink rivers, create more lakes and canals were just discussing, discussing and jettisoning scheme after scheme.

The National Water Development Agency is only collecting the data offered by various state governments and compiling them into reports. Even to do that the agency seems to be having 2010 as the time frame to complete all feasibility studies. Moreover like previous proposals this may also be jettisoned citing similar reasons or fresh excuses. Thereafter, after 2010, India may go for global tenders to make a study of this project and all will be back to square one in 21st century too. In view of this, petitioner had to pray before the court to find out what was going to ultimately happen to this project. While rivers that flow through different countries had benefited such countries with mutual agreements on water sharing, some Indian states prefer waters to be emptied in oceans instead of benefiting fellow Indians.

Following this petition the Apex Court of the country had set a new deadline for networking of rivers. The Center accordingly prepared a blueprint for linking all major rivers in the country under the National Riverine Network (NRN) project (the details of the project has been discussed later). A full bench of the Supreme Court had asked the Government to ensure that the network would be in place by 2012 pushing the earlier deadline of 2043 ahead by 31 years.

Existing Projects in India

The initiative for inter linking of rivers has given rise to certain apprehensions: few real, some hypothetical and others just misplaced. While local issues on the doctrine of riparian rights and appropriation of water for priority uses within a basin or sub-basin are important, a broad national perspective is of no less significance and required to be considered against unreasonable use and or non-use of this vital resource.

Projects for inter-basin transfer of water are in existence and are also being planned in countries such as the United States, Canada, Mexico, Spain, China, etc., For India, long distance water transfer is not a new concept. The Western Yamuna canal and Agra canal have been carrying waters from the Himalayas to the distant plains of Punjab, Uttar Pradesh and Rajasthan for the last a few centuries. The Kurnool—Cuddapah Canal and the Periyar—Vaigal Canal executed in the 19th century—all stand testimony to the sense of cooperation and goodwill exhibited by the people of yore to remove the suffering of the water starved fellow human beings.

The water resources of the Sutlej, the Beas and the Ravi have been optimally developed after construction of Bhakra, Nangal, Pong, Pandoh and Ranjit Sagar multipurpose projects. The installed hydro-electric power generation capacity of these dams taken together is 3,420 MW. Irrigation benefits have been provided to 54 Mha; most of it in Punjab, some extended to Haryana and further to desert districts of Bikaner and Jaisalmer in Rajasthan. Several towns and villages in Punjab, Haryana, Western Rajasthan, Union Territory of Chandigarh and also the national capital Delhi in the Yamuna basin are provided with 28,368 million cubic-meter (MCM) water for drinking purpose from Bhakra—Beas system.

Except Nangal having a negligible storage in the balancing reservoir, the other dams have live storage capacity of 16,843 MCM to cater to various demands and protecting Punjab from flood. The area protected from flood is yet to be properly quantified. The submergence in most cases is within Himachal Pradesh affecting about 500 sq.km area.

The Indus basin has two classic cases of interlinking its rivers. The oldest is a link canal of 283 cusec capacity constructed in 1955 for diverting surplus water of the Ravi at Madhopur headworks to the Beas near Pathankot. Transfer of 4,716 MCM Beas waters into the Sutlej has been made at Pandoh in Himachal, about 21 km upstream of Mandi town through a 13.1-km long tunnel of 255 cusec capacity. The link tunnel outfalls into the Sutlej at Slapper. These have proved useful augmenting water for generation of power and other uses.

The interlinking of the basin's three rivers has also facilitated better control in regulating river flows through integrated operation of the storage reservoir at Bhakra, Pandoh, Pong and Ranjit Sagar. The system feeds the largest irrigation net work of 900-km long canal from Harike to Gadra, also known as the Indira Gandhi Canal, its 8,800-km distributaries and 6,500-km field channels by long distance transfer of the Ravi-Beas waters. The Rajasthan desert is still thirsty and needs more water to reclaim the moving sand dunes for productive use. The ongoing Sutlej-Yamuna Link (SYL) proposal for interbasin transfer of water to Haryana is however troubled by political posturing notwithstanding the Supreme Court directive.

The number of dams, weirs, barrages, river-interlinks, large network of canals and distributaries have been constructed on these three rivers with massive investment after independence. Punjab has been the main beneficiary and became the granary of India. Haryana carved out of erstwhile Punjab, western Rajasthan and Delhi, though outside the Indus basin, have shared some of the benefits in accordance with old agreements, awards and the Punjab Reorganisation Act, 1966.

Himachal Pradesh, an upper riparian basin state of these three rivers contributing major flow from perennial snow cover, higher rainfall and large drainage area of 51,358 sq. km has received a raw deal. Himachal did not have enough political clout for hard bargaining. Its administration displayed mute generosity while signing the agreements and accepting the awards. Sadly, Himachal cannot draw water from Bhakra and Pong reservoirs even for drinking purposes. As per the National Water Policy, it should receive the first charge from the nearest available source. The villages and towns situated along the periphery of these reservoirs in Bilaspur, Ghumarwin, Paragpur and Dehra areas have to depend upon tanker supplies in summer months. Similar is the situation of Mandi town and hill top villages around Pandoh reservoir. Drawal of only 0.57 cusec water out of canal's capacity of 255 cusec is stipulated from Sungernagar hydel channel of BSL for irrigating the fertile Balh valley. The local farmers and small towns in proximity of BSL channels are depending upon nearby seasonal streams or on groundwater to meet their additional demand. Its share of hydro-electric power from these major dams is a pittance when compared to benefits accruing in the downstream. Jammu and Kashmir too has not been able to utilize its share of the Ravi waters due to delay in the construction of the Shahpur-Kandi barrage.

Indiscriminate and unscientific irrigation has disturbed the equilibrium between inflow and outflow of water in command areas of main beneficiary states leading to excess salt concentration. The area affected by salinity in Punjab is about 4,90,000 ha. and in Haryana 1,97,200 ha. The over-irrigated command of the Indira Gandhi Canal is also suffering from water logging and salinity. Misuse and quality degradation are bound to affect the productivity of land and cause shortage of water.

Efficient water management practices by reducing the water allowance and irrigation frequency are necessary to check these problems. The water thus saved can provide critical irrigation to more desert areas in Rajasthan. Despite optimum development, the principles of easy access, fair use, justice and social equity have not been adequately addressed in all adjoining states.

Harnessing the water resources of the Indus, the Jhelum and the Chenab within the Indian catchment of 1,78,406 sq. km is negligible though the Indo-Pak Treaty affords the use of some flow for existing and specified future purposes. This is generally viewed with abundant caution. Only a couple of hydro electric projects in the Chenab basin at Salal in Jammu and Kashmir and Hurot in Himachal have been completed. Work on Dulhasi power project in Jammu and Kashmir is in progress by the NHPC.

The Jhelum has no project worth the name although a flood storage dam is needed to control inundation in the rainy season and to regulate inflow to the Wular Lake for inland navigation.

The main Indus river flowing through the Ladakh region enters into Pakistan except for small abstractions to meet day-to-day water requirements of thinly populated local habitations. These snowfed rivers have untapped hydro-electric potential which could be exploited for evacuating power to Pakistan and Afghanistan through the sub-continental power grid after meeting the local demand. The Chenab is the largest river in terms of water originating in Himachal Pradesh. Its upstream tributary, the Chandra, could be linked with the Beas through a tunnel parallel to the one already planned for road traffic near Manali to supplement the SYL and the Indira Gandhi Canal supplies via the Beas-Sutlej link for the benefit of Haryana and Rajasthan. Similar possibility exists in the downstream for diversion of still larger quantity of Chandra Bhaga waters to Budhil nala near Kugti and its further transfer to the indira Gandhi Canal through the Ravi-Beas Link. Himachal can be the facilitator for such a project. The feasibility and economics of the two are a matter of detailed study.

The water requirement for diverse uses is greatly influenced by social behaviour, population dynamics and economic growth of the communities. None of the three parameters is static. Therefore, the increasing demand for water cannot be tied down forever to an old and one-time treaty. The Supreme Court has given a broadbased interpretation to Article 21 of the Constitution guaranteeing fundamental right to life to include right to water as well. The old water sharing instruments should be redrawn in terms of basic human right and relative importance of water for various uses.

The Pakistani Experience

As India has decided to undertake an ambitious programme of linking its rivers at an estimated cost of Rs. 5,60,000 crore in a period of ten years, it would be beneficial to examine how other countries have handled such projects. Pakistan, for instance, built a network of river links as a part of the Indus Treaty works which functions as replacement links to irrigate those areas which got deprived of irrigation when three eastern rivers of the Indus System were allocated to India. The experience of Pakistan on this front will be of utmost relevance to Punjab as the soil and environment conditions are similar.

Pakistan built ten links, six barrages and two dams during the post-treaty period of 1960-70. Its Indus Plains host most of its population and the world's largest irrigation network commanding an area of 345 lakh acres which contribute to 90 per cent of its national agricultural production.

At the time of partition, the barrages of Madhopur and Hussenawala fell to the share of India whereas six barrages of Kalabagh, Sukkur, Trimmu, Baloki, Sulemanki and Islam went to Pakistan. There was no storage dam built then.

After Partiion, the irrigation system conceived originally got divided between India and Pakistan without any regard to irrigation boundaries which resulted in disruption of irrigation. This triggered a dispute which was resolved by the Indus Water Treaty in 1960. This treaty covered the regulation of the river system of the two countries.

The treaty assigned three eastern rivers of the Sutlej, the Beas and the Ravi to India and the other three rivers to Pakistan. It provided for construction of replacement links by Pakistan for transfer of irrigation supplies from the western rivers of the Indus, the Jhelum and the Chenab to the areas served earlier by the three eastern rivers. It also provided for building of two storages dams—Mangla on the Jhelum and Tarbela on the Indus. The time fixed for the completion of treaty works was 10 years and if there was any time overrun, Pakistan was required to pay to India for the water utilized.

Consequently, Pakistan built ten links and some other associated works which provided on short-term basis, the losses incurred on account of partition of the irrigation system. The system built resulted in meticulous water management and optimum utilization of the allocated water resources to Pakistan.

The links built by Pakistan are:

1. Marala-Ravi link, (capacity—22,000 cusecs, 100 km long) to transfer water from the Chenab to the Ravi.
2. Bombanwal-Ravi-Bedian-Ripalpur link (2,200-4,200 cusecs, 158 km) to transfer water of the Chenab which was brought by the Marala-Ravi link to the Sutlej.
3. Rasul-Qadriabad link (19,000 cusec,40 km) to transfer water from the Jhelum to the Chenab at the Qadriabada barrage, built some distance below the old Marala Barrage.
4. Qadribad-Balloki link (18,600 cusecs, 125 km) to transfer water brought by the link above the Balloki Head-works,
5. Balloki-Sulemanki II link (15,400,84 km) to deliver water to the Sutlej above the Sulemanki headworks.
6. Balloki-Sulemanki II link (6,500 cusecs, 58 km) to transfer additional supplies to the Sutlej above Sulemanki Headworks.

PROPOSED INTER BASIN WATER TRANSFER LINKS

HIMALAYAN COMPONENT NAME OF LINK

1. Brahmaputra-Ganga (MSTG)
2. Kosi-Ghagra
3. Gandak-Ganga
4. Ghagra-Yamuna
5. Sarda-Yamuna
6. Yamuna-Rajasthan
7. Rajasthan-Sabarmati
8. Chunar-Sone Barrage
9. Sone Dam-Southern Tributaries of Ganga
10. Ganga-Damodar-Subernarekha
11. Subernarekha-Mahanadi
12. Kosi-Mechi
13. Farakka-Sunderbans
14. Brahmaputra-Ganga (JTF) (ALT)

PENINSULAR COMPONENT NAME OF THE LINKS

15. Mahanadi (Manibhadra) -Godavari (Dowlaiswaram)
16. Godavari (Inchampalli Low Dam)-Krishna (Nagarjunasagar Tail Pond)
17. Godavari (Inchampalli) - Krishna (Nagarjunasagar)
18. Godavari (Polavaram) - Krishna (Vijaywada)
19. Krishna (Almatti) - Pennar
20. Krishna (Srisailam) - Pennar
21. Krishna (Nagarjunasagar) - Pennar (Somasila)
22. Pennar (Somasila) Cauvery (Grand Anicut)
23. Cauvery (Kattalai) - Vaigai - Gundar
24. Ken - Betwa
25. Parbati - Kalisindh - Chambal
26. Par - Tapi - Narmada
27. Damanganga - Pinjal
28. Bedti - Varda
29. Netravati - Hemavati
30. Pamba - Achankovil - Vaippar

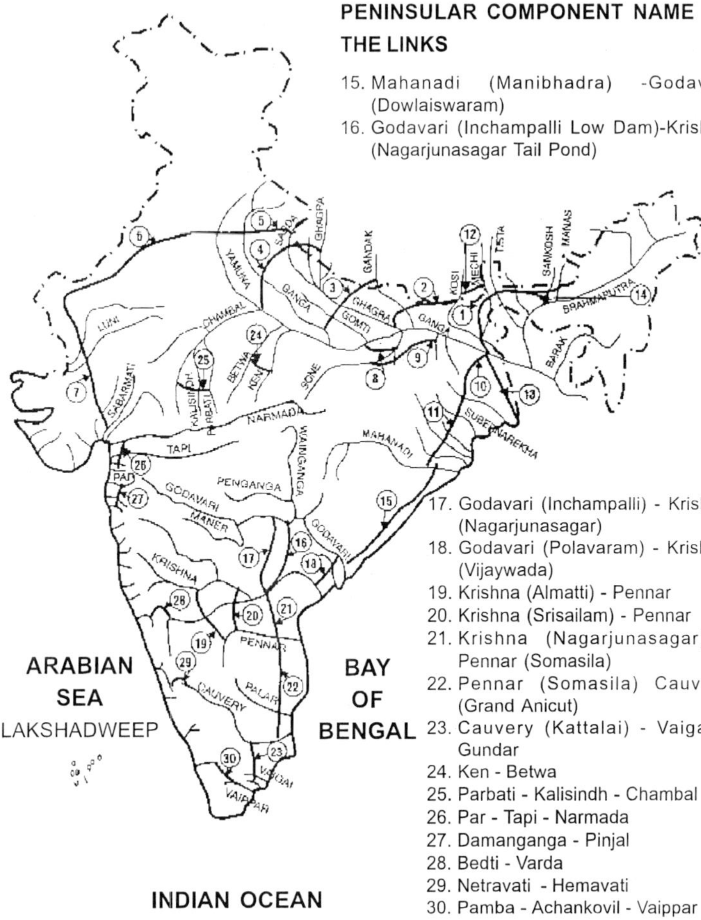

7. Chasma Jhelum link (21,700 cusecs, 94 km) to transfer water from the Indus to the Jhelum.
8. Trimmu—Sidhnao link (11,000 cusecs, 66 km) to transfer from the Chenab after Syphon crossing.
9. Sidhnai-Mailsi link (10,100 cusecs, 198 km) has been built.
10. Taunsa-Panjnad link (12,000 cusecs, 58 km) to transfer water from the Jhelum to the Chenab.

Most of the links built are unlined channels except a part of the Sidhani-Mailsi link. The total length of the links works out to be 899 km, built at a cost of Rs. 400 crore. These links have the aggregate capacity of 140, 500 cusecs and in a day transfer some 2,81 MAF of water.

Five barrages were built as part of the Indus Treaty Works. These were the Chasma Barrage on the Indus below the Tarbela Dam, the Rasult Barrage built on the Jehlum; the Marala Barrage; the Qadribad Barrage which works like a level crossing on the Chenab; and the Sidhani Barrage on the Ravi.

In addition, a unique structure of the Mailsi Siphon has been built, 29 km downstream of the Islam Barrage. It enables irrigation water supplies of the Sidhnai-Mailsi-Bahawl Link across the Sutlej to Bahawal Canal. This structure works as a siphon-cum-weir, a gated control structure provided to handle a flood of 4.29 lakh cusecs. The siphon measures 160 ft between the Inlet Portal and the Outlet Portal. It is located a few kilometers downstream of the Islam Barrage. The Mailsi-Bahawal link constructed at the downstream end of the Mailsi-Syphon into the Bahawal Canal at RD 15,000. It is a lined channel with full supply of 4,000 cusecs and is 21 km long.

Dams built as a part of the Indus Treaty Works are those of the Mangla Dam on the Jhelum having a storage capacity of 5.34 MAF and the Tarbela Dam on the river Indus (Capacity 9.7 MAF). Thus, a storage of 15.04 MAF got created. The Mangla Dam was completed in just five years.

THE PRESENT PROJECT

As depicted earlier, the concept of inter-basin transfer of water per se is by no means new to this country or others that have implemented schemes of similar nature. Inter-basin transfer of water is considered as an outstanding example of effective and efficient management of water resources on the basis of need and equity though there are many strong points against such a venture. The government established a national water development agency (NWDA) in July 1982 to explore the scientific, technological, socio-economic and environmental feasibility of transferring water from water surplus areas to the deficit ones. The gigantic technical task was divided into two components, the Himalayan rivers development and the peninsula river development. NWDA has already conducted scientific studies dealing with water balances in various basins/ sub-basins. Before establishing the feasibility of any link, the agency has meticulously studied all relevant ecological and environmental concerns and to include all ameliorative and mitigative measures to keep the sustainability of the system intact. No doubt, the task envisages some large dams and the consequent displacement, rehabilitation and resettlement of the project-affected people not only pose a challenge but also provide an opportunity that would be countered with liberal and enlightened package.

According to the Director-General, NWDA, Radha Singh, the studies are scientific and incorporate relevant data on surveys and investigations, geological, geophysical and geo-technical, soil surveys of the command areas, appropriate cropping patterns., socio and ecological compensatory afforestation. The project aims to link 37 rivers to transfer water from water-surplus areas to relatively dry regions through 30 link canals spread over 9,600 kms. The network will interlink 32 dams, using 56 million tonnes of cement and two million tonnes of steel. As stated earlier, the Government of India proposed to complete the linking job by 2043. However, the Supreme Court directed to accomplish the plan by 2012, *i.e.*, within a period of ten years, to effectively control flood and drought situations. However, according to the Ministry of Water Resources, a number of issues have to be resolved and agreements forged for different projects. Yet, efforts will be made to complete the programme by 2016.

The Cost

The estimated cost of the whole scheme at 2002 price level has been worked out to be mind boggling Rs. 5,60,000 ($12 billion). This is nearly 50 times the total allocation for the ongoing water resource development project in the 10th Plan—and 300 per cent of the government's total resources in 2003-04 and about ten times the amount of Rs. 58,000 crore—spent for irrigation in the past 52 years. Also, it will have to spend 2 per cent of the current GDP for the next 10 years. Though the project cost has been tentatively put at Rs. 5.6 lakh crore, the estimated cost of the river linking programme for all acceptable links will be based on development and planning report (Feasibility Report) to be prepared by the Task Force appointed for the purpose. The Task Force has now suggested imposition of cess, taxes and duties on certain items to raise funds for the project, estimated to cost more than Rs. 5.6 lakh crores.

In its Action Plan II submitted to the Union Government, the Task Force has stated that the cost of inter-linking rivers will be much higher than the estimated Rs. 5.6 lakh crores. It has recommended fund mobilization through a combination of private funding, public—private partnership and pure public funding. The recovery of costs will be from farmers (irrigated agriculture, hydropower), consumers (drinking water, navigation) and industry (industrial water).

The Task Force has observated that the initial estimate of Rs. 5.6 lakh crores by the National Water Development Agency under the Union Water Resources Ministry does not envisage yearly inflation and costs relating to ecology, environment, wildlife, resettlement and rehabilitation of displaced people. More than five lakh people are likely to be displaced by various river inter-linking projects under which 32 new dams would be constructed to reduce flooding by about 30 per cent. The area to be submerged would be 3.5 lakh hectares of which 1.2 lakh hectares would be forest land. The Task Force has suggested that projects that generate direct revenue, such as hydro power projects, should be handed over totally to private parties while the Irrigation Infrastructure, which may not generate adequate revenue, could be developed in the form of public-private partnership. Hydropower projects (34,000 MW) would constitute about 25 per cent of the total investment needed for the programme. For this, the debt requirement would be 70 per cent of the project cost and the balance 30 per cent would be promoters' contribution.

Funding for irrigation infrastructure would be on "annuity model" wherein the private developer would finance, construct and operate, while the market risk would be borne by the government and annual revenues would be assured to the developers. Significantly, the Task Force has not put any figures to its assessments and has sought nine months to come out with a detailed Action Plan II on the funding pattern and the manner of cost recovery. For evolving a political consensus, it has suggested integration of master plans of water resources development of various States, harmonization of water tribunal awards and inter-State agreements with the Rivers Inter-Linking programme.

Expected Outcome

The implementation of the above programme of inter-basin water transfer is expected to provide additional irrigation benefits of 35 million hectares *i.e.* 25 million hectares from surface waters and 10 million hectares from increased use of ground waters which will be over and above the ultimate irrigation potential of about 140 million hectare envisaged from major, medium and minor irrigation projects and would generate 34,000 MW of power, apart from the benefits of flood control, navigation, water supply, fisheries, salinity and pollution control etc.

Project Components

The project aims to transfer 'surplus' waters of the eastern rivers—especially the Brahmaputra and various tributaries of the Ganga-to the drought-affected western and southern parts of India. Various links between rivers are being considered for the purpose. The project will, therefore, entail the construction of several major dams and lengthy canals cutting across various river basins.

It has been stated above that the proposed scheme has two components[2]. The Himalayan component proposes to have 14 canal links (Table 14.1) while the Peninsular component has 10 links (Table 14.2).

(i) Himalayan Component

The Himalayan Rivers Development Component of National Perspective for water resources development envisages construction of storage reservoirs on the principal tributaries of Ganga and the Brahmaputra in India, Nepal and Bhutan along with interlinking canal systems to transfer surplus flows of the eastern tributaries of the Ganga to the west, apart from linking of the main Brahmaputra and its tributaries with the Ganga and Ganga with Mahanadi. It would also provide the necessary discharge for augmentation of flows at Farakka to, inter-alia, flush the Calcutta Port and facilitate inland navigation across the country.

Table 14.1: Links Identified in Himalayan Component

1. Kosi-Mechi
2. Kosi-Ghagra
3. Gandak-Ganga
4. Ghagra-Yamuna
5. Sarda-Yamuna
6. Yamuna-Rajasthan
7. Rajasthan-Sabarmati
8. Chunnar-Sone barrage
9. Sone dam-Southern tributaries of Ganga
10. Brahmaputra—Ganga (Manas-Sankosh-Tista-Ganga)
11. Brahmaputra-Ganga (Jogighopa-Tista-Farakka)
12. Farakka-Sunderbans
13. Farakka-Damodar-Subarnarekha
14. Subarnarekha-Mahanadi

Studies Undertaken

The work under this component comprises of water balance studies of 19 diversion points, toposheet studies of 16 reservoirs, toposheet studies of 19 water transfer links, preparation of pre-feasibility report of 14 water transfer links and survey and investigations of 14 water transfer links for preparation of feasibility reports. All these studies have since been completed. Further, during the year 2001-02, topographical survey for the three links namely, Sarda—Yamuna, Ghagra—Yamuna (Indian territory) and Chunar—Sone Barrage were completed. Besides, topographical surveys for preparation of feasibility—reports in respect of 4 links namely, Manas—Sankosh—Teesta—Ganga, Ganga—Damodar—Subarnarekha, Yamuna—Rajasthan and Sone Dam, Southern Tributaries of Ganga were under progress.

Peninsular Component

The Peninsular Rivers Development is divided into following four parts:

(i) Interlinking of the Mahanadi—Godavari-Krishna-Pennar-Cauvery rivers.

(ii) Interlinking of the west flowing rivers north of Bombay and south of Tapi.

(iii) Interlinking of the river Ken with Chambal.

(iv) Diversion of the west flowing rivers of Kerala and Karnataka to the east.

Under the component, surplus waters of Mahanadi and Godavari rivers are proposed to be transferred to the deficit basins of Krishna, Pennar, Cauvery and Vaigai. In the transfer of waters from Godavari to Krishna, a lift of 1,200 cusecs of water over about 116m is essential. In addition, water is proposed to be transferred from Ken River to Betwa river to benefit Madhya Pradesh and Andhra Pradesh and interlinking of Parbati, Kalisindh and Chambal rivers to benefit MP and Rajasthan. Peninsular component aims at benefiting Orissa, Karnataka, Tamilnadu, Pondicherry, Maharashtra and Gujarat.

The work under this component comprises of collection of data for 137 basins/sub-basins, water balance studies of 137 basins/sub-basins and at 49 identified diversion points, toposheet & storage capacity studies of 58 identified reservoirs, toposheet studies of 17 links including identifications of command area enroute, preparation of prefeasibility reports of 17 links and survey and investigations of 16 water transfer links for preparing feasibility reports.

Table 14.2: Links Identified in Peninsular Component

1. Mahanadi (Manibhadra)-Godavari (Dowlaiswaram)
2. Godavari (Inchampalli)-Krishna (Nagarjunasagar)
3. Godavari (Inchampalli Low dam)-Krishna (Nagarjuna Tail pond)
4. Godavari (Polavaram)-Krishna (Vijaywada)
5. Krishna (Almatti)-Pennar
6. Krishna (Srisailam)—Pennar (Prodattur)
7. Krishna (Nagarjunasagar)-Pennar (Somasila)
8. Pennar (Somasila)-Cauvery (Grand Anicut)
9. Cauvery (Kattaiai)—Vaigai—Gundar
10. Ken-Betwa-Link
11. Parbati_Kalisindh—Chambal
12. Par-Tapi-Narmada
13. Damanganga_Pinjal
14. Bedti-Varda
15. Netravati-Hemavati
16. Pamba-Achankovil-Vaippar

The **main features of the scheme** thus may be summarized as follows:

1.	Estimated Cost (At 2002 price leve)	Rs. 5,60,000 Crore
2.	Number of Links	30
3.	Interdependent Links	21
4.	Independent Links	9
5.	Time Period	10 Years
6.	Annual Financial Outlay	Rs. 56,000 Crore
7.	Countries involved	India, Nepal, Bhutan

Steps for Implementation of the Link Proposals

Mandate of NWDA is to complete the feasibility reports of link schemes. The steps involved in the implementation of link schemes, after completion of the feasibility reports, are:

- Negotiations and interstate agreements amongst the concerned states to arrive at consensus regarding sharing of surplus water and other project related agreements.
- Preparation of Detailed Project Reports (DPRs)
- Techno-economic and Environmental approvals and investment clearance by Planning Commission.
- Financing arrangements and mode of funding.
- Execution of Link Projects

The whole programme of implementation depends on the cooperation extended by the State Govts. Negotiations and interstate agreements are pre-requisite for taking up the work of DPRs. Thus, states have to enter into serious discussion amongst themselves or if required under the aegis of Central Govt. to arrive at consensus.

Perspective Plan for Implementation of Link Proposals

In order to speed up the process of implementation of link schemes which are not involving major inter-state issues or which are intra-state, NWDA has identified following five links (out of total 30 links) for preparation of DPRs:

1.	Ken-Betwa link	(involving M.P. & U.P.)
2.	Par-Tapi—Narmata link	(involving Gujarat & Maharashtra)
3.	Damanganga—Pinjal link	(involving Gujarat & Maharashtra)

4. Debti—Varda link (Within Karnataka)
5. Netravati—Hemavati link (Within Karnataka)

Task Force

A Task Force under the Chairmanship of former Power Minister Shri Suresh Prabhu had been constituted in 2002. The terms of reference of the Task Force are as follows:

(i) Provide guidance on norms of appraisal of individual projects in respect of economic viability, socio-economic impacts, environmental impacts and preparation of resettlement plans;
(ii) Devise suitable mechanism for bringing about speedy consensus amongst the States;
(iii) Prioritize the different project components for preparation of Detail Project Reports and Implementation;
(iv) Propose suitable organizational structure for implementing the project;
(v) Consider various modalities for project funding; and
(vi) Consider International dimensions that may be involved in some project components

The milestone/ time table for achieving the goal of interlinking of rivers as indicated in the above resolution is in Table 14.3:

Table 14.3: Milestonnes Dates/Time Table for Interlinking of Rivers

i)	Notification of the Task Force	By 16.12.2002
ii)	Preparation of Action Plan—I giving an outline of the time schedules for the completion of the feasibility studies, detailed project reports, estimated cost, implementation schedule, concrete benefits and advantages of the project etc.	30.4.2003
iii)	Preparation of Action Plan—II, giving alternative Options for funding and execution of the project and also suggested methods for cost recovery	31.7.2003
iv)	Meeting with the Chief Ministers to deliberate over the project and to elicit their cooperation	May/June 2003
v)	Completion of Feasibility Studies (already in progress)	31.12.2005
vi)	Completion of Detailed Project Reports (Preparation of DPRs will start simultaneously since FRs in respect of six river links have already been completed)	31.12.2006
vii)	Implementation of the Project (10 years)	31.12.2016

Major Hurdles in Implementation

i) *Apprehensions of States*: While the donee states agree to the link proposals, donor states do not because they have apprehensions about such surplus in their states even though all the future water requirements have been taken care of. Some states feel that the existing Tribunal awards will get disturbed and thus no water can be taken out of the basin due to the Tribunal awards. The requirement of power to lift waters as proposed in the link schemes may not be available. The link proposals may have adverse effects on the existing irrigation and power generation. The reliability and adequacy of the water transfer from distant sources is also being questioned.

ii) Consensus amongst the States regarding sharing of surplus waters and entering into interstate and project related agreements may not be easy.

iii) Necessity for construction of large storage dams both within India and in the neighboring countries of Nepal and Bhutan which would require international agreements with these countries.

iv) Large dams involve environmental issues like submergence of forest and cultivated areas in habitation, necessitating compensatory afforestation and rehabilitation and resettlement of people. In this connection, it may be mentioned that topographical survey of Manas—Sankosh-Tista—Ganga link could not be completed for want of clearance from Ministry of Environment & Forests, as this link passes through the reserve forests & wild life sanctuaries, namely, Raidak & Buxa Tiger reserve forests.

v) Acquisition of land for the long and wide link canals could pose a problem.

vi) For some of the link canals water has to be pumped. Brahmaputra water after reaching Ganga has to be lifted by 60 m to reach Subernarekha and by about 58 m for further transfer to Mahanadi—Godavari link. Similarly for transferring water from Inchampalli dam on Godavari to Krishna, it has to be lifted by about 116m. These three lifts would require about 3,700 megawatt of power.

CRITICS' VIEWS

Shri Ramaswamy R Iyer[3] opines that "we entertain grand visions of long-distance water transfers from one basin to another, when we cannot even persuade neighbouring states within a basin to agree upon a sharing of waters (eg. Ravi-Beas, Cauvery). The National Water Development Agency has been studying the possibilities of linking the peninsular rivers (Mahandi-Godavari-Krishna—Pennar—Cauvery), but Orissa does not agree that there is a surplus in the Mahanadi, and Andhra Pradesh does not agree that there is surplus in the Godavari." Iyer further points out that every inter-basin transfer must necessarily involve carrying of water across the natural barrier between basins (which is what makes them basins) by lifting, or by traveling through, or by a long circuitous routing around the mountains if such a possibility exists in a given case. Exceptionally heavy capital investments and continuing energy costs (in operation) are always likely to be involved. In addition, big dams, reservoirs and conveyance systems will need to be built, involving not merely large investment but also substantial environmental impacts and displacement/rehabilitation problems. All this will need to be looked into very closely in every case.

Dr. Bharat Singh[4] Professor Emeritus at the Water Resources Development Training Center at IIT, Roorkee, finds it odd that the present initiative for inter—linking of rivers comes not from the Planning Commission nor from National Water Resources Development Council but from the Supreme Court. In fact, any water resource engineer will immediately discard inter-linking of rivers as a flood control measure. Yet, Dr. Singh laments, the Government of India almost immediately endorsed the Supreme Court directive and gave a time span of 10 years for implementation of the project. In view of the fact that we have been taking more than 20 years for completion of a single high dam project, the time seems to be totally unrealistic. He also believes that looking at the on-going disputes between states on sharing of river waters, inter-state agreements would be extremely difficult to achieve. According to him, there is no convincing agreement on vital national interest which can justify undertaking this mammoth undertaking in its entirety. Yet in the opinion of few other critics, the project is a mirage which the country can ill afford to chase. In support of their contention, they highlight the following issues[5]:

One, the very inter-State disputes over water that have breathed fresh air into the linking proposal stand in the way of its execution. If water is to be moved from the Mahanadi to the Godavari and then to the Cauvery basin, Orissa and Andhra Pradesh must agree to the transfers. But neither State agrees that there is a surplus in the two rivers, so how then can there be any sharing of waters? More generally, the States and even regions within a State are loath to share water with one another. The need to appropriate water is a natural consequence of an attitude, nurtured for decades, that the use of more and more water for cultivation of water-intensive crops is the only way to cope with rainfall uncertainty and the only recipe for agricultural development.

Two, incomplete projects which have already swallowed huge resources bedevil the irrigation sector. According to the NCIWRDP, it will take Rs. 180,000 crores over the next 10 years to complete these projects. Should the nation's priority be to complete these schemes or abandon them and start on a new and gigantic one? In fact, in a situation of severe resource scarcity, the question of the relative priorities to be accorded to the improvement of existing facilities and the expeditious completion of viable projects on hand as against major projects based on questionable premises is particularly relevant. This issue ought to be debated seriously.

Three, the river basin is the natural unit of planning which we are supposed to be following. While in the past there have been inter-basin projects that have been implemented, a countrywide inter-basin transfer of water will be a fundamental violation of this very basic principle. Will it be worth in economic terms? Four, can we have even a reasonable estimate of capital and operating costs? Depending on whichever proposal is now going to be drawn up, it will be a mind-boggling project. It will require the construction of a network of large reservoirs, excavation of canals that are thousands of kilometers in length, consumption of enormous amounts of energy to lift water across mountain ranges (requiring 10,000 MW of capacity, according to one estimate) and construction of even longer canals if lifting of water is to be avoided by taking a circuitous route around ranges. A figure that the Government mentioned in the Supreme Court is the stupendous one of Rs. 5,60,000 crores. If the project is to be completed in a decade, then it will mean an expenditure of Rs. 56,000 crores a year. What does expenditure of this magnitude imply? This annual outlay would consume 30 per cent of the Center's total tax revenue. It would be more than the Centre's current level of Plan capital outlay on all social and economic projects. And it would be equivalent to about 80 per cent of the Centre's annual capital expenditure in its entirety on defence, power, irrigation, roads, health, education and everything the Government is involved in. In short, this one project will devour Government revenues and push every single social and economic priority off the agenda for years to come.

Five, if this what execution of the project implies, can we have a cold-headed and reasonable assessment of the costs and benefits before committing ourselves to the grand idea? Six, can we also look at the larger human, social and environmental costs at this attempt to redraw India's geography? How many million people will be uprooted by the reservoirs that will be created and the canals that will cut through the country? Can we even think of rehabilitating them and how? What permanent changes will the project bring to the environment?

Seven, this gigantic project will kill any dream we may have of an integrated approach to water use in the form of pursuing major and minor irrigation projects, watershed development and efficiency in water use. Two years ago, Parliament was informed that "linking of rivers is only a part of the solution to flood, drought and drinking water problems. The solution lies in a conjunctive, holistic and integrated water management". Exactly, yet, now the political establishment is prepared to invest the country's energies and finances in one single project.

Eight, is the Supreme Court time-frame of execution in a decade even remotely feasible? The Government itself says it will take at least four decades to complete. Or, rather can a project of this kind be ever completed or will it be abandoned mid-way when the human and financial costs become too much to bear? It should be just as worrying if the mirage is given up-mid-stream, abandoned after an expenditure of perhaps hundreds of thousands of crores.

As for environmental issues, it may be pointed out that each river is an ecosystem and any diversion causes imbalance. In the case of the Ganga, it will even lose religious connotations and we will have a social problem too. The ministry of environment and forests has also pointed out the loss of bio-diversity, loss of traditional knowledge linked to it, and loss of migratory path and corridor for birds and animals. It has stressed on unknown possibilities like climate change, changes in hydrological cycle, impact on aquatic life like fish, spread of contaminants (fluorides, sulphides, nitrates, chlorides, toxic chemicals and heavy metals) from one basin to another. Since resettlement and rehabilitation would also impact environment, the ministry has pushed for an inter-disciplinary approach to the matter in partnership with civil society.

There are innumerable other economic, hydrological and social issues that need to be addressed. History will not forgive us if we do not take a decision after a careful examination of the costs and benefits of linking rivers. A decision to take up what will be the largest project ever executed in India cannot be based on the whims and fancies of groups looking for a quick-fix solution to India's water challenges.

Shri S.K. Lahiri, retired Secretary, Irrigation and Waterways, Government of West Bengal states that the idea of tackling the flood problem in Assam's Brahmaputra basin to the Ganga in West Bengal as a part of the ambitious river-link project is preposterous[6]. Excess water of the mighty Brahmaputra cannot be canalized into the Ganga through the existing irrigation canals because they cannot be used as flood flow canals. In the dry season, however, the problem is quite different. As the Brahmaputra and the Ganga and their major tributaries flow into Bangladesh, the sharing of dry season water requires special international commitment and agreement. There is a great demand for the dry season flow because it helps cultivate rabi crops and ensures navigational facilities. Bangladesh has to bear the brunt of the floods in the Brahmaputra and the Ganga, so it would naturally demand the lion's share of the two rivers' and the Teesta's waters in the dry season as well. If only dry season discharge is to be canalized, perhaps there will be nothing left of the Brahmaputra for Assam after Bangladesh gets its share.

Feeding other states with Ganga's water from Farakka during the dry season is considered to be an impossibility. If there is not enough dry season flow in the Brahmaputra either, it would be absolutely futile to interlink them at a huge cost and against a strong public opposition.

Attention is also drawn on the contentions of the water rich states who assert that they do not have any surplus water to transfer. Given the requisite finances they insist that it would be possible for them to utilize all their water resources. Hence they do not agree to the basic philosophy of inter-basin transfer as a method of correcting the natural imbalance of inequitable distribution of water.

An international group consisting of scientists from the U.K., Nepal and India conducting research on Himalayan glacier hydrology has concluded that the Himalayan glacier system is melting at a rapid rate as a result of global warming. If the process continues, nearly 80 per cent of the glaciers would melt in about 25 years. Then the river Ganga would also be monsoon-fed like the Cauvery. This means that the South Indian rivers would not get water from the North during summer. A glaciologist attached to the above research group has expressed surprise at many eminent scientists speaking up in support of the national river grid project even though scientific evidence like "database on the Ganga headwater glaciers and discharge patterns" is not available.

There has also been criticism against the plan from some groups involved in the sector. According to them, if the proposals are meant only for meeting future irrigation needs, there are options to improve water availability. Better water management is economically more attractive then bringing water through a costly grid. Demand management by improving water use efficiency, evaporation control, recycling and reuse of water, etc., can also achieve the requisite objectives. Some of them even feel that investments in water transfers being economically less efficient compared to industrial and commercial investments for such ventures, deficiency in food can be met from imports at a cheaper cost. There are also groups advocating small water harvesting structures to reduce flood effects; improve soil moisture and recharging groundwater as an alternative to long distance water transfers.

Shri Subrata Sinha, former Deputy Director General, Geological Survey of India and now member, Rajiv Gandhi Watershed Mission, Government of Madhya Pradesh[7] opines that the arguments against river linking, stressed at conventions, published in newspapers and even discussed within internet groups, leave little doubt that any proposal for river linking is repugnant to all natural and economic logic. It may be worthwhile to consider some of the proposed link clusters to understand the fraud on the people.

The links involving Brahmaputra, Ganga, Sundarbans, Kosi and Gandak will traverse the Himalayan foothills and the Ganga basin, richly, endowed with fertility. It is an admitted fact that these areas are already over-irrigated; and incremental irrigation would only aggravate floods and soil degradation. The linking system would also be ravaged by excessive erosion/deposition originating in the fragile Himalayas. Worse, the links would allow only an insignificant fraction of

Brahmaputra's stupendous monsoon discharge to be offloaded while affecting the natural equilibrium.

Even otherwise, the effectiveness of large dams is considered as a matter of controversy in more than one way. Silt formation reduces the reservoir depth, causing reduced power generation. Regarding irrigation, some impressive gains have, of course, been made. But this has drawbacks as well. The main drawback of irrigation systems is that it tends to concentrate salts at the surface, ultimately causing soil infertility. This problem is now adversely affecting about one-third of the world's irrigated land. Holding back the sediments also lowers the flow of silicates into the sea leading to a change in the types of algae that live in the sea, favouring species that form toxic blooms, which can kill fish and other marine life.

The creation of reservoirs and construction of roads and buildings associated with big dams affect the quantity of rain and seepage of water in the catchment area in addition to threat to wild life. This affects the flow of water through the river in terms of quantity, velocity and contents. The drying-up of wells and ponds along the banks has necessitated digging of bore-wells. Thus the cost on drinking water is increased while the quality is reduced. Losses and expenditure on agricultural operations along the riverbanks have increased. There is considerable reduction in the quantity of sand available for mining along the river. With very low amount of river water flowing into the sea, saline water enters deep into the river. This causes more losses in agriculture, fish cultivation and thermal power generation, if any, downstream.

The dams cause many environmental and social problems such as the loss of habitat and forcible removal of local communities. A long-term study by the University of Hamburg and the National Institute of Oceanography, Goa, has brought out a more disastrous effect. The study says that dams and other diversions from rivers might adversely affect the ability of the sea to absorb atmospheric carbon, aiding what is called the 'greenhouse effect'. This is a reason for warming of earth's atmosphere and very high sea-level rise and climatic changes. Even flood water, in spite of its damaging effects, has favourable aspects in the areas of agriculture. The advantages of dams and other diversions from rivers in matters of electricity generation and irrigation have to be weighed with the various losses and other problems accompanying them.

Government's Rejoinder

The Ministry of Water Resources, Government of India have examined carefully all the questions raised by critics and are of the view that though there are many serious impediments in implementation of the Inter Linking of Rivers (ILR) programme, these are not insurmountable with the available state-of-the-art technologies. The Task Force which has been set up for the purpose will suggest modalities for arriving at speedy consensus amongst the States for sharing and transfer of surplus water to deficit areas, providing guidance on norms of appraisal of individual projects in respect of economic viability, socio-economic impacts, environmental impacts and preparation of resettlement plans. The Task Force will also suggest the prioritization of different project components for preparation of Detailed Project Reports and its implementation, a suitable organizational structure for implementing the project and modalities for project funding. In fact, the Ministry has answered all the relevant questions raised on the subject and the important of them are indicated below (Table 14.4).

Table 14.4: Frequently Asked Questions about ILR Programme and replies thereon

Sl. No.	*Questions*	*Reply*
1.	Is the interlinking of rivers desirable? If yes, is it feasible?	Yes, the interlinking of rivers is necessary to overcome anomaly of uneven temporal and spatial distribution of the rainfall in the country. The interlinking of rivers is also necessary to remove the regional imbalances. The studies carried out by the National Water Development Agency for interlinking of rivers of the country, prima facie, indicate that the linking is technically feasible.
2.	What are the challenges of the project?	The consensus on implementation of project among the co-basin states, international issues involved, resettlement and rehabilitation of the project affected people, timely clearance of the concerned agencies and project funding are the major challenges likely to be faced.

Sl. No.	*Questions*	*Reply*
3.	Is river-linking a viable proposition from environmental hazards point of view?	Yes, it is a viable proposition as construction of storage reservoirs and canals rather improve the environment. In interlinking proposals a provision of the minimum lean season flow will be safeguarded to maintain the ecology and the river regime. The general prosperity that will follow the implementation of the project will help alleviate poverty, which today is the biggest polluter. Nevertheless each project will be studied for its environmental impact and all measures necessary to mitigate any adverse impact will be taken.
4.	Ambitious projects like this involve mass movement and rehabilitation. Highlight the initiatives that Task Force is likely to take in the event of mass uprooting.	It is estimated that a few lakhs of people may get affected due to creation of reservoirs. The cost of rehabilitation will be included in the project cost preferably up front. The displaced persons will be provided proper rehabilitation & resettlement packages to ensure that their socio-economic conditions improves after resettlement at new places. In order to ensure that the package of R & R well handled, measures like 'Special Purpose Vehicle' (SPV) associating NGO's sensitive to humane problems will be suggested.
5.	How competent task force technical expertise to undertake the project?	Fully competent. Dr. C.C. Patel, Vice-Chairman and Dr. C.D. Thatte, Member-Secretary, Task Force both are Ex. Secretaries of Ministry of Water Resources and eminent engineers having a vast experience of various aspects of water resources development projects such as planning, execution and running & maintenance. Shri M. Gopalakrishnan, Ex. Member CWC has been appointed by Ministry of Water Resources as Coordinator, Shri G.C.Sahu, Ex. E-in-C., Shri. A. Mohankrishnan, Ex. E-in-C and Member Task Force are having the experience of similar nature. In addition to these persons, it is also proposed to utilize the expertise of various eminent engineers and Institutions like IITs etc. as resource persons/ Institutions for the Task Force.

Sl. No.	*Questions*	*Reply*
6.	Total approximate cost	The overall costs is yet to be established based on Detailed Project Reports which are being commissioned for individual links. After firming up the cost of each of these links, the overall cost will be established.
7.	How many rivers included into the project	Almost all the major rivers of the country are getting linked.
8.	How to collect fund for the river project?	This is being looked into by the Task Force through the specialized Institutions and Agencies.
9.	How to deal with various mountain ranges like Aravalli etc.?	The present day tunneling technology is sufficient to handle these issues.
10.	Some parts of the country are reeling with water logging problems. How we can assure that this project will not create problem like this?	Inadequate drainage & improper application of irrigation had caused the problems of water logging in certain cases in the past. The project will be planned to minimize such problems like water logging by efficient water planning, lining of canal systems, improved method of drainage & efficient irrigation application.
11.	Will the water treaties with Bangladesh and Pakistan be affected due to this programme?	The ILR Programme will respect all the existing Treaties with the neighbouring countries.
12.	Have multinationals/ private companies any interest in the project at all?	The initial response of industry is encouraging.
13.	Is the GOI in a hurry to start the project? Any particular reason for that?	The Projject will be launched only after establishing its viability based on Detailed Project Reports (DPR) which will be addressing all concerns like Social issues, Environment & Ecology, Economic and other issues.
14.	How much energy will be required for this programme? And how will it be made available?	About 4,000 MW will be required for operationalizing the lifts involved in some of the links. The requirement being small, it can be met with from the available generation. In addition, the programme will be generating 34,000 MW.
15.	Will this programme avail of foreign technical assistance?	There are enough indigenous technologies to handle the task of ILR Programme. However, if required there will be no hesitation to avail foreign technologies.

Sl. No.	*Questions*	*Reply*
16.	Considering the country's present economic condition, how do you justify this "Mega budget project"? Instead, is it not possible to go in for a "refined" irrigation system, which will be cost effective and will also reach the deprived rural population?	The Interlinking of Rivers proposal is not one project, it consists of a series of link projects which are to be implemented in a phased manner over a period of time. While studying the proposals of individual link projects at the Pre-feasibility stage, its financial and economic viability based on existing norms for irrigation projects has been studied. Notwithstanding this each link will be further examined by the Task Force on Interlinking of Rivers set up by Govt. of India and Detailed Project Reports prepared, after which only the links will be implemented. Refined irrigation systems are also to form part in the comprehensive programme to ensure efficiency in water management.
17.	Realistically speaking will it ever be possible to complete the project, let alone by 2016? It is a huge task needing a lot of money and involving many states. Just look at the Sardar Sarovar Project in Gujarat a very small task compared to the Interlinking of Rivers and yet, not complete even decades later?	Project implementation is dependant of several factors: financial, technical, social and environmental. The modalities for implementing the scheme within the stipulated time frame of year 2016 are to be suggested by the Task Force. The implementation of multipurpose projects often gets delayed due to non-availability of timely clearance from Environment and Forest angle, R & R, Inter-state issues and others. However, for the ILR, special attention is being given to these aspects so as to resolve them up front in a time bound manner, to avoid delays during implementation phase.
18.	India's forest cover is already less than the required 23%. ILR Programme is estimated to submerge 72,202 hectares of forest land. What TF is going to do about it?	Sufficient provisions exist in Forest Conservation Act of Government of India for the protection of forest by means of raising compensatory afforestation, water shed treatment and other land and water management programmes. Such norms will also be applied in ILR to avoid reduction in the forest cover of the country.

Sl. No.	*Questions*	*Reply*
19.	Is there a plan to interlink the Tapi and Narmada rivers—especially the Ukai Dam and Sardar Sarovar Dam—on priority as the water in Ukai dam is very inadequate to meet the full demand for irrigation water in South Gujarat and for the existing hydro-electricity project as well?	Yes, the studies have been carried out by NWDA to interlink the west flowing rivers north of Mumbai and South of Tapi. One such Link under this component is Par—Tapi—Narmada link. This link although passing through Ukai reservoir but does not envisage diverting any amount of water of Tapi basin. Similarly the link canal crosses the river Narmada, but no water of Narmada is proposed to be diverted in this link.

CONCLUSION

Inter-linking of rivers which was first dreamt by Dr. K.L. Rao, the then Power and Irrigation Minister and then by Captain Dastur, remained only as a dream for nearly four decades. But as the Supreme Court decreed that the rivers of India shall be linked within 10 years, the project is being presented as a major initiative, and as the definitive answer to the country's future needs and problems, though in the past, the Indian Scientists and Engineers always rejected the idea on account of technical unfeasibility or socio-economic issues. This ambitious project at a mind boggling cost of Rs. 5,60,000 crore has been given a go ahead allegedly without any detailed agreement of its financial viability, technological capability, ecological sustainability and political feasibility. Some express the extreme view that such directive of the Supreme Court tantamounts to 'Judicial Activism' for a wrong cause. They feel that there should have been a national debate on an issue of such national importance. The first requisite for such a debate is that the government should make its detailed plans public for engineers, geographers, environmentalists, economists, agronomists, soil scientists, sociologists, social anthropologists, financial analysts and others to examine and offer their comments. They assert that this massive project is too important a matter to be left entirely to the internal processes of politicians and the bureaucrats.

The general feeling of the people at large in India is that the political scenario of the country greatly affects the implementation of the laid down policies which may undergo changes with the change in governments. This is a crucial issue which need to be tackled parallel to the studies for technical viability. The proposal may get stuck if the water-surplus states are not duly compensated and if they do not agree even if the scheme is found as technically sound and financially viable. Further, inter-linking of northern rivers itself has international ramifications. Indeed, in addition to arrival of consensus among the co-basin states, negotiations need to be held with neighbouring countries, Bhutan and Nepal, as the scheme demands construction of storage dams in these two countries. It is a fact that we can never do anything comprehensive without their cooperation and the government expect to get it from these two upper-riparian countries. In so far as Bangladesh, a lower riparian country, is concerned, it may not have any reason to worry. The inter-linkage scheme starts with the assumption that Ganga is a water deficit river right from Hardwar. So our country is proposing to transfer water from the Brahmaputra's tributaries such as Teesta and Mahananda to the Ganga. That will augment the flow in Ganga and Bangladesh will have more water after the completion of the river-linking scheme than they have today. So will Pakistan.

The time—frame of 10 years ordered by the Supreme Court is also reasonably being questioned. In view of the fact that we have been taking more than 20 years for completion single high dam projects, the time seems totally unrealistic. Serious apprehensions are also being expressed on the ecological front, considering the fall out of 200 large water storage reservoirs and extensive networks of canals. Linking these rivers and storage reservoirs would eat into the natural habitats of wildlife and re-shape the ecology of the country with unknown consequences. There are not even any estimates regarding the number of people who would be dislocated, 'guestimates' vary from a few lakhs to a few crores! This would surely add to the political as well as economic costs of the project.

Voices are also heard that we are wasting a good deal of time in pursuing such grandiose and unpromising propositions, and distracting ourselves from finding time and money for more modest, worthwhile and urgent activities, such as extensive water harvesting all over the country, wherever feasible, and the massive task of rehabilitation of tanks in the south and other similar traditional systems ('dying wisdom') elsewhere. Even more important is effective demand management through improved efficiency and economy in water whether in agriculture or in industry or in domestic and municipal uses so as to minimize the need for supply-simulating solutions.

Hence, keeping in view all the points raised by eminent scientists, engineers, social activists, environmentalists, the concerned state governments and various central government departments and others, the proposal needs to be examined in all its facets before actual execution. Since long distance transfer of water is not considered an end by itself, but a means to end the human sufferings from frequent drought and flood, it is all the more necessary to have a National debate on the proposal so that instead of ameliorating sufferings, we create new irreversible problems for the Indian milieu.

REFERENCES

1. Government of India: Planning Commission: Ninth Five Year Plan 1997-2002: Thematic Issues and Sectoral Programmes: Volume II, p.492
2. Government of India, Ministry of Water Resources; Annual Report 2001-02.
3. Iyer, Ramaswamy R: Linking Rivers: Judicial Activism or Error? *Economic and Political Weekly*, November 16, 2002.
4. Sing, Dr. Bharat; A big dream of little logic.
5. Reddy, C. Rammonohar: Chasing a mirage of water.
6. Lahiri, S. K.: Link to certain disaster; *The Statesman*, New Delhi Edition. May 2003.
7. Sinha, Subrata: River Linking Plan, *The Statesman*, New Delhi, August, 20, 2003.

15

Innovative Approaches to Irrigation

"Hunger is not just a manifestation of poverty; it perpetuates poverty, and has been the major stumbling block to efforts aimed at eliminating poverty. Food security is therefore an essential factor for breaking the vicious circle of poverty and malnutrition. In order to eliminate hunger, concerted efforts are required to accelerate agriculture and rural development in countries where poverty and under-nutrition exist".

—FAO

Introduction

As the water is widely perceived to becoming increasingly scarce with the passage of time, it needs to be considered whether the usage of water irrigation is efficient in our country. It is estimated that in the next fifty years, India's irrigation capacity requirements will increase by 50 per cent, and the demand to domestic and industrial water will rise at an even faster rate. Bulk of the increase in demand for irrigation water in the future may have to be met by improving the usage of water, apart from increasing the supply. A global water usage model developed by the International Water Management Institute (IWMI) shows that 50 per cent of the world's additional water needs by 2050 can be met by increasing the efficiency of water use. Most of this gain would come from the countries with large areas of irrigated rice, which singled out India and China. Together, these two countries account for half the water savings effected in their model.

The importance of increasing efficiency in water usage can hardly be overstated. While the decline in public investment and inability to levy user charges on surface irrigation has already taken its toll on water management, the over-exploitation of ground water has created its own set of problems. The Planning Commission estimated that the number of "dark" (ground-water deficient) areas increased by 70 per cent between 1985 and 2000. In order to avert the impending water crisis, the overuse and misuse of fresh water must end as soon as possible. It is no longer possible to keep wasting and fouling our precious resource of freshwater. Increasingly, human activities are altering the flow of water and drawing down freshwater supplies faster than they can be replenished. Not only in India, but throughout the world enormous amounts of water are wasted due to inappropriate agricultural subsidies, inefficient irrigation system, leaky municipal pipes, improper pricing of water, poor watershed management and other imprudent practices. It is high time for widespread conservation measures, effective water management policies and growing attention to assure fresh water supplies.

Most irrigation systems waste water. Typically, only between 35 per cent and 50 per cent of water withdrawn for irrigated agriculture even reaches the crops. Most soaks into unlined canals, leaks out of pipes, or evaporates before reaching the fields. Poorly planned and poorly built irrigation system limit the yields on irrigated land. Paradoxically, even when sufficient water reaches agricultural fields, it can spoil much of the lands unless drained properly. Also, poorly drained irrigation water can raise the groundwater table until it reaches the root zone, waterlogging and drowning crops. In view of this, there is an urgent need for improving irrigation efficiencies for saving water and protecting the land. A few innovative methods of irrigation are now in vogue which are described in the chapter.

Artificial Recharge and Retrieval

Besides full utilisation of the potential already created, new technologies to make full use of rain water is also important. Result of artificial recharge and retrieval (ARR) experiments carried out show that ARR wells could be used to mitigate the seasonal drinking water scarcity felt in many villages in hard rock areas of India. The government of Gujarat has engaged an Israeli company to recharge the underground water by methods of recharging, recycling and inter-basin transfer. These activities are underway in Saurashtra in North Gujarat.

The experiment, the first of its kind in India, conducted by the National Geophysical Research Institute (NGRI) in the Jam river basin located near Nagpur reveals that ARR technology could be simultaneously applied to improve quality of drinking water in certain situations. Feasibility of artificial recharge through injection of bore wells in the basalts was evaluated through field experiments in Karol area of Nagpur district. The intake capacity of nine borewells was measured by injecting water under gravity. Electrical logging, packer and pumping tests were carried out. The effects of hydrofracturing was evaluated in some wells.

ARR technology, which meant storage of water in a suitable aquifer during times of wet weather and pumping it out from the same borewell during dry weather, could be applied to mitigate the drinking water scarcity faced by villagers during summer. Part of rain water percolates through the solid joined groundwater table every year. But NGRI studies on this process of natural recharge indicated that only five to 10 per cent of annual rainfall replenished groundwater in hard rock areas. Rest flowed off, joined major rivers and finally the sea and a portion was lost due to evaporation.

Encouragement of 'irrigation dry' is another innovation worth experimenting Supplying water in periodical way (2 or 3 wets) to the crop is called irrigated dry. This method is to be encouraged more in coastal districts instead of intensive supply of water. By all means, irrigation dry will cover more farmers, more land under guarantee of crop. In this way, the states can reduce poverty and promote more production of farm goods.

A New Integrated Water Harvesting Technique

A new integrated water harvesting technique developed and used at Central Arid Zone Research Institute, Jodhpur has shown promising results besides being practicable and economically feasible. Pits of 50 cm diameter and 60 cm depth are dug manually or with the help of a mechanical auger. A 5 mm layer of locally available fine pond sediments, bentonite clay or locally available clay is put at the bottom of the pit and compacted with the help of a disc compacter to a bulk density of 1.60 g/cc for checking the deep percolation losses of water and nutrients. The excavated soil is mixed with pond sediments or manure at the rate of 10 or 5 kg/ pit respectively. The pit is refilled and a circular catchment of 1.5 m radius and 5 per cent slope are made around the pit for water harvesting. Tree seeding is done at the centre of the filled pit.

The results of the study show that after rainfall initial soil moisture storage of 60 cm. profile increased from 73 mm (control) to 91 mm.

Due to increase in profile moisture storage there was marked increase in height, canopy diameter and dry weight of pruned material after four months of planting. The integrated water harvesting technique does not cost anything except farmers own labour in the digging of pits, making catchments and the use of locally available materials like pond sediments and manure. The use of bentonite clay at the rate of 2 kg/pit, however, costs about Rs. 1.50 and this could be used for the establishment of orchard crops. The run-off of rain water to the sea can be reduced by creating biotic and engineering impediments, and letting it to percolate into the soil and sub-soil strata. The restoration of permanent vegetable cover, construction of innumerable small weirs, check-dams and small tanks across all drainage lines in micro-catchments and the treatment of all erosion-prone agricultural lands for conservation of both soil and water are elements of this strategy. Since around 150 million hectares (or nearly half of the country's land surface) suffer from soil erosion, this treatment should be successful in ameliorating this problems as well. Such a programme would solve many problems: of waste-land development, rainfed agriculture, drought prone and desert areas, poverty and rural unemployment. The Ministry of Water Resources should contribute money for the priority treatment of the catchments of flood prone rivers as well as of reservoirs that are vulnerable.

MICRO IRRIGATION

Micro Irrigation is a generic term used for water management technologies deployed at the individual farm level. The two most important micro-irrigation methods are: drip irrigation and sprinkler irrigation. Plastic mulching and green house cultivation are also becoming popular now. Inconsistencies in rainfall, and the sinking of the water-table in most parts of the country, has made water-conservation a national priority. Farmers are increasingly accepting technologies which provide the right amount of water to their crops at the right time.

Drip and sprinkler irrigation systems were originally developed for arid and semi-arid regions of the world. The drip system uses only 30-50 per cent of the water used by conventional flood irrigation methods, while sprinkler systems use about 70 per cent. The farmers can irrigate about 2/3 times the farmland with the water saved by micro-irrigation systems. Over the last three decades, however, these technologies have come to be valued the world over for water conservation, as well as for their contribution to increasing yield levels. Additional benefits include economies in the utilisation of power and labour, and ensuring a reduction in weeds and plant diseases which thrive under excessively humid conditions of flood-irrigation.

The principle under Drip Irrigation System is to apply the required amount of water for a plant directly in the root zone. This is achieved by conveying the water/nutrient solution from the source to a plant with the help of a network of mains, sub mains and laterals and then applying this water to the root zone through the dripper or microtube. By knowing the quantity of water to be applied and the rate at which the water comes out through a dripper, time for the operation of the system can be determined. The effort is to keep only the root zone wet and, thus, the infiltration losses are completely avoided. The water is not lost in wetting the extra soil between two plants; the practice also controls weed growth. Fertilizer could be mixed at the source and applied at the root zone through the drip irrigation system. This practice results into significant savings (estimated at around 30 per cent) in the quantity of fertilizers to be applied per unit area. Since the root zone is maintained at favourable moisture contact at all times, the plants do not suffer any moisture stress and, therefore, the productivity of the crop and the quality of the produce have been found to improve. Table 15.1 shows the impact of drip irrigation practices on the yield and water savings of different crops. It is evident that a given quantity of water can command a larger area under drip irrigation system as compared to the traditional flood irrigation system.

Table 15.1: Crop Yields under Drip Irrigation Practices.

Crop	*% Water Saving*	*Yield Per Drip*	*Hectare Conventional*
Pomegranate	45	109000 nos	75000 nos
Grapes	48	32500 kg	26400 kg
Sugarcane	50	100000 kg	72000 kg
Banana	70	82500 kg	40000 kg
Tomato	39	48000 kg	32000 kg
Chillies	62	6068 kg	4233 kg

Source: Annual Reports of PDCs, Ministry of Agriculture, GOI.

Productivity Differences

Productivity of a crop is directly related with the amount of use of yield increasing inputs besides source of irrigation. Productivity of canal irrigated crops is higher than the tank irrigated crops. Similarly, productivity of crops, which are cultivated using groundwater irrigation, is much higher than canal and tank irrigated crops. Most of the time yield is affected because of moisture stress faced by crops. It is difficult to maintain the water supply constantly for crops by surface method of irrigation due to various reasons. Studies related to drip method irrigation have confirmed that problem of moisture stress is completely reduced by providing irrigation through drip as it supplies water at the root zone of the crops at a required frequency and quantity. As a result, the yield of crops cultivated under drip method of irrigation as shown in Mahasrashtra in the case of sugarcane (Table 15.2 & 15.3)

Table 15.2: Water Consumption by Drip and Non-drip Irrigated Sugarcane

District	*Water Consumption*		*Water Saving Over FMI*	
	DMI	*FMI*	*In Percent*	*In Quality*
Pune District	1740.08	2934.08	40.69	1194.00
Ahmednagar District	1793.92	3452.88	47.63	1631.96
Two Districts Average	1767.00	3179.98	44.43	1412.98

Table 15.3: Productivity of Sugarcane Under Drip and Flood Irrigated Condition

Method of Irrigation	*Pune District*	*Ahmednagar District*	*Two District Average*
Drip Method	141.32 (22.17)	135.40 (17.33)	138.36 (19.92)
Flood Method	116.80 (12.90)	108.08 (10.48)	112.44 (12.44)
Yield Increase Over FMI			
(i) Percentage	20.99	25.27	23.05
(ii) Quantity	24.52	27.32	25.92

Note: Figures in brackets are standard deviation.

Source: Field Survey data

With drip irrigation, poor quality water can be used safely since the continuous supply of water prevents accumulation of salts associated with other methods of irrigation. With 'furrow irrigation' differential salt accumulation occurs on the top of the ridges where plants are located. So, this method cannot be combined with use of saline waters. In case of irrigation through' flooding' with border method, fewer salinity problems are experienced particularly during the germination stage of plants, but as this method results in greater loss of water, this should be combined with the Precision Land Leveling (PLL). Some of the major benefits of drip irrigation are stated below:

- **Increased Yields:** The system facilitates water application at regular interval thereby maintaining optimum moisture level at the root zone for longer period thus preventing moisture stress or shock associated with other methods of irrigation. This promotes optimum plant performance resulting in higher yield and better quality produce.
- **Improved Quality and Early Maturity:** Crop Quality improvement and early maturity are results of even growth and ripening. In addition to this benefit, damage and losses due to water contact with fruit or foliage are eliminated.
- **Water Saving:** Water saving experienced upto 70 per cent in various crops as all percolation and evaporation losses are eliminated.
- **Control weed growth, saving on fertilizers and labour costs:** Water is applied directly in the root zone, wetting only a fraction of the soil; inter-space between the row of plants is not allowed to go beyond the root zone. Saving in fertilizer use upto 50 per cent is effected.
- **Provision of application of fertilizer and chemical through the System:** Drip Systems comprise of a chemical application device to enable application of soluble fertilizers, plant protection chemicals through the system, directly near the root zone of the plant.
- **Improved disease control:** Disease control is enhanced under micro irrigation system, because the soil moisture and chemical additive levels can be closely controlled. In addition, spread of any disease organism by wetting the leaf and surface water runoff movement is eliminated.

- **Ideally suited for difficult terrain and problematic soils & water:** Drip irrigation can convert a vast area of waste land into productive purpose. And at the same time it eliminates the need of land leveling and removal of productive top soil. Drip irrigation can be used on saline soil even with brackish water.

A typical micro-irrigation system has the following components: Dripper/micro sprinkler/sprayer, distribution lines and fittings, control head system, fertilizer tank and fittings.

The drip system can be used for all row crops other than bunch crops like rice, jute, wheat, millets etc. Though capital intensive, the system has become cost-efficient primarily in plantation crops/cash-crops such as sugarcane, coconut, arecanut, spices, and also cotton, rubber, tea & coffee and horticultural crops such as grapes, bananas, mangoes & other fruits, flowers and vegetables.

The principle of sprinkler irrigation is to nearly simulate the rain. It is achieved by forcing water at high pressure through the nozzles and breaking it into fine droplets which pass through the crop canopy and go on to wet the soil surface. The quantity of water to be applied through the sprinkler system is so determined as to avoid any infiltration losses through the soil. Water could be applied more frequently than the traditional flood irrigation systems and, thus, minimizing the risks of moisture stress. This, not only the water applied per unit area is reduced but also the yield is increased.

Water-Salt Balance: To avoid setting of 'negative' or secondary effects of introduction of irrigation to an area, proper watch need to be kept over the water-salt balance of the irrigation command area. This will need frequent salinity surveys covering the land return flows from the irrigated or drainage systems.[1]

Current Scenario

Presently micro irrigation is being practiced in as many as thirty five countries in the world, out of which India ranks seventh in terms of coverage of area. The countries, which have brought substantial area under drip irrigation include USA, Spain, Australia, South Africa, Israel, France and Italy. This system has gained wide popularity in areas of acute water scarcity and in areas where horticultural and commercial crops are grown. In the world scenario, almost 54 per cent of the drip irrigation is applied to orchard crops with citrus as a major crop.

In India some crude forms of drip irrigation was prevalent in various States like Maharashtra, Rajasthan, Haryana and Meghalaya. Scientific development of drip irrigation, however, took place only during the eighties. About 27 million ha. in the country is siuitable for drip irrigation (Table 15.4).

Water scarcity has compelled the Indian farmers to go in for advanced methods of irrigation, in order to increase the production and farm income and to utilise the water more efficiently for high value crops. It is reported that farmers making use of drip irrigation in

Table 15.4: Potential for Micro Irrigation in India

Crop	*Area M.H.A.*	*Area Suitable for M.I.*
Cereals & Millets	100.40	0.00
Pulses	22.50	0.00
Sugarcane	4.10	4.10
Spices	2.19	1.40
Fruits	3.40	3.40
Vegetables	5.30	5.30
Coconut	1.90	1.90
Oilseeds	26.20	1.90
Cotton	9.00	9.00
Others	1.40	0.00
Total	176.39	27.00

Maharashtra State, are able to get a net profit of Rs. 1,25,000 to Rs. 2,50,000 from one ha by growing grapes, orange, pomegranate, tomato and other fruit and vegetable crops. Similarly the yield of tea crop has increased by about 30 per cent by providing drip irrigation during summer. In spite of these advantages, the area under drip irrigation in only about 75,000 hectares which is considered a meagre achievement. Therefore, it needs no emphasis that planning and implementation of the increased programme of drip irrigation in future should be such as to achieve a target of 1 m. ha by the end of the first decade of 21st century.

Andhra Pradesh Experiment

Andhra Pradesh, left at the receiving end of the contentious war between its neighbours Karnataka and Tamil Nadu over Cauvery waters, is striving to spread a 'drip revolution' in this rocky region located near the trijunction of the three states. In Kuppam the "revolution is being carried out with help from a US-Israeli multinational. The drip irrigation system has been successfully implemented along with advanced farming technology, provided by the Israeli firm BHC International in Kuppam, which has erratic rainfall and virtually no permanent sources of water. The project is a principal political trajectory of not only making the farmers better off, but also exporting it to other districts of the state. After eight years effort covering 3,800 acres, the take-off has now been reached. Around 1,000 acres are being added each month and the target here is 20,000 acres. What is called "Bangalore Climate" helps Kuppam's farmers, although there is water shortage and check dams need to be constructed to conserve water. The state would have 6.5 lakh acres under drip eventually. The best part of market-driven success story is that the entire produce is exported to the US, Canada, Europe and West Asia. Besides the potatoes, tomatoes and chillies Kuppam produces gherkin, paprika and baby corn. Three Israeli experts have been brought who in turn give proper advice.

The task force on drip irrigation set up by the Central Government has now decided a nation-wide replication of the model adopted by the Andhra Pradesh Government in drip irrigation, agricultural practices and water management. The task force would confabulate with experts to fine-tune this model. It may also provide alternatives to various areas depending upon specific needs and climatic zones. Apart from providing a road map for long-term drip irrigation policy, the task force is expected to draw up a policy framework for water management. It will complement the work initiated by another task force on river inter-linking. The task force is also expected to recommend measures to not only expand the coverage of irrigation based on drip and sprinklers, but also suggest safeguards to ensure that the benefits reach the targeted groups.

Evaluation

Though not many evaluation studies on the impact of drip irrigation are available, a few studies carried out by agencies like Agricultural Finance Corporation Ltd. and NABARD are revealing. The drip irrigation component of the scheme was evaluated through Agricultural Finance Corporation Ltd. during 1997-98. The study covered about 3,900 farmers spread over 26 districts in the states of Andhra Pradesh,

Haryana, Karnataka, Maharashtra, Orissa and Tamil Nadu. The study revealed that the area under drip irrigation went up from 1,500 ha. in 1985 to 1.86 lakh/ha. by April, 1998. In all, about three lakh ha have been brought under drip irrigation including the area covered under the state sector scheme on drip irrigation in Maharashtra. The study also revealed that the beneficiary farmers invariably introduced high value horticultural crops like grapes, banana, mango, cashewnut and coconut after installing the drip system. There was yield increment for crops like banana, grapes, citrus, pomegranate ranging between 41 per cent (grapes) to 141 per cent (pomegranate) over the state average yield. Economic analysis of 695 beneficiary farmers and 76 non-beneficiary (who installed drip system without any Govt. subsidy) farmers indicated that the cost (inclusive of subsidy) was recovered in a period of less than three seasons in majority of the cases. The study further indicated that in Andhra Pradesh the coverage of area under drip irrigation was concentrated in the districts of Ananthapur, Chittoor, Rangareddy, Medak, Maahboob Nagar and Nalgonda. In the case of Karnataka the coverage is largely in the districts of Shimoga, Chitradurga, Tumkur and Bijapur while in Tamil Nadu almost 65 per cent of the coverage under drip irrigation was in the districts of Chengalpattu MGR, Dindigul, Coimbatore and Periyar.

The study also revealed that the majority of the farmers are able to recover their investments on drip system within a period of three seasons; the benefit: cost ratio being more than 2.5:1 in most of the cases. In another recent study by NABARD, it emerged that introduction of drip irrigation contributed towards bringing additional area under irrigation to the extent of 24 per cent in Chitradurga district of Karnataka. The study also revealed the financial viability of drip irrigation in yielding arecanut gardens. In his research paper entitled—Paradox of Deficit and Surplus Water: Impact on Food Security, published in Intensive Agriculture of September-October, 2002 issue, P. Raghu Ram Reddy of Central Research Institute for Dryland Agriculture, Hyderabad points out that more water use resulting in higher production may not always be true. He finds out from his field experiments that there is a distinct advantage of drip system over the traditional flood irrigation for certain fruit and vegetable crops. His findings are presented in Table 15.5. The productivity increase was 14 to 52 per cent. The water saved was from 39 to 62 per cent. The water use efficiency was only 30 to 35 per cent, while in drip it was 90 to 95 per cent.

Constraints in Promoting Micro Irrigation

Even though the advantages of drip irrigation have been scientifically proved, there remain many problems requiring further research so as to increase their applicability to horticultural crops. Realistic methods for predicting the temporal and spatial variation of soil moisture, the minimal and optimal fraction of the soil volume to be wetted, and management system in different horticultural crops under varying weather and soil conditions are not known. Application of nutrients along with water through drip irrigation can help in enhancing the production and productivity of horticultural crops. However, fertigation is yet to be adopted in large scale due to non-availability of soluble fertilizers, apart from high cost of imported fertilizers. The unbalanced growth of drip irrigation limiting mainly to the Southern Peninsular states is yet another area of concern. Some of the reasons for the uneven development in the drip irrigation are:

Table 15.5: Comparison of yield per ha. and water used in mm under flood & drip

Crops	*Yield Mt/ha.*			*Water Used (mm)*		
	Traditional Method (Flood)	*Drip System*	*Increase in Yield*	*Traditional Method (Flood)*	*Drip System*	*% Water Saving*
Fruit Crops						
Banana	57.50	87.50	52	1,760	970	45
Grapes	26.40	32.50	23	0532	278	48
Citrus	100.00	150.00	50	1,660	640	61
Vegetables						
Tomato	32.00	48.00	50	0300	184	39
Brinjal	28.00	32.00	14	0090	042	53
Chilli	04.20	06.10	44	0100	042	62

(i) Inadequate awareness about the advantages of micro irrigation.
(ii) Sufficient availability of surface and ground water in some of the states, particularly in Northern and Eastern India.
(iii) Lack of trained manpower.
(iv) Inadequate credit facilities for the farmer.
(v) Uneven distribution of drip manufacturers. Although there are 74 firms involved in supplying drip irrigation systems, only one is a large-scale manufacturer having complete

infrastructure for manufacture, design and after-sales service to the farmers.

Future Programme for Promoting Micro Irrigation

The tempo generated during the 8th plan in covering large areas under drip irrigation was continued during the ninth plan also. However, in comparison to the available potential of 27 million ha and the increasing pressure on water resources, the coverage under drip irrigation is only about 0.4 million ha. The working group on horticulture for tenth plan has recommended the need to bring one million ha under micro irrigation during the tenth plan.

Drip irrigation technology has been recognized as an answer to meet the increasing demands of water for irrigating the horticultural crops. It ensures increase in crop yield, higher quality of crop, less water and energy consumption, less chemicals and fertilizer use, reduced leaching and run-off, less weeds and less soil compaction. There has been significant increase in the area under drip irrigation in the country, brought about mainly due to governmental intervention. Investments to the tune of Rs. 400 crore have been made by the Government for promoting micro irrigation. The financial assistance provided by the Government along with the infrastructure created by the drip manufactures and the concerted efforts of the farmers have helped to bring about 4.0 lakh hectares till 2000-01 under micro irrigation in the country. In general about 22 per cent of the area covered is under coconut. Other crops covered are mango, grapes, banana, pomegranate, arecanut, etc. Presently, more than 80 per cent of coverage is restricted in four states, viz. Andhra Pradesh, Karnataka, Maharashtra and Tamil Nadu where water is scarce. Increased yield, reduced harvesting time and economy in water use have been the factors which promoted the adoption of this system particularly in high value horticultural crops.

Different states give importance to different aspects of irrigation methods. In the north, drainage and reclamation of land are given priority. In Madhya Pradesh the Government is considering large scale sprinkler irrigation system in their canal command areas. In Maharashtra, the Government and farmers work together to grow fruit and vegetable crops and to introduce drip irrigation in areas irrigated by wells. In Tamil Nadu, since water is scarce, importance is given to command area development programmes in canal/tank irrigated area and to drip irrigation in areas irrigated by wells. In the north and eastern part of Uttar Pradesh and Bihar tubewells are constructed to bring more area under irrigation.

Low—energy Precision Application (LEPA)

Another provision conservation method—low-energy precision application (LEPA)-offers substantial improvements over conventional sprinkler systems that spray water into the air. Instead, the LEPA method delivers water to the crops from drop tubes that extend from the sprinkler's arm. When applied together with appropriate water-saving farming techniques, this method also can achieve efficiencies as high as 95 per cent. Since this method operates at low pressure, energy costs drop by 20 per cent to 50 per cent compared with conventional systems. Farmers in Texas (USA) who have fitted conventional sprinkler systems with LEPA have reported that their yields have increased by as much as 20 per cent and the their investment costs have been recouped within one or two years (source: Population Reports, Series M, Number 14, February 1999).

In fine, the advantages in micro irrigation systems as discussed above are as follows:

1. Water efficiency 90-95 per cent as against 30-35 under flooding system.
2. Suitable in sloppy, hilly and undulated land areas.
3. Productivity increases by 25 to 50 per cent
4. Only system where soil, water and plant relationship can be completely combined and balanced with climatological variations and physical conditions of the crop at any stage.
5. MIS is a system which is designed taking into consideration water quality, hydraulics, agronomy, climatology and soil characteristics of a given farm plot.

Plastic Mulching

This involves covering the soil around plants to make conditions for crop growth more favourable. Plastic films of low density polyethylene and linear low density polyethylene are quite popular as mulches. Normally, films of 15 to 100 micron thickness are employed for this purpose depending upon the crop duration. Smaller duration crops require thinner film. The plastic film creates almost an impervious barrier between the soil and the atmosphere so that the evaporation from the soil is minimised. A transparent mulch film on the other hand can be used for raising soil temperatures significantly. If the objective is to reduce the soil temperature then a white coloured or reflective film is used for mulching. Plastic mulches also result into better root development, insect/pest control and productivity increase. Some results of the plastic mulching are presented in Table 15.6. By reducing the evaporation losses from the soil, plastic mulches increase the water utilization efficiency of the crop production.

Table 15.6: Performance of Crops (Preliminary) Mulched with Plastic Films in India

Crop	*Percent yield increase over control treatment*	*Remarks*
Tomato	20-50	Use of drip irrigation with mulches improves the yield.
Sunflower	23	
Banana	12	With drop irrigation system.
Groundnut	46	Rainfed
Maize	30	
Sesamum	136	
Potato	20	
Cabbage	58	
Chilli	32	

Source: Annual Reports of Plasticulture Development Centres. National Committee on the use of the Plastics in Agriculture. Ministry of Agriculture, Department of Horticulture, New Delhi.

Plastics in Agriculture

Companies in the micro-irrigation business design, install, and train in the maintenance of irrigation application systems. Success in this industry requires companies to design custom-made solutions for farmer's problems through a network reaching taluka levels and undertake continuous research to reduce the costs and improve the performance of its patented technologies and equipment. Against a potential coverage of 27 million hectares, the current coverage of micro-irrigation systems is even less than 2 per cent in India. Of this 87 per cent is in three states of Maharashtra, Tamil Nadu and Karnataka. As against this, China has already more than a million hectare under the system.

In order to speed up the acceptance of micro-irrigation technologies, state and central governments are subsidising installations. Subsidies towards the capital cost range from Rs. 12,000 to Rs. 60,000 per farmer in the case of drip irrigation systems, and Rs. 5,000 to Rs. 10,000 in the case of sprinkler systems.

Scientists at the Central Electronics Engineering Research Institute (CEERI) at Pilani (Rajasthan), have developed an advanced automatic drip irrigation system which is more efficient than the Israeli-made ones available in the market. The system can automatically control irrigation over four fields simultaneously round-the-clock and the savings in water is substantial. The CEERI system uses a highly-sensitive ceramic-based sensor to monitor moisture availability near the roots. As soon as the moisture content decreases below a certain value, the sensor sends a message to switch on a valve to resume water supply. CEERI has applied for a patent for the sensor.

The heart of the system is an electronic control equipment which can be programmed to monitor signals from four different fields simultaneously. Depending on the crop, it is sufficient to install five to 10 sensors for each acre of cropped land. The entire system, including the control panel, will cost approximately Rs. 15,000.

With the market potential from additional acreages growing at over 50 per cent per year during the present decade, many manufacturers of pipes can be expected to bid for a share of profits generated by the micro-irrigation industry. However, sales of micro irrigation are service oriented whereas the pipe manufacturers are commodity merchants. It is therefore also reasonable to assume that very few amongst today's pipe manufacturers will be able to acquire the total irrigation management knowhow and the ability to generate profits from the design and service component of the installations. As a result, many companies who do not make the transition will be forced to revert to the status of piping and components manufacturers.

Today's leaders can be expected to emerge as total farm-management consultants, not only providing the water-management technology, but also providing soluble fertilisers and micro-nutrients which the water will carry in order to obtain the best farm-yields. Along with the speedy commercialisation of agriculture, India will see the evolution of a few large and competent farm-management companies which generate profits by assisting farmers with total productivity management.

Reusing Urban Waste Water

A number of countries channel treated urban wastewater from towns and cities onto nearby farms growing vegetables and fruits. Today at least half a million hectares in 15 countries are being irrigated with treated urban wastewater, often referred to as "brown water". Israel has the most ambitious brown-water programme of any country. Most of Israel's sewage is purified and reused to irrigate 20,000 hectares of farm land. Some developing countries also use this technique, such as:

- In Mexico City, treated urban wastewater irrigates and fertilizes alfalfa fields. The alfalfa in turn is sold as feed to small-scale farmers who raise guinea pigs and rabbits.
- One-third of the vegetables grown in Asmara, Eritrea, are irrigated with treated urban wastewater.
- In Lusaka, Zambia, one of the city's biggest squatter settlements irrigates its vegetable crops with sewage water from nearby settling ponds.

> *If every drop of water that is drained or wasted is recycled, it is enough to irrigate a million hectares of cultivable land. After preliminary treatment, drainage water can be used to recharge ground-water (as happens in Los Angeles in the USA), pumped to industries for cooling purposes (as most western nations do or diverted to farmers to raise crops (as Israel does) or pumped into rivers to pollute them dangerously as India does!*

In other countries also the same concept is applied differently. For example, Kolkata, India, channels much of its raw sewage into a system of natural lagoons, where fish are raised. The city's 3,000 hectares of lagoons produce about 6,000 metric tons of fish a year for urban consumers. The fish are safe to eat because the lagoons soak up and clean the sewage. Unless urban wastewater receives some pretreatment, either from natural wetlands or sewage treatment plants, it can transfer disease organisms to vegetables and fruits and endanger human health.

Such natural water treatment technologies as using wetlands often can be an alternative to modern water treatment systems that are too costly for poor urban areas of developing countries. Recycling waste for agricultural purposes by using oxidation ponds and aerated lagoons does not require as much land as is often assumed. Moreover, it decreases pollution, reduces the need for fertilizers, and often can be accomplished with small-scale, low-cost technology that is based on local traditions, decentralized, and ecologically sound.

Greenhouse

A greenhouse is a framed or inflated structure covered with transparent or translucent glazing materials where crop environment could be controlled to some extent and which is large enough to permit human entry to carry out cultural operations. Traditionally, glass had been most popular glazing material until the invention of plastic films and sheets in the fiftees. Now more than 90 per cent of the new greenhouse constructions use plastics as the glazing material. A plastic greenhouse is a light structure and is a considerably low cost construction.

A greenhouse encloses the crop and creates a favourable microclimate for the crop growth. The transpired and the evaporated moisture from the greenhouse is not allowed to escape freely. As a result the humidity in the greenhouse is normally higher than that in the open fields and, therefore, the evapo-transpiration process is reduced. Obviously if the evapotranspiration losses are reduced, net irrigation requirements are also reduced. In this way a given quantity of water can meet the requirements of a larger crop stand. This attribute of greenhouse permits efficient crop production under arid conditions.

It is possible to combine the above mentioned technologies to maximise the water use efficiency for crop production. For example, it is quite practical to conceive of a greenhouse irrigated with either a drip or a sprinkler system and employing plastic mulching for crop production. If the soils are very sandy then some amount of infiltration is inevitable even under drip and sprinkler systems. It is possible to think of a plastic film placed at some depth in the soil under sandy soil condition so that the water and nutrients applied through drip or sprinkler system are not lost. Such a system is more practical in a greenhouse type of situation.

The Co-Efficient of Uniformity

A key to evaluation of all systems is the co-efficient of uniformity, or the CU. This is a calculated figure that represents the uniformity of water distribution over an area. In fact, the higher the CU, the greater the uniformity and the more effective the irrigation. Drip methods typically have CUs of about 90, sprinklers around 80 and gravity methods from 50 to 75. The impact of different CU levels on crop yields can be significant, as indicated in Table 15.7.

Table 15.7: Crop Sensitivity to Water Distribution

CU	*Yield (Kgs./ha.)*		
	Banana	*Potatoes*	*Maize*
60	23.4	26.6	4.2
62	25.4	28.8	4.5
64	27.2	31.0	4.8
66	28.9	32.9	5.2
68	30.6	34.9	5.4
70	32.1	36.6	5.7
72	33.6	38.2	6.0
74	34.9	39.7	6.2
76	36.1	41.2	6.4
78	37.3	42.5	6.6
80	38.3	43.7	6.8

Source: Rain Bird International.

It is obvious that high CUs are important for irrigation to result in high crop yields. In the case of maize, for example, increasing the CU from 70 to 80 results in an 18 per cent increase in yield.

In the conventional methods of furrow irrigation, for example, only a quarter of each 100 cubic metres of moisture reaches the roots of cotton. Sub-soil irrigation represents tubes buried in the soil to a definite depth. Through the openings in the tubes, water moistens the root layer. One variety of this method is drop-let irrigation, where a special dropping bottle is installed near each plant from the polythene irrigating main line. The whole of the water coming by small doses is assimilated by the roots. Nine years of experimentation in Russian Federation has shown that the yield of raw cotton was 41.3 metric centers per hectare in the case of furrow irrigation against 50.5 metric centers in the case of sub-soil irrigation with the distance of 60 centimetres between the tubes. The loss of irrigation water per metric center of raw cotton was reduced by 33 per cent.

Lining of Reservoirs and Canals

As far as unlined canals and distributaries are concerned, a study by the Central Water and Power Commission in 1967 revealed that about 71 per cent of the water is lost in transit from the reservoir to the field. This is broken down as shown below:

	Per cent
Canals	15
Distributaries	7
Water Courses	22
Field Losses	27

Source: IR: NCPA

It is apparent from the above that prima facie there is a case for considering the conservation of water by reducing seepage losses through the use of an impervious medium between the water and the porous soil below. Conventional lining materials such as clay, single tiles, double tiles and cement have been used for lining of canals with varying costs and benefits, while clay offers only limited protection, tiles and cement can develop cracks with time leading to some seepage loss. Plastic film with soil cover in suitable circumstances or with tile or cement cover elsewhere can provide improved protection against seepage, often at lower cost than a similar level of protection with other materials.

The supplementary Report of the C & AG for the year 1975-76 had drawn attention to the excessive loss of water during transmission and distribution. The data furnished by the Ministry in this regard indicates that in the case of unlined canals, losses are shown and assumed in the design at 8 cusecs per sq. ft. while in the case of lined canals the figure is taken as 2 cusecs per million sq. ft. based on the previous experience in the country. Actual measurements have, however, shown that the losses have been much more than estimated. For example, the losses observed in the Mahanadi Canal System have been as high as 39.7 cusecs/million sq. ft., in the case of Mula Right Bank Canal, these are of the order of 24-25 cusecs/million sq. ft. and for Tawa Project 22.8 cusecs/million sq. ft. Losses in the case of ten other projects, studied by the Public Accounts Committee were also significant.

The Economic Survey, 1982-83, also pointed out that conveyance losses in the canal systems were very high. According to an estimate made in 1960, over six million hectares of additional land could be irrigated by lining the canal systems. The colossal loss occurring to the country in such large scale wastage of the precious water resources can be easily imagined. The situation does not seem to have improved much since then. Similarly live storage capacity and withdrawal of water from reservoirs could be optimised through control on siltation which would also enhance the economic life-span of the project. It is necessary to strengthen the infrastructure to monitor the performance of each major/medium project instead of limiting important elements like annual inflow of water and location of state deposits and the annual draw of water conveyance losses.

Losses can often be reduced by relatively cheap improvements in irrigation design. In Maharashtra, the distribution system traditionally consisted of unlined canals serving forty-hectare blocks. Where they have been replaced by lined tertiary canals serving eight-hectare blocks, water available at the field has increased by 40 per cent, and net returns on the investment by some 160 per cent. In Madhya Pradesh, combining underground water supplies with dams and canals has raised the rate of return on investment from 10 to 29 per cent.

INTERNATIONAL EXPERIENCE

In order to introduce modern techniques and establish a proper system for an efficient use of water resources, it would be of interest to study the experience of other countries which have done well in these areas. We depict here details of Israel and France-two countries which have successfully introduced modern tools and established model systems.

Example of Israel

The example of Israeli system which is regarded as one of the most advanced in agriculture because of its most efficient and maximum utilisation of its water resources is of immense interest. The country measures its agricultural production as per cubic meter of water, which has been increased from 0.0548 in 1950 to 2.084 in 1990. They have developed a science oriented agriculture. It is really high tech. The agricultural schools there have the most sophisticated equipment and Israel could increase its yield by as much as 17 times through high tech. They export millions of dollars worth of foodstuffs every year. Israel has been hawking its agricultural expertise to the world. It is running two experimental farms in Egypt, one in China, one in Russia and two in Kazhakistan. A notable feature of Israeli expertise is the use of brackish, sea water for irrigation.

Israel does not get more than 1/2 inch rainfall per annum and two thirds of the country is desert. Its dry tropical climate is akin to that of Indian Jodhpur and Jaisalmer regions. Yet, every one in Israel gets high-quality drinking water, exactly as per needs. Further it is a wonder that even with such low rainfall Israel exports vegetables to Europe. In contrast, India has over 45 inches rainfall per annum, yet, not even 50 per cent of the people in India get even average fair quality water for drinking-what to talk of vegetables?. We must learn from Israel where every litre of water (regardless of the purpose for which it is used) is metered. It is priced realistically according to its quality and the use to which it is put. The water is recycled there as a matter of course and incentives operate for economical use, and disincentives are imposed for wastage.

Possible Israeli Contribution

Tech vs. *Resources*

Two factors explain Indian agri-businesses' growing fascination with Israel. First, the similarity between the two countries so far as landscape (semi-arid), water resources (limited) and natural vagaries (unpredictable) go. And next, Israel's amazing feats in agriculture with the help of innovations in drip irrigation, fertigation, greenhouses, pest control, climate control and farm mechanization. It generates inputs of over $40,0000 per farmer and achieves some of the highest yields in the world, including 500 tonnes of tomatoes and 3 million roses per greenhouse hectare. India's agricultural ties with Israel began with a December 1993 agreement. But the results of the deals in irrigation and water management techniques were nothing to write home about.

Second Wind

The last couple of years, however, have seen a gradual change in the entire Indian approach to agritech. "Politicians and farmers are realizing that old standards of water and power supply cannot continue" says Zvi Feler, MD of Netafim, the largest drip irrigation company in the world, which has seen its Indian business grow by 60 per cent since 2001. "Now governments appreciate the fact that free hardware does not help farmers".

Companies such as Netafim have had a presence in India since

1993, but so long, they were dependent on government subsidies and short-term clients. But if 10 years ago, Maharashtra subsidized nearly 90-100 per cent of the input, Andhra Pradesh is looking to subsidise just 50 per cent of his newest baby, the Andhra Pradesh Micro Irrigation Programme. The state even boasts of a 30-member association of drip irrigation manufacturers. Maharashtra, Gujarat and Tamil Nadu, too, are signing up to the 50 per cent subsidy plan. Gujarat, plagued by power theft, is even packaging a legitimate Gujarat Electricity Board connection with free drip irrigation equipment.

Big Returns

With three Israeli companies shortlisted for the large Andhra contract, and more possibilities opening up, the foreign firms are now focusing on spruced up after-sales service and package deals, a combination of fertigation, computerized farming plans, five-year after-sales service and drip irrigation hardware. But will the technology be cost-effective for the Indian farmer? Officials put the investments for fruits and flowers at between Rs. 20,000—50,000/hectare. And Feler, for one, is confident that the input will be worth it. "The cost in sugarcane, vegetables, fruits can be recovered within a couple of years". The advanced agriculture solution is the provision of drip irrigation which is a method by which just adequate amount of water is dripped to each plant that brings about a saving in irrigated water usage and enhances the yield as well as quality. Eighty per cent of Israel's cultivation is through the drip method.

The high-end solution Netafim offers includes survey, design, installation, agreement and after-sales support for five years. The cost, which admittedly is about 20 per cent higher than offered by Indian competitors, is around $300 to $600 per acre of sugarcane, for example, with a promised increase in yield from 25-30 tonnes per acre to 70 to 80. The list of Israeli innovations in farming is:

- *Drip Irrigation*: one of the first technologies to come from Israel, it ensures a continuous flow of water to the roots of the plant in drops with the help of a network of pipes. Saves 60 per cent water compared to flood irrigation. Israel has 80 per cent of its cultivation under drip irrigation
- *Fertigation*: This involves mixing fertilizer with irrigation water so as to minimize environmental pollution. Many Israeli

companies make the compatible highly soluble and liquid fertilizer.

- *Greenhouses*: Plastic-covered ones have been found to be cost-effective. Used as a filter to screen or admit UV rays and to refract and distribute light to maximize benefit to plants.
- *Brackish Water*: Israeli experience has shown some crops—cotton, tomatoes, melons—to be salinity-tolerant, improving quality but pulling down yield. Reclaimed water is also being used after the Soil Aquifer Treatment.
- *Irrigation Regime*: A countrywide network of agro meteorological stations that deliver real-time weather data to farmers. Parameters such as leaf water potential, fruit growth rate and soil-moisture monitoring device are utilized for precision in water application.
- *Drought-Resistant Seeds*: They have made hybrid cotton with stronger and longer fibre that also requires less water. Other achievements include onions with the highest number of dry layers, tomatoes with a 40 per cent longer shelf life, seedless and fridge-friendly watermelons.
- *Advanced Machinery*: They include multi-purpose carriers, which can be loaded or off-loaded from trucks, airblast sprays for use in citrus fruit-farming, hydraulic lifts for picking fruits from trees.
- *Drip Emitter Tech*: The revolutionary drip irrigation system developed by Israeli engineers and agriculturists reduced water consumption by 50-70 per cent compared with gravity irrigation and by 10-20 per cent compared to sprinkler irrigation. Recently, though, growers have been introducing the first generation of ultra low application rate (minute irrigation) drip emitters, with 200 cc/h flow rates. Considered even more advanced than the drip system, the emitters create optimal air-water relationships in the plant's root zones and, being more efficient, save yet more water. By use of drip emitters, the yield for tomatoes, for instance, has gone up to 30 tonnes for every 1/10th hectare as compared to only three tonnes just a few years ago. There are also plans by the government to introduce specialized drought resistant crops developed by Israel. Water saving (top priority in Israel) and recycling technology and infrastructure, irrigation technology, urban waste water recycling and bio technological advances,

including pooling genetic resources, and use of Israeli technology to boost floriculture and aquaculture will also figure prominently in the agreements being reached between the two countries.

France's "Model" System

Finally, it deserves to be mentioned about the French system of water resources management, adopted after many years of study and debate. This could well serve as a model for developed and developing countries, as they look for the best way to put the new "demand"-focused approach into action. Key elements of the system are:

- *Well-defined laws and regulations:* The water acts of 1964 and 1992 are the foundation of the French system. The earlier law established specific quality objectives and regulations for pollution control, while the later act is designed in part to meet stricter European directives on water management.
- *Hydrographic basin management*: The system is organized around six major hydrographic basins, with appropriate national policy oversight. These correspond to the country's four main catchment areas and to two areas of dense population and intense industrial activity.
- *Comprehensive management, decentralization, and participation:* Each of the six basins has a basin committee and a corresponding executing agency, a water board. The basin committee, also known as a "water parliament" because of its representation and powers, reflects regional-rather than central-government control and is designed to promote the role and responsibility of different interest groups in the basin. The water boards, while executing the committee's directives, are also responsible to the central government for certain technical matters (e.g., upholding national standards). Water and sewerage services are provided by either public or private firms (increasingly through competitive bidding) and are chosen by communities.
- *Cost recovery and incentives*: The companies and entities operating water services deliver a portion of the charges they collect to the basin agencies. In addition, a "pollution fee" (*i.e.* a penalty) is collected by the basin agency. Most of these

revenues are reinjected to provide technical assistance and help the public or private sector ensuring safe, purified water.

Conclusion

Demand for the world's increasingly scarce water supply is rising rapidly, challenging its availability for food production and putting global food security at risk. Agriculture upon which a burgeoning population depends for food, is competing with industrial, household, and environmental uses for this scarce water supply. Even as demand for water by all users grows, groundwater is being depleted, other water ecosystems are becoming polluted and degraded, and developing new sources of water is getting more costly. In India, the utilizable groundwater resource is estimated at only 41.19 million hectare meters per year. The present groundwater use, for irrigation is just 30 per cent of the available utilizable groundwater and this 30 per cent use again is not uniform throughout India. Intensive groundwater extraction in many regions has led to critical situations like depletion of water and premature failure of irrigation wells. In this situation, there is an urgent need for improving irrigation efficiencies for saving water and protecting the land. Micro irrigation system like drip and sprinkler, which is now being practiced in many countries may be more widely adopted in India.

There is now a need for paradigm shift in our approach to irrigation in view of the rapid depletion of ground water, particularly in the arid and semi-arid zones where the recharge potential is very low. However, despite many advantages of the micro irrigation systems, there are a number of constraints requiring further research so as to increase their applicability. Israel, a desert country; has made a spectacular progress in their agriculture by extensive adoption of micro irrigation systems. Apart from adopting micro irrigation systems, other innovative methods of using water for irrigation may be practiced. As many countries use channel treated wastewater from towns and cities for growing vegetables and fruits, this technique may also be adopted in India. Simultaneously there is a need for adopting fertigation in large scale due to non-availability of soluble fertilizers, apart from high cost of imported fertilizers. Low-energy Precision Application (LEPA), Plastic moulding and lining of reservoirs and canals may go a long way in mitigating the water requirement in India's agriculture.

Part IV

CONCLUSIONS

16

Water Policy in India

"Fire and Water pair are the soul and body that enables life to become dynamic"

—Rig Veda

Prologue

It has been well said that water is life's matter and matrix, mother and medium: nay, water is life itself. But guaranteed sustainable supply of water, the quality of which is also good enough to meet requirements of people and environment, is an enormous challenge to the government. Caught between increasing freshwater consumption for agriculture, industry and municipalities on the one hand and increasingly polluted water supplies on the other, India has been facing a difficult choice. Yet there is no more water available what it was there, say, 2000 years ago when the population was a fraction of the present size. India has 2.45 per cent of the world's land and 4 per cent of its freshwater resources[1]. Rising demands for water for various uses are forcing stiff competition over the allocation of scarce resources among both areas and types of uses. There are also striking differences in water availability within the country. Moreover, the freshwater supply in India comes in the form of seasonal rains. For instance, the country gets 90 per cent of its rainfall during summer monsoon season which lasts from June to September. For the other eight months the country gets very little rainfall. Such monsoon rains run off too quickly for efficient use. Further, the rainfall in the country is very much skewed geographically, with the average ranging from 100 mm in the west to 1,170 mm in the east.

Because of the skewed distribution of rainfall, every summer, on an average 91 districts are hit by drought and 40 million hectares in 83 districts are submerged in flood waters. While India receives a surface water flow of 1869 bcm., it can hardly store only 180 bcm. Hence, the hydrological cycle comes with no guarantee for the Indian population. Notwithstanding the utilizable water flow of 690 bcm., the storage and irrigation potential are grossly underused[2]. The solution is obvious. Stress should be laid on building water storages, increasing capacity in basins and working on inter-basin transfer wherever possible.

Instead of focusing on long-term solutions, every government has found it easier to allow exploitation of groundwater exceeding supply in the medium and long term. While for the government it meant less investment, for the farmer it is free water, even power where available, to irrigate his land. True, India's food security was propelled by the "tube-well revolution", but it led to long-term damage as the pump culture has wrought havoc on the hydrological cycle. It needs no reiteration that overuse leads to lower groundwater levels and increasing costs of supply, and, under certain circumstances also to subsidence and landfalls. The latter usually results in irreversible damage to the water-bearing bed. In nearshore areas it is not unusual for the sinking of the water level to allow saline seawater into water-bearing bed, (as it has happened in many parts of Tamil Nadu State), thus leading to its degradation. In fact, according to the Central Groundwater Board, groundwater levels have plunged in 206 districts in the country. With nearly 3.5 million hand-pumps and 56 lakh tubewells in operation, pumping of underground water is now nearly double the rate of aquifer recharge from rainfall.

The demand for irrigation water in future in the country is destined to multiply with increased crop intensity of improved planting materials, but the escalating and fierce competition from industrial and urbanized domestic use will drastically reduce the very availability of irrigation water for agriculture. Therefore, if agriculture is to remain sustainable, conservation and management of land and water resources and improvement in their per unit productivity and usage assume unparallel significance.

Contamination of water with harmful and pathogenic substances has become an additional problem aggravating further the fresh water scarcity in the country. The contributing factors include untreated or inadequately treated waste water from settlements, industrial waste, diffused contamination from agriculture and the usually isolated contamination by mining. The unsatisfactory amount of drinking water available in combination with inadequate sanitary conditions at the domestic and municipal levels is one of the main causes for the high incidences of diseases, especially among children in India. Thus, the importance of the sustainability of water supplies of sufficient quality for development can hardly over-emphasized.

As in other developing countries, in India too, exploitation of new water resources, was concentrated primarily on the surface waters by building dams. As the possibilities of their utilization are largely exhausted and have become too expensive, there has been large-scale exploitation of other water resources, especially groundwater. Normally, until the possibility of using renewable water resources as a whole are largely exhausted that recourse is taken to exploiting non-conventional water resources, for example by reprocessing human and industrial waste water, water transfer, the desalinization of sea and brackish water. But in supply management, reuse of waste water after treatment is still trivial in the country.

However, maintaining the balance between supply and demand by means of supply management does not always depend on exploiting new water resources. It can be done in many cases by improved management of the water resources made available. Since situations of shortages are not only a consequence of general water shortage but are all too often the consequences of inadequate operation and maintenance planning, it is possible to overcome shortage situations by proper planning and application of planning[3]. Indeed, water shortage and drastically increased supply costs call for changes in thinking more and more urgently necessary. In view of the usually inefficient water use the need today is for a greater orientation to demand management. More efficient use of water is expected to produce not only more economic treatment of water as a resource but also less environmental impact and thus the creation of one of the important preconditions for sustainable development.

It is widely conceived that a rational water policy is a crucial precondition for the framework of supply and demand management. However, water policy will certainly differ from one country to another and it depends on, *inter alia*, degree of economic development, urbanization, institutional capacity, and on the degree of water shortage and the intensity of agricultural production. Taking these influencing factors into account, a specific and practicable water policy should be worked out. According to Rosegrant[4], the key elements of the comprehensive water policy reform should be;

- legal and social safeguarding of water rights and if need be the introduction of trading in these rights;
- management of the irrigation systems by water users and/or user associations;
- creation of price incentives for saving water;
- introductions of suitable technologies;
- environmental safeguards;
- encouragement of international cross-border cooperation.

The Food and Agricultural Organization of the UN[5] also sees the reform of water resource policies as one of the fundamental preconditions for sustainable use of water resources and has, of late, begun a number of initiatives to encourage the necessary reforms.

Keeping in view the urgency of managing efficiently supply and demand of quality water for various uses and conserving this precious resource, availability of which is limited by nature, India had formulated its water policy for the first time in 1987. However, as the problem of availability of freshwater aggravated further over time and space, the government felt the need for reforming the policy and a new water policy has been adopted in 2002. This chapter highlights the backgrounds of adoption of Water Policies and critically examines the efficacies of these policies in overcoming the existing scarcity of water in the different regions of the country as also the measures suggested for conservation of water resource and augmentation of its supply.

The Requisites for a National Water Policy

The compelling needs of formulating the first National Water Policy in 1987 by the Ministry of Water Resources, Government of India were well documented. At the outset it was realized that water was a prime finite natural resource and was an increasingly scarce and extremely valuable resource without which human survival and sustainable development were impossible. Hence, there was a call for doing everything humanly possible to guarantee a sustainable use of fresh water.

Water Vision

- Sustainable Development
- Maintenance of Quality
- Efficient Use of Country's Water Resources to Match the Growing Demands
- Active Involvement of all Stakeholders in order to
- Achieve Accelerated Economic Development of the Country

At the outset, an assessment of availability was made by the Ministry. It was pointed out that out of the total precipitation of around 400 million hectare meters in the country, the surface water availability was about 178 million hectare meters. Out of this only about 50 per cent could be put to beneficial use and rest could not be due to topographical and other constraints. In addition, it was then assessed that there was a groundwater potential of about 42 million hectare meters. Further, as mentioned earlier, the disparity in the availability of water over time and space was also highlighted.

The water does not respect state boundaries. Not merely rivers but even underground aquifers often cut across state boundaries. Water as a resource is one and indivisible: rainfall, river waters, surface ponds and lakes, and groundwater are all parts of one system, and water itself is also a part of a larger ecological system. Flood and drought affect vast areas of the country transcending state boundaries. A third of the country is drought prone. Floods affect an average area of around 9 million hectares every year. According to the National Commission on Floods, the area susceptible to floods is around 40 million hectares. Considering their formidable situation the policy framers were of the opinion that the approach to the management of drought and floods needed to be coordinated and guided at the national level.

Even the planning and implementation of individual irrigation or multi-purpose projects, though done at the state level, involved a number of aspects and issues such as environmental protection, rehabilitation of project-affected people and livestock, public health, consequences of water impoundment, dam safety, etc. On these matters common approaches and guidelines were felt necessary. Moreover, certain problems and weaknesses affected a large number of projects all over the country. There were also substantial time and cost overruns on projects. In some irrigation commands, problems of waterlogging and soil salinity emerged alarmingly leading to the degradation of good agricultural land. There were also complex problems of equity and social justice in regard to water distribution. The development and exploitation of the country's groundwater resources also gave rise to questions of judicious and scientific resources management and conservation. All these questions needed to be tackled on the basis of common policies and strategies.

Identified Objectives

- To Help Meet the Feed and Fibre Requirement of a Projected Population of 1,330 Million by 2025
- To Make Available Safe Drinking Water for All
- To Safeguard Existing Water Sources From Pollution and Over Exploitation
- To Maximize the Hydropower Development
- To Ensure Safety and Serviceability of Existing Infrastructure and
- Improve Efficiency of Irrigation Water Systems and Its Application

The production of foodgrains had increased from around 50 million tonnes in the fifties to about 150 million tonnes at the time of framing the first Water Policy (1987) and it was planned to raise the production to around 240 million tonnes by the year 2000 AD. The drinking water needs of people and livestock were also required to be met. In keeping with the objectives of the International Drinking Water Supply and Sanitation Decade Programme (1981-91), adequate drinking water facilities were to be provided to the entire population in both urban and rural areas and sanitation facilities to 80 per cent of the urban population and 25 per cent of the rural population by the end of the 90s decade. Domestic and industrial water needs were largely concentrated in or near the principal cities, but the demand from rural society was expected to increase sharply as the development programmes resulted in improvement of economic conditions in the rural areas. The demand for water for hydro and thermal power generation and for other industrial uses was also anticipated to increase substantially. As a result, water, which was already a scarce resource, was apprehended to become even scarcer in future. This underscored the need for the utmost efficiency in water utilization and a public awareness of the importance of its conservation.

As stated earlier, an important aspect of water is its quality. Improvements in existing strategies and the innovation of strong science and technology based new techniques would be needed to eliminate the pollution of surface and groundwater resources, and improve water quality. This would also step up recycling and re-use of water. Science and technology, and training were also considered to play an important role in water resources development in general.

Keeping in view all these issues, the national government felt the need for a comprehensive water policy and as the country was preparing itself to enter the 21st Century, it was felt necessary to make sincere efforts to develop, conserve, utilize and manage this important resource according to the national perspective.

The National Water Policy, 1987 and 2002

The National Policy of 1987 recognized water as a precious national resource and it talked of its development based on national perspectives. It called for conjunctive use of surface and ground water, the need for planning water use on hydrological basis and inter-basin transfer of water from surplus to deficit. The pride of place in water use had been accorded to drinking water followed by irrigation, hydro-power, industrial uses, etc. It championed the need for periodic assessment of ground water potential and its regulated exploitation with reference to recharge possibilities and considerations of social equity. It also called for close integration of water use and land use policies and the need for water conservation efforts through education, regulation, incentives and disincentives. It advocated revision of water rates to motivate economy in use and also to cover maintenance and operational charges and a part of fixed costs as well. For bringing farmers to the forefront in the management of irrigation system, it talked of participatory irrigation management and watershed management.

National Water Policy-2002

The National Water Policy was adopted by National Water Resources Commission (NWRC) with a major consensus in its 5th meeting held on 1st April 2002; It will Guide the Nation in Realizing it's Vision with regard to Water in a holistic manner.

After almost fifteen years of the announcement of the first National Water Policy, the country has achieved precious little in the fulfillment of the objectives laid down in the Policy. In fact, the situation has further worsened whether in respect of augmentation of supply or conservation or maintenance of quality or unregulated groundwater exploitation. We have in fact discussed in this and in other chapters the seriousness of the problems. It is now widely believed that the problems may further aggravate and the country may soon fall in the category of water stress. Thus, there was an imperative to revise the National Water Policy of 1987 and a new Policy was adopted by the National Water Resources Council on 1st April, 2002.

The National Water Policy—2002 also recognized water as a precious national asset. It embodies the Nation's resolve that planning, development and management of water resources would be governed by National Perspectives. The new Policy emphasizes integrated water resources development and management for optimal and sustainable utilization of the available surface and groundwater, creation of well-developed information system, use of traditional methods of water conservation, non-conventional method for water utilization and demand management. The revised policy aimed at integrating quantity and quality aspects as well as environmental considerations for water through adequate institutional arrangements. Besides, 'ecological needs' have been assigned due priority in water allocation. Greater emphasis has been laid on water quality aspects. The policy also stresses involvement of people in project planning and participatory approach in water resources management[6].

However, a comparative study of the provisions of the 1987—Policy and 2002—Policy reveals that there are very little differences in the two Policies including their language. In fact, most of the provisions of the earlier policy have been repeated verbatim. The two Policies are presented below:

1987—Policy	2002—Policy
Information System	
2. The prime requisite for resource planning is a well developed information system. A standardized national information system should be established with a network of data banks and data bases, integrating and strengthening the existing Central and State level agencies and improving the quality of data and the processing capabilities. There should be free exchange of data among the various agencies and duplication in data collection should be avoided. Apart from the data regarding water availability and actual water use, the system should also include comprehensive and reasonably reliable projections of future demands for water for diverse purposes.	Repetition with the following additional section: **2.2.** Standards for coding, classification, processing of data and methods/procedures for its collection should be adopted. Advances in information technology must be introduced to create a modern information system promoting free exchange of data among various agencies. Special efforts should be made to develop and continuously upgrade technological capability to collect, process and disseminate reliable data in the desired time frame.
Maximizing Availability/Water Resource Planning	
3.1. The water resources available to the country should be brought within the category of utilizable resources to the maximum possible extent. Resources should be conserved and the availability augmented by measures for maximizing retention and minimizing losses. **3.2.** Resources planning in the case of water has to be done for a hydrological unit such as a drainage basis as a whole or for a sub-basin. All individual developmental projects and proposals should be formulated by the States and considered within the framework of such an overall plan for a basin or sub-basin, so that the best possible combination of options can be made. **3.3.** Appropriate organization should be established for the planned development and management of a river basin as a whole, special multi-disciplinary units should be set up in each state to prepare comprehensive plans taking into account not only the needs of irrigation but also harmonizing various other water uses, so that the available water resources are determined and put to optimum use having under the relevant laws.	Repetition except section 3.2 which has been replaced by the following and addition of a new clause: **3.2.** Non-conventional methods for utilization of water such as through inter-basin transfers, artificial recharge of ground water and desalination of brackish or sea water as well as traditional water conservation practices like rainwater harvesting, including roof-top rainwater harvesting, need to be practiced to further increase the utilizable water resource. Promotion of frontier research and development, in a focused manner, for these techniques is necessary. **3.4.** Watershed management through extensive soil conservation, catchment-area treatment, preservation of forests and increasing the forest cover and the construction of check-dams should be promoted. Efforts shall be to conserve the water in the catchment.

<table>
<tr><td>3.4. Water should be made available to water short areas by transfer from other areas including transfers from one river basin to another based on a national perspective, after taking into account the requirements of the areas/basins.

3.5. Recycling and re-use of water should be an integral part of water resource development.</td><td></td></tr>
<tr><td colspan="2" align="center">Project Planning</td></tr>
<tr><td>4.1. Water resource development projects should be as far as possible be planned and developed as multipurpose projects. Provision for drinking water should be a primary consideration. The projects should provide for irrigation, flood mitigation, hydro-electric power generation, navigation, pisciculture and recreation wherever possible.

4.2. The study of the impact of a project during construction and later on human lives, settlements, occupations, economic and other aspects should be an essential component of project planning.

4.3. In the planning, implementation and operation of projects, the preservation of the quality of environment and the ecological balances should be a primary consideration. The adverse impact, if any, on the environment should be minimized and should be off-set by adequate compensatory measures.

4.4. There should be an integrated and multi-disciplinary approach to the planning, formulation, clearance and implementation of projects, including catchment treatment and management, environmental and ecological aspects, the rehabilitation of affected people and command area development.

4.5. Special efforts should be made to investigate and formulate projects either in, or for the benefit of areas inhabited by tribal or other specially disadvantaged groups such as Scheduled Castes and Scheduled Tribes. In other areas also, project planning should pay special attention to the needs of Scheduled Castes</td><td>Repetition with the additional provisions of the following:

6.4. The planning of projects in hilly areas should take into account the need to provide assured drinking water, possibilities of hydro-power development and the proper approach to irrigation in such areas, in the context of physical features and constraints of the basin such as steep slopes, rapid run-off and the incidence of soil erosion. The economic evaluation of projects in such areas should also take these factors into account.

6.6. The drainage system should form an integral part of any irrigation project right from the planning stage.</td></tr>
</table>

and Scheduled Tribes and other weaker sections of society. **4.6.** The planning of projects in hilly areas should take into account the need to provide assured drinking water, possibilities of hydro-power development and the proper approach to irrigation in such areas, in the context of physical features and constraints such as steep slopes, rapid run-off and the incidence of soil erosion. The economic evaluation of projects in such areas should also take these factors into account. **4.7.** Time and cost overruns and deficient realization of benefits characterizing most irrigation projects should be overcome by upgrading the quality of project preparation and management. The under-funding of projects should be obviated by an optimal allocation of resources having regard to the early completion of on-going projects as well as the need to reduce regional imbalances.	
Maintenance and Modernization	
5.1. Structures and systems created through massive investments should be properly maintained in good health. Appropriate annual provisions should be made for this purpose in the budgets. **5.2.** There should be a regular monitoring of structures and systems and necessary rehabilitation and modernization programmes should be undertaken.	No mention in this policy
Safety of Structure	
6. There should be proper organizational arrangements at the national and state levels for ensuring the safety of storage dams and other water-related structures. The Central guidelines on the subject should be kept under constant review and periodically updated and reformulated. There should be system of continuous surveillance and regular visits by experts.	**24.** There should be proper organizational arrangements at the national and state levels for ensuring the safety of storage dams and other water-related structures consisting of specialists in investigation, design, construction, hydrology, geology, etc. A dam safety legislation may be enacted to ensure proper inspection, maintenance and surveillance of existing dams and also to ensure proper planning, investigation, design and constructions for safety of new dams. The Guidelines on the subject should be periodically updated and reformulated. There should be a system of

	continuous surveillance and regular visits by experts.
Ground Water Development	
7.1. There should be a periodical reassessment on a scientific basis of the groundwater potential, taking into consideration the quality of the water available and economic viability. **7.2.** Exploitation of groundwater resources should be so regulated as not to exceed the recharging possibilities, as also to ensure social equity. Groundwater recharge projects should be developed and implemented for augmenting the available supplies. **7.3.** Integrated and co-ordinated development of surface water and groundwater and their conjunctive use, should be envisaged right from the project planning stage and should form an essential part of the project. **7.4.** Over-exploitation of groundwater should be avoided near the coast to prevent ingress of sea water into sweet water aquifers.	Repetition
Water Allocation Priorities	
8. In the planning and operation of systems, water allocation priorities should be broadly as follows:- - Drinking Water - Irrigation - Hydro-power - Navigation - Industrial and other uses However, these priorities might be modified, if necessary, in particular regions with reference to area specific considerations.	Repetition
Drinking Water	
9. Adequate drinking water facilities should be provided to the entire population both in urban and in rural areas by 1991. Irrigation and multipurpose projects should invariably include a drinking water component, wherever there is no alternative source of drinking water.	Repetition except deletion of 'by 1991'.

Drinking water needs of human beings and animals should be the first charge on any available water.	
Irrigation	
10.1. Irrigation planning, either in an individual project or in a basin as a whole, should take into account the irrigability of land, cost-effective irrigation options possible from all available sources of water and appropriate irrigation techniques. The irrigation intensity should be such as to extend the benefits of irrigation to as large a number of farm families as possible, keeping in view the need to maximize production. **10.2.** There should be a close integration of water-use and land-use policies. **10.3.** Water allocation in an irrigation system should be done with due regard to equity and social justice. Disparities in the availability of water between head-reach and tail-end farms and between large and small farms should be obviated by adoption of a rotational water distribution system and supply of water on a volumetric basis subject to certain ceilings. **10.4.** Concerted efforts should be made to ensure that the irrigation potential created is fully utilized and the gap between the potential created and its utilization is removed. For this purpose, the command area development approach should be adopted in all irrigation projects.	Repetition with the additions of the following sections **9.5.** Irrigation being the largest consumer of fresh water, the aim should be to get optimal productivity per unit of water. Scientific water management, farm practices and sprinkler and drip system of irrigation should be adopted wherever feasible. **9.6.** Reclamation of water logged/saline affected land by scientific and cost-effective methods should form a part of command area development programme.
Water Rates	
11. Water rates should be such as to convey the scarcity value of the resource to the users and to foster the motivation for economy in water-use. They should be adequate to cover the annual maintenance and operation charges and a part of the fixed costs. Efforts should be made to reach this ideal over a period, while ensuring the assumed and timely supplies of irrigation water. The water rates for surface water and groundwater should be rationalized with due regard to the interests of small and marginal farmers.	No mention in this policy

Participation of Farmers and Voluntary Agency	
12. Efforts should be made to involve farmers progressively in various aspects of management of irrigation systems, particularly in water distribution and collection of water rates. Assistance of voluntary agencies should be enlisted in educating the farmers in efficient water-use and water management.	**12.** Management of the water resources for diverse uses should incorporate a participatory approach; by involving not only the various governmental agencies but also the users and other stakeholders, in an effective and decisive manner, in various aspects of planning, design, development and management of the water resources schemes. Necessary legal and institutional changes should be made at various levels for the purpose, duly ensuring appropriate role for women. Water Users' Associations and the local bodies such as municipalities and *gram panchayats* should particularly be involved in the operation, maintenance and management of water infrastructures/ facilities at appropriate levels progressively, with a view to eventually transfer the management of such facilities to the user groups/local bodies.
Water Quality	
13. Both surface water and groundwater should be regularly monitored for quality. A phased programme should be undertaken for improvements in water quality.	Repetition with the additions of the following sections: **14.2.** Effluents should be treated to acceptable levels and standards before discharging them into natural streams. **14.3.** Minimum flow should be ensured in the perennial streams for maintaining ecology and social considerations **14.4.** Principle of 'polluter pays' should be followed in management of polluted water. **14.5.** Necessary legislation is to be made for preservation of existing water bodies by preventing encroachment and deterioration of water quality.
Water Zoning	
14. Economic development and activities including agricultural, industrial and urban development, should be planned with due regard to the constraints imposed by the configuration of water availability. There should be a water zoning of the country and the economic activities should be guided and regulated in accordance with such zoning.	Repetition

Conservation of Water	
15. The efficiency of utilization in all the diverse uses of water should be improved and an awareness of water as a scarce resource should be fostered. Conservation consciousness should be promoted through education, regulation, incentives and disincentives.	Repetition with the addition of the following section: **16.2.** The resources should be conserved and the availability augmented by maximizing retention, eliminating pollution and minimizing losses. For this, measures like selective linings in the conveyance system, modernization and rehabilitation of existing systems including tanks, recycling and re-use of treated effluents and adoption of traditional techniques like mulching or pitcher irrigation and new techniques like drip and sprinkler may be promoted, wherever feasible.
Flood Control and Management	
16. There should be a master plan for flood control and management for each flood-prone basin. Sound watershed management through extensive soil conservation, catchment-area treatment, preservation of forests and increasing the forest area and the construction of check dams should be promoted to reduce the intensity of floods. Adequate flood-cushion should be provided in water storage projects wherever feasible to facilitate better flood management. An extensive network for flood forecasting should be established for timely warning to the settlements and economic activity in the flood plain zones, to minimize the loss of life and property on account of floods. While physical flood protection works like embankments dykes will continue to be necessary, the emphasis should be on non-structural measures for the minimization of losses, such as flood forecasting and warning and flood plain zoning, so as to reduce the recurring expenditure on flood relief.	**17.1.** There should be a master plan for flood control and management for each flood prone basin. **17.2.** Adequate flood-cushion should be provided in water storage projects, wherever feasible, to facilitate better flood management. In highly flood prone areas, flood control should be given overriding consideration in reservoir regulation policy even at the cost of sacrificing some irrigation or power benefits. **17.3.** While physical flood protection works like embankments and dykes will continue to be necessary, increased emphasis should be laid on non-structural measures such as flood forecasting and warning, flood plain zoning and flood proofing for the minimization of losses and to reduce the recurring expenditure on flood relief. **17.4.** There should be strict regulation of settlements and economic activity in the flood plain zones with flood proofing, to minimize the loss of life and property on account of floods. **17.5.** The flood forecasting activities should be modernized, value added and extended to other uncovered areas. Inflow forecasting to reservoirs should be instituted for their effective regulation.

<table>
<tr><th colspan="2">Land Erosion by Sea or River</th></tr>
<tr><td>17. The erosion of land, whether by the sea in coastal areas or by river waters inland, should be minimized by suitable cost-effective measures. The States and Union Territories should also undertake all requisite steps to ensure that indiscriminate occupation and exploitation of coastal strips of land are discouraged and that the location of economic activities in areas adjacent to the sea is regulated.</td><td>Repetition</td></tr>
<tr><th colspan="2">Drought Management</th></tr>
<tr><td>18.1. Drought-prone areas should be made less vulnerable to drought-associated problems through soil-moisture conservation measures, water harvesting practices, the minimization of evaporation losses, the development of the groundwater potential and the transfer of surface water from surplus areas where feasible and appropriate. Pastures, forestry or other modes of development which are relatively less water-demanding should be encouraged. In planning water resource development projects, the needs of drought-prone areas should be given priority.

18.2. Relief works undertaken for providing employment to drought-stricken populations should preferably be for drought proofing.</td><td>Repetition</td></tr>
<tr><th colspan="2">Science and Technology</th></tr>
<tr><td>19. For effective and economical management of our water resources, the frontiers of knowledge need to be pushed forward in several directions by intensifying research efforts in various areas, including the following:-
- hydrometeorology;
- assessment of water resources;
- snow and lake hydrology
- groundwater hydrology and recharge;
- prevention of salinity ingress;
- water-harvesting;
- evaporation and seepage losses;
- economical designs for water resources projects
- crops and cropping systems;</td><td>Repetition with the following additional items (Section 25):
- Water quality
- Water conservation
- New construction material and technology (with particular reference to roller compacted concrete, fiber reinforced concrete, new methodologies in tunneling technologies, instrumentation, advanced numerical analysis in structures and back analysis)
- Seismology and seismic design structures
- Use of remote sensing techniques in development and management</td></tr>
</table>

- sedimentation of reservoirs; - the safety and longevity of water-related structures; - river morphology and hydraulics; - soils and material research; - better water management practices and improvements in operational technology; - recycling and re-use; - use of sea water resources.	- Use of static ground water resources as a crisis management measure - Prevention of water logging and soil salinity - Reclamation of water logged and saline lands - Environmental impact - Regional equity.
Training	
20. A perspective plan for standardized training should be an integral part of water resources development. It should cover training in information systems, sectoral planning, project planning and formulation, project management, operation of projects and their physical structures and systems, and the management of the water distribution systems. The training should extend to all the categories of personnel involved in these activities as also the farmers.	Repetition
Conclusion	
21. In view of the vital importance of water for human and animal life, for maintaining ecological balance and for economic and developmental activities of all kinds, and considering its increasing scarcity, the planning and management of this resource and its optimal, economical and equitable use has become a matter of utmost urgency. The success of the national water policy will depend entirely on the developments to its underlying principles and objectives.	**27.** In view of the vital importance of water for human and animal life, for maintaining ecological balance and for economic and developmental activities of all kinds, and considering its increasing scarcity, the planning and management of this resource and its optimal, economical and equitable use has become a matter of the utmost urgency. Concerns of the community needs to be taken into account for water resources development and management. The success of the National Water Policy will depend entirely on evolving and maintaining a national consensus and commitment to its underlying principles and objectives. To achieve the desired objectives, State Water Policy backed with an operational action plan shall be formulated in a time bound manner say in two years. National Water Policy may be revised periodically as and when need arises.

In addition to the above National Water Policy—2002 also contains certain provisions under the following heads:

Institutional Mechanism

4.1. With a view to give effect to the planning, development and management of the water resources on a hydrological unit basis, along with a multi-sectoral, multi-disciplinary and participatory approach as well as integrating quality, quantity and the environmental aspects, the existing institutions at various levels under the water resources sector will have to be appropriately reoriented/reorganized and even created, wherever necessary. As maintenance of water resource schemes is under non-plan budget, it is generally being neglected. The institutional arrangements should be such that this vital aspect is given importance equal or even more than that of new constructions.

4.2. Appropriate river basin organizations should be established for the planned development and management of a river basin as a whole or sub-basins, wherever necessary. Special multi-disciplinary units should be set up to prepare comprehensive plans taking into account not only the needs of irrigation but also harmonizing various other water uses, so that the available water resources are determined and put to optimum use having regard to existing agreements or awards of Tribunals under the relevant laws. The scope and powers of the river basin organizations shall be decided by the basin states themselves.

Resettlement and Rehabilitation

10. Optimal use of water resources necessitates construction of storages and the consequent resettlement and rehabilitation of population. A skeletal national policy in this regard needs to be formulated so that the project affected persons share the benefits through proper rehabilitation. States should accordingly evolve their own detailed resettlement and rehabilitation policies for the sector, taking into account the local conditions. Careful planning is necessary to ensure that the construction and rehabilitation activities proceed simultaneously and smoothly.

Financial and Physical Sustainability

11. Besides creating additional water resources facilities for various uses, adequate emphasis needs to be given to the physical and financial sustainability of existing facilities. There is, therefore, a need to ensure that the water charges for various uses should be fixed in such a way that they cover at least the operation and maintenance charges of providing the service initially and a part of the capital costs subsequently. These rates should be linked directly to the quality of service provided. The subsidy on water rates to the disadvantaged and poorer sections of the society should be well targeted and transparent.

Private Sector Participation

13. Private sector participation should be encouraged in planning, development and management of water resources projects for diverse uses, wherever feasible. Private sector participation may help in introducing innovative ideas, generating financial resources and introducing corporate management and improving service efficiency and accountability to users. Depending upon the specific situations, various combinations of private sector participation, in building, owning, operating, leasing and transferring of water resources facilities, may be considered.

Monitoring of Projects **20.1.** A close monitoring of projects to identify bottlenecks and to adopt timely measures to obviate time and cost overrun should form part of project planning and execution. **20.2.** There should be a system to monitor and evaluate the performance and socio-economic impact of the project.
Water Sharing/Distribution amongst the States **21.2.** The water sharing/distribution amongst the states should be guided by a national perspective with due regard to water resources availability and needs within the river basin. Necessary guidelines, including for water short states even outside the basin, need to be evolved for facilitating future agreements amongst the basin states.
Performance Improvement **22.** There is an urgent need of paradigm shift in the emphasis in the management of water resources sector. From the present emphasis on the creation and expansion of water resources infrastructures for diverse uses, there is now a need to give greater emphasis on the improvement of the performance of the existing water resources facilities. Therefore, allocation of funds under the water resources sector should be re-prioritized to ensure that the needs for development as well as operation and maintenance of the facilities are met.
Maintenance and Modernization **23.1.** Structures and systems created through massive investments should be properly maintained in good health. Appropriate annual provisions should be made for this purpose in the budgets. **23.2.** There should be a regular monitoring of structures and systems and necessary rehabilitation and modernization programmes should be undertaken. **23.3.** Formation of Water Users' Association with authority and responsibility should be encouraged to facilitate the management including maintenance of irrigation system in a time bound manner.

State Water Policy

The National Water Policy-2002 envisages that each State shall formulate its own State Water Policy backed with an Operational Action Plan in a time bound manner say in two years to achieve the desired objectives of the Policy. Prior to adoption of the National Water Policy-2002 various Governments had formulated a State Water Policy or initiated action in this direction. The State Water Policy is either to be revised in line with the National Water Policy, wherever the same exist, or to be formulated by the States not having such policy.

- The success for the National Water Policy will depend entirely on evolving and maintaining a national consensus and commitment to its underlying principles and objectives
- To achieve the desired objectives, State Water policy backed with an operational action plan shall be formulated.
- NWP may be revised periodically as & when need arises.

A Critical Review of the Water Policy-2002

> **Sustainable Planning, Development and Management of Water Resources Governed by National Perspectives through**
> - Planning with Basin as a Hydrological Unit
> - Establishment of Appropriate River Basin Organizations.
> - Transfer of Water from Surplus to Deficit Basins
> - Watershed Management
> - Prioritization for Optimal Allocation of Resources

The looming water crisis in the different parts of the country, particularly in years of drought, is a cause of concern calling for a rational and scientific national water policy. Unfortunately our country has not yet woken up sufficiently to the impending water crisis. The need for conservation and better utilization of the limited water resources is, in fact, a global concern. However, "more than ever before in human history, the need for sustainable development, properly balanced with environmental protection, has engaged the attention of governments, NGOs, scientists, entrepreneurs and common people all over the world"[7]. The prevailing water crisis, no doubt, can be tackled to a great extent only through concerted efforts, both at the government and at people's level to regulate the over-exploitation of groundwater and rain water harvesting. There should be optimization of water conservation structures and regulation of groundwater exploitation is as essential as its conservation, particularly in plateaus. Education and training of the persons associated with groundwater conservation is also required for improving their understanding of groundwater.

The role of voluntary organizations in taking initiatives for water development projects can hardly be over-emphasized. The need of the hour is to have a synergy between the government and the non-governmental organizations in accomplishing adequate availability of quality water and its conservation.

The National Water Policy-2002 has been, as we have seen, by and large repetition of the first Policy of 1987 except a few new provisions. In fact, the new policy belies the peoples' expectations of decentralization of water development and management and encouragement to community-based water harnessing. The policy has left many loopholes for the market forces to take advantage. The policy does not have any punitive measures against wrong doers on a clause for encouraging those who harnessed water.

Promoting Beneficiaries' Participation

- Participation in various aspects of Planning, Design, Development and Management
- Necessary Legal and Institutional Changes
- Ensuring Appropriate Role for Women
- Transfer of the Management of Water Resources Facilities to People

The document is criticized as a hotchpotch of contradictory perspectives, completely lacking in integrity. It is alleged that a policy made in the backdrop of repeated droughts affecting large parts of the country in recent years could have been reasonably expected to give a central thrust to long-term drought-proofing. The country needs urgently to recognize that so long as we insist on spreading green revolution-type agricultural development to all regions, there will be no relief from the growing water crisis.

More than 60 per cent of India's landmass is underlain by hard rock types that are not conducive to irrigation through deep drilled tubewells. But tubewells have now become India's single largest source of irrigation in virtually all areas. This is the consequence of a policy that seeks to blindly replicate the success of Punjab, Haryana and other alluvial tracts in expanding tubewell irrigation. Even in areas where the rate of groundwater recharge is very low, in such hard rock areas, water takes a long time to accumulate below the ground. It has been estimated that in the Malwa region of Madhya Pradesh, for example, water that took over 10,000 years to accumulate in underground aquifers has been mined dry like coal in the last 30 years, creating a crisis in an area fabled for its plentiful water.

What the new policy should have emphasized is that tubewll irrigation needs to be undertaken with the greatest of care and under severe social regulation in hard rock regions. What is more, we need to learn lessons from the alluvial tracts themselves where after initial years of apparent success, problems caused by over-extraction of groundwater are beginning to surface. The key issue that needs attention here is that water policy has an integral relation to agricultural policy.

One could almost say that wrong agricultural priorities have created the water crisis in large parts of the country. India is a land of great hydro-geological and agro-bio diversity. Cropping patterns in each area must intimately respect and reflect this diversity. The move towards deep drilling of tubewells in hard rock areas is a consequence of the change in cropping patterns towards highly water-intensive crops and seed varieties in the vain hope of a Green Revolution in yields, without any consideration for the critical questions of sustainability.

And questions of sustainability cannot be addressed without recognizing the common property resource character of groundwater. Effective regulation of groundwater could emerge as the single most critical issue of rural governance in the 21st century. This entails mobilization of local communities to check private excesses and to specially protect the interests of the poor, dalits, adivasis and women. These efforts need legal and administrative backing from the state. They also require a range of technical inputs to study various aquifers for their transmission and storage characteristics. Such studies would enable communities to collectively plan the sequence and intensity of pumping of their wells.

The clause under section 10 which has been incorporated at the very last stage of the new policy deals with resettlement and rehabilitation of the dam ousters. The clause is considered as most dangerous and retrogressive since this nullifies the success achieved by the displaced families of such projects during the last 15 years forcing governments to formulate policies that are more liberal in their provisions, at least on paper. One of the most important achievements has been the principle of "land for land"—loss of land due to submergence would be compensated by other land to be mandatorily made available by the government. The old practice of bulldozing people out after a mere cash compensation would be stopped.

The new clause seeks to loosen this strict provision by speaking only of what it calls a "skeletal national policy" for resettlement and rehabilitation. The state will be allowed to "evolve their own detailed re-settlement and rehabilitation policies taking into the local conditions". So if, "local conditions" include difficulty in the government finding land they may revert back to old practice of cash compensation. Not only that, the policy also goes back on the condition reiterated recently by the Supreme Court that 'resettlement and rehabilitation' must precede dam construction. The new policy document just speaks of ensuring "that construction and rehabilitation activities proceed simultaneously and smoothly".

Whereas in 1987-policy document there was a section (No.11) dealing with 'water rates', but in the new policy this provision does not find place at all. There is a lot of misuse of water in all sectors whether agriculture or industry or domestic due to wrong pricing policy. In fact, despite the already existing supply bottlenecks and the many varied challenges, unfortunately fresh water is still not regarded as a scarce resource and both urban and rural water users, particularly the farmers, are supplied with highly subsidized water. Irrigation water is provided largely free of charge. In urban areas the prices charged for water fall far short of the cost of providing it. Because water is provided very cheaply or even free of charge by public water utilities, users have no incentive to use water as a resource economically. The new policy document should have dealt with this aspect and provided guidance for pricing of water for various uses.

Water Policy Blind to Local Situation

The Magsaysay Award winner Rajendra Singh, better known as "Waterman", has suggested to the States to prepare their respective water policies after taking into consideration the agro-climatic conditions in their areas. "Adequate thought is not being given to the formulation of water policies by the States even when water has the potential to become the most crucial issue in the days to come," he said addressing a seminar on Water Policy at Jaipur, Rajasthan.

Mr. Singh, who heads the Rashtriya Jal Biradari, a forum of conservationists and harvesters of water, said that the stakeholders should be consulted before the States went about finalizing their water policies. The governments should fix priorities for water use and the rights of the citizens on water. There was a need to decide on the ownership of water, its consumption and the agricultural patterns to be followed to ensure maximization of water use. The citizens should oppose any move by the governments to take away the community's right on water.

Jal Biradari, which is opposing the provision for privatization of water in the National Water Policy as well as the theme of linking of rivers has been seeking a review of the national policy. Now those in the organization are apprehensive of the likelihood of the States blindly following the pattern of the National Water Policy without giving any thought to the local situations.

"The responsibility and role of stakeholders should be spelt out in the water Policy," said S.S. Acharya, Director of the Institute of Development Studies (IDS). The present policy at the State level did not carry any reference to the use of ground water as well as sharing of water between those living in rural areas and urban centers.

Making a specific reference to the draft Water Policy of Rajasthan, M.S. Rathore of IDS said it was being formulated without holding consultations with the stakeholders in the State. "What we have with us is not even a water policy. It is only an irrigation policy" he remarked. "In economic planning, water is the basic component and this aspect is being overlooked". Dr. Rathore suggested formulation of water management plans on the basis of availability of the commodity. In a vast State like Rajasthan, the water policy should deal with various climatic zones for a more realistic approach.

Epilogue

There has been elaborate discussions on the needs of a national water policy both in 1987 and in 2002, the various clauses and provision of the two policies and the drawbacks of the new policy. The Ministry of Water Resources have also formulated a Draft Action Plan for Implementation of the National Water Policy-2002. The Draft Action Plan was reported to have been prepared in consultation with the State Governments and Union Territories and with inputs from the Non Governmental/ Voluntary Organizations and Experts through deliberations in a series of workshops and thereafter it was adopted on 5th February 2003. The policy is claimed to have been designed to meet the needs of the changing times. The government also asserts that the policy provides an opportunity for the private sector's participation in managing water resources like dams and pumping houses and does not give them any rights over water sources. In fact, the NGOs need to play an important role particularly in executing the water policy and there should be people's participation at the implementation and monitoring level as well. But the plan has allocated most of the tasks to ministry itself. Needless to say, individuals, institutions, NGOs and Government, each has a role to play to manage supply and demand and conserve water resources. Despite many shortcomings, the National Water Policy-2002 is a well thought out document carrying the message that water is an increasingly scarce and extremely valuable resource, without which sustainable development is impossible and all efforts should be made to guarantee sustainable use of fresh water.

REFERENCES

1. Aiyar, Shankar: Thirst India, Why Water has become Problem No.1 in cities and villages; *India Today*, June 9, 2003.
2. Ibid
3. Wolff, Peter: On the Sustainability of Water Use; National Resources and Development Focus: Water—the Life line of Our Future; Institute for Scientific Co-operation, Tubingen.
4. Rossgrant, M.W.: Water Resources in the 21st Century. Increasing Scarcity, Declining Quality, and Implications for Action—Paper presented at the Conference on the Sustainable Future of the Global System, Tokyo, Organized by the United Nations University and the National Institute for Environment Studies, Japan, 1995.
5. Food and Agricultural Organization of the United Nations (FAO): Sustainable Development and the Environment: FAO Policies and Actions, 1995.
6. Annual Report 2002-03, Government of India, Ministry of Water Resources.
7. Vajpayee, Atal Bihari, Prime Minister of India: PM Voices Concern over water crisis; *The Tribune*, September 4, 2002.

17

Conclusions and Recommendations

"Water is likely to become one of the limiting resources of the next century, as well as one with multiple, often conflicting uses."

—UN Commission for Sustainable Development, Second Session, New York, 1994

Total water resource on earth is estimated to be about 1,360 million cubic kilometers which is 0.25 per cent of the planet's mass and if spread evenly over its surface, it would cover the planet to a height of 2.7 kilometers. But more than 97 per cent is in the form of ocean and seas and only 3 per cent constitutes fresh water. Of this fresh water, nearly 77 per cent is in glaciers and ice caps, 22 per cent is ground water, of which only 11 per cent is available upto a depth of 800 meters that can be extracted for use. India—which has 16 per cent of the world's population, 2.45 per cent of the world's land area and 4 per cent of the world's water resources—is already heading towards a state of water crisis. Water is, literally, the source of life on earth. The human body is 70 per cent water. People begin to feel thirsty after a loss of only 1 per cent of body fluids and risk death if fluid loss nears 10 per cent. Human beings can survive for only a few days without fresh-water. Yet in a growing number of places people are withdrawing water from rivers, lakes and underground sources faster than they can be recharged.

The magnitude of the problem of scarcity of freshwater in India can be realized from the fact that some 200 million Indians do not have access to safe and clean water. India may be one of the major water-stressed country in the world by 2025 when nearly half the world population will experience water shortage. The stress on water resources is the consequence of a multitude of factors such as rapidly rising population and changing lifestyles that have increased the need for freshwater and intense competition among agriculture, industry and the domestic sector that is pushing the groundwater deeper. Several perennial flows like the Ganga-Brahmaputra—Barak are gradually becoming seasonal and many other rivers are drying up or declining in volume. There are vast stretches of Indian rivers that are unable to support fish due to high levels of pollution caused by discharge of untreated effluents. Many basins in India already have a low per capita availability and the water stress is being felt through shortages and deteriorating quality. The annual water withdrawal requirement, currently at about 650 billion cubic meters (BCM), is likely to increase to around 700 BCM by 2010 and to around 810 BCM by 2025. Although the agriculture/irrigation use would be predominant, domestic and industrial water supply would become increasingly important. While domestic use generally attracts high priority, waste water treatment gets neglected.

The population explosion had hit the cities and towns in the '70s and '80s but the authorities were not bothered to augment civic services. Tapping the groundwater reserve became the most preferred theme as water crisis became worse with each passing year in big cities. And in rural areas it was the lack of infrastructure, with no continued investments coming in, which made matters worse. According to a study, the annual per capita water availability in the country is going to fall by 67 per cent. In close to 300 districts, water levels have fallen by more than four meters in the past 20 years. Getting water for an hour twice a week or for an hour once a week are stories that have been in circulation for long. In some states more than 40 per cent of the population have no access to water, as a result, most of them have to depend on water brought on tankers. But this is not sufficient even to meet their half demand.

Availability of water in the country varies widely from one region to another. While eastern and north-eastern regions have by and large adequate availability of water both for domestic and agricultural uses, the states in western and southern regions suffer from acute scarcity of water, particularly during the summer months. Mismanagement and misuse of water further aggravates the situation. For instance, volume and depth of groundwater extraction in agriculture (the major consumer) is governed by the cropping pattern.

In India, nearly 90 per cent of the available fresh water is used for irrigation purpose, six per cent for industries and three per cent for domestic use. With further industrialization and increasing foodgrains demand to feed the growing population, the competition for freshwater use may increase alarmingly in the coming decades accentuating the water stress at an unimaginable level. Judicious use of water for irrigation is imperative in view of its limited availability and competing demand for industrial and municipal uses. The race for industrialization ignores checks on wasteful technology and pollution of water resources. Considering the fast decline of irrigation water potential and low water use efficiency under flood (conventional) method of irrigation, drip· and sprinkler methods of irrigation may be widely practiced. These methods, besides assisting land degradation, save substantial amount of water and help increasing the productivity of crops, that too with reduced cost of cultivation.

To date, most increases in food crop yields took place in irrigated land. Progress in rainfed agriculture associated with new crops, supplemental irrigation, deficit irrigation, watershed management, rainwater harvesting and some forms of precision irrigation, could lead to fast yield increases in rainfed lands. This would, no doubt, generate an increase in food production without a concomitant increase in water irrigation. However, continued emphasis on national food security may lead to overuse of water resources and hasten attaining the status of water-stress country. Increased availability of water through creation of irrigation potential has resulted in a major change in the cropping system in the various Indian states. As a result, signs of crisis in agrarian economy are visible in a few states following switching over to the water consuming crops like paddy, wheat and sugarcane. Water management problems all the same are very location specific, and thus need a special attention for their solution on individual farm units. Criteria most suitable for scheduling irrigation may vary in different situations, but in all cases, it should permit favourable crop yields, optimum use of water, and proper attention to other factors (irrigation cost, control of groundwater level, leaching of salts, accommodation to other farm operations) involved. The efficiency of irrigation water use is as important, indeed more so, than the quantum of water delivered or area irrigated. The potential gains to effort and resources spent on improving would be both substantial and more assured than pursuit of irrigation water supply augmentation. There is much scope for improving the efficiency of water use in the existing systems, which if properly exploited, can ensure wider, more equitable and productive use of water.

Our experience is that the administered prices have played an important role in changing the cropping system in many parts of the country. The crops suited to the dry lands of India have suffered negiect not merely in poor state-sponsored research & development, they have also been discriminated against in the procurement and price support mechanisms. Farmers tended to grow water-intensive crops, as these have been the ones favoured by state procurement and pricing policies. Over-extraction of groundwater for irrigation purposes and its unsustainable use are the major factors intensifying droughts suffered by the hard rock regions of the country. As the water level goes down deeper, the quality of water too deteriorates with the presence of fluorides, nitrates and other minerals.

The indiscriminate drawing of water through tubewells and the excessive watering of soil for crops like sugarcane has resulted in unsustainable agriculture on both counts. Wells are drying up due to excessive drawing of water, and at the same time excessive watering has led to the soil becoming saline, and thus unusable for agriculture. However, of late, the national government have been laying emphasis on Participatory Irrigation Management (PIM) and asserting that participation of beneficiaries would help in the proper upkeep of irrigation system and the optimal utilization of irrigation water. The PIM is expected to foster a sense of ownership, indisputable right to water, self-management, institutional access to irrigation information, choice of crops, improved water distribution system, better system maintenance and more disciplined and efficient use of water. The need of the hour is Integrated Water Resource Management (IWRM) which is a process to promote coordinated development and management of water to maximize economic and social welfare in an equitable manner, without compromising the sustainability of vital eco-system. Conjunctive use of surface and groundwater is also an essential aspect of water resource management.

While stating the importance of irrigation in accelerating growth in agricultural output, surface irrigation projects have also direct and indirect effects (indirect being defined specifically as those operating through the concomitant exploitation of groundwater) on cropping patterns and cropping intensities. Cropping pattern changes affect the use of water relatively to land and yields, and cropping intensity in the absence of cropping pattern changes, constitutes a kind of scale effect which is land augmenting. Of late, a related aspect, that of land degradation has come to be recognized as a significant consequence. The degradation caused by over-irrigation is further accelerated by intensive use of fertilizers. In fact, land degradation is a major threat to the country's food and environmental security and much of the irrigated crop land has degraded significantly over the decades through water logging and salinization. Widespread erosion in the hilly catchments is also causing excessive siltation of multipurpose reservoirs and other water bodies of the county at rates higher than the designed capacity.

Though irrigation is the major consumer of freshwater, the importance of requirements of water for domestic and industrial uses can hardly be over emphasized. The problems of water supply for domestic uses in urban areas and in rural areas are not similar and hence provision of water in these two areas and their solutions are also different.

It is very unfortunate that despite more than five decades of planned development, safe drinking water remains a problem area for the country and nearly 45 per cent of India's rural population does not have access to potable water in adequate quantity. There are two main issues in respect of rural drinking water problem. The first is the dependence only on the limited resources of the states for funding drinking water systems, and its consequent uncertainties. The second is the sustainability of drinking water systems created. Sustainability problems arise because the present policy is almost entirely 'supply driven'. The user community is rarely involved in the planning, implementation, operation and maintenance of the drinking water system created for them. Hence, demand-based policies—*i.e.* in terms of the type of system desired by the community or the level of service demanded are rarely adopted.

The gargantuan water problem has further been accentuating following rapid urbanization accompanied by faster population growth. There are stresses on the natural environment which are extending far beyond the land that urban areas actually occupy to affect the land that provides the resources to sustain urban life. Urban areas claim the ecological output and life-support functions of both nearby areas and distant regions. In fact, urbanization dramatically increases per capita freshwater use, as million of households gain access to piped water, as industry increases, and as large-scale irrigated agriculture replaces subsistence farming. India needs a Blue Revolution in water management, just as we need another Green Revolution in agriculture. Time is of essence. Dwindling freshwater supplies per capita are threatening the health and living standards of thousands of denizens of the country's cities and towns. Achieving a blue revolution will require coordinated policies and responses to problems at international, national and local levels.

In most parts of the country the groundwater resources have been over-exploited causing alarming environmental problems. Rainwater harvesting appears to be a major solution. High urban usage of water requires steps to augment supply by ensuring roof-top rainwater harvesting compulsorily. Further, as urban water supply is heavily subsidized, there is little compulsion of conservation of scanty freshwater. Proper pricing of water may only reverse the situation. Unless people are made to shell out appropriate rates for the water they use, advocating conservation is almost crying in the wilderness. The most vital part of a long-term solution to urban water supply problem lies, however, on recognition of the links between rapidly growing populations and shrinking freshwater supplies.

In fact, in much of India, water is basically a free good. But those who have the best access to it are the wealthy farmers and urban dwellers, because they are the ones who are connected to the water system. Poor people in rural areas or urban slums, who are not connected to a piped water supply, factually pay a higher price for their water, in terms of time spent in collecting it or money spent buying it from small-scale private water vendors. So the government pays to supply water to the well off, whereas the poor pay the price for their water themselves. For example, in Delhi city, the cost of supplying water is around Rs. 15 per 1,000 litres, whereas, the consumers pay only Rs. 1.50. Low water tariff, thus encourages wasteful utilization of water. As in other developing countries, in India too, obtaining water for domestic use is usually the responsibility of women and girls, and the hours that they spend fetching water is time that is not available for child care, productive activity and schooling. Pricing would be an effective mechanism in putting a curb on its unlimited use-and this option has to be exercised if a change has to be seen on the ground. Besides, industry and agriculture sectors have to exercise control over the use of water.

Needless to say, water pricing policy has the potential to mitigate water scarcity. Because of its key role in managing water demand and augmenting water supply, water pricing is an important policy instrument for creating incentives to conserve and allocate water efficiently. Low water charges and poor cost recovery lead to declining funds available for investment in water infrastructure, poor maintenance of existing systems, inefficient water allocation, and growing conflicts between those with or without access to water. Water rates are however still subsidized even in developed countries like Australia, Israel and the United States. This is rooted in the political economy of water as the powerful state and user interests often oppose charging the full cost of water. As a result, the gap is wide between the observed water rates and the ideal prices of water as reflected by its scarcity value and opportunity costs.

Though household water use is small, the requirement is in the details of ensuring that water is safe and getting it to the people of both urban and rural areas who need it in an affordable manner. In fact, the health problems associated with unclean water are so enormous that globally, nearly three and half million people are dying each year by water-related diseases, more than 2 million, mostly children, die from diarrheal diseases stemming from lack of access to clean water, and the issue of clean drinking water is inextricably linked to sanitation and hygiene. An estimated 90 per cent of the country's water sources are polluted with untreated industrial and domestic waste, pesticides and fertilizers run-off from fields and nearly 1.5 million children under five die each year from water-borne diseases. Thus, there is a need to integrate water supply, sanitation and hygiene projects.

The problem is so serious in most of the developing countries that according to UNICEF, globally one child dies every eight seconds from lack of access to safe water. In India, according to the World Bank, nearly 21 per cent of all communicable diseases are water borne. Diseases like hepatitis, diarrhea, and roundworm, hookworm and guinea worm injections amongst the population are commonplace. The increased deterioration of water quality due to industrial, agricultural and urban waste and to insufficient investments in domestic water supply infrastructure is a matter of serious concern. In most of the countries, policy documents on integrated management of water resources give highest priority to the drinking water supply in water allocation decisions. The global concern for water quality reflects the high standards imposed on drinking-water quality by institutions and professionals in the western world, where the quality of drinking water ranks above all and where the direct use of surface water from irrigation systems for drinking seems unacceptable.

The main strategy for improving rural drinking water has been to install low-cost hand pumps that draw groundwater uncontaminated by disease-causing microbes. But recently the high concentrations of arsenic in many tube-wells, particularly in the eastern states of India have become apparent leading to arsenic poisoning of millions of people. Another problem is high content of fluoride in the groundwater. While arsenic is toxic and carcinogenic, excess fluoride leads to serious dental and skeletal deformities and other health problems. The freshwater resources are not only getting exhausted day by day, it is also increasingly becoming contaminated.

The debate veers to this point of view that water management is the key to the entire issue of getting the best out of available resources. Urban and rural areas can do their bit in getting over this crisis. While in the urban areas it is leaking pipelines and poor distribution that has been responsible for much of the problem, in rural areas it is the outdated irrigation system that makes us see the worst side of this issue. In urban areas, the shift to rain water harvesting has made a steady progress. In this endeavour, the corporate sector's initiatives have been noteworthy. Financial policies which improve their combined viability and attract private/community participation also become important. Decentralized treatment of domestic sewage can save water, create useful bi-products and reduce pollution loads. An important step has been the formation of a body consisting of IIT alumni, which will be deliberating at length on infrastructural problems like water and suggest ways to get out of such problem areas.

Water resource planning is a scientific tool aimed at not only optimizing the utilization and management of water resources, but also to make the developmentalists aware of environmental and social equity considerations as well as to make environmentalists aware of economic and development needs. But unfortunately, a scientific approach to water management has not yet received the attention it deserves. There is a general agreement that water resource development has been far too preoccupied with increased exploitation of this resource in the name of development without paying much, if any, attention to its environmental impact; its efficient and equitable use; and long-term sustainability.

Needless to reiterate that in the existing situation of business as usual, the problem of adequate availability of freshwater may further aggravate in the near future. However, growing scarcity and inter-sectoral competition for water along with rapid depletion of groundwater and the problem associated with major surface-water control system, have raised interest in revitalizing water harvesting systems that capture rainwater wherever it falls. The rainwater harvesting in India is an ancient concept. But slowly with the advent of tap water, rainwater harvesting lost its pristine importance. Now our country is in a situation where efficient management of water resources has become an urgent necessity. India receives an annual precipitation of 4,000 billion cubic meters (bcm). However, the usable water resources from this precipitation are only about 1,100 bcm, *i.e.* roughly 28 per cent. Even, if this proportion could be raised by 5 per cent, it would mean an additional 200 bcm of usable water, which is 70 bcm more than the total reservoir capacity in the country.

Water harvesting requires collective action in the densely populated areas where more than one person uses catchment and command areas for multiple purposes. The government has to work with the people and modify policies to facilitate community participation in developing and maintaining rainwater harvesting systems. The government should make it compulsory by framing law to prepare area-wise water harvesting structures so that not a drop of water is wasted.

Our focus of discussion all through has been about the limited availability of freshwater and the need for judicious management of this resource. Janel Abramovitz of the Worldwatch Institute rightly observed that we were required to consider a river or lake alongwith its watershed and all its physical, chemical, and biological elements, as a part of a complex integrated system. But the people living in most of the watershed areas have radically altered their natural surroundings and such tempering with watershed has proved ruinous almost irreversibly. An imperative need therefore is being felt to implement productive, environmentally sustainable land and watershed management systems, more particularly, in the arid and semi-arid areas. Integrated watershed management is a multidisciplinary approach for rational utilization of natural resources existing in the watershed for optimum and sustained production with minimum hazards, considering holistic development for all the users of watershed. In fact, watershed management conceptually takes into account social, economical, environmental sustainability and institutional factors operating within and outside the watershed for effective management of natural resources. To make successful watershed management practice, full participation of project beneficiaries and careful attention of social organizations are imperative.

It is more than two decades, watershed programmes are being implemented in the various states and union territories in the country. The programme has attained popularity among development planners and agricultural scientists as it promises a win-win situation in which natural resource conservation and agricultural productivity are complementary. Unfortunately, much of the hype surrounding the country's watershed development draws on a small number of extraordinary success stories that may be difficult or impossible to replicate on a large scale. Various studies indicate that those projects which take a more participatory approach and are managed by the NGOs have performed better than their technocratic, top-down counterparts. However, such success has often come at the expense of the poorest people in watershed areas following restriction to the natural resources base on which they have been depending on.

India, nay the world, needs sustainable water management. To meet increasing water demands, storage of monsoon flows building storage reservoirs on our rivers are essential. Many basins in the country are surplus in water resource even in the ultimate stage of development while others face serious shortages. Creation of storages and inter-basin transfer of water from surplus to deficit regions could therefore be an option for achieving more equitable distribution of our water wealth and its optimal utilization. The planners now argue that if the resource abundant rivers like the Brahmaputra, Ganga, etc. could be linked with other rivers and a National Water Grid created, the miseries of recurring droughts and floods could be contained to a great extent.

Inter-linking of rivers first dreamt by Dr. K.L. Rao, the then Power and Irrigation Minister and then by Captain Dastur, remained only as a dream for nearly three decades. But as the supreme court decreed on a writ petition that the rivers of India should be linked within next 10 years, the project is now being presented as a major initiative and as the definitive answer to the country's future needs and problems. This, ambitious project has been given a go ahead though there are still no detailed agreements of its financial viability, technological capability, ecological sustainability and political feasibility. Critics point out that prima facie there are serious doubts about the feasibility and wisdom of pursuing this route to solving the country's water problem. The idea of inter basin transfers may have some merit if seasonal patterns of rainfall and stream flow in different parts of the country differed significantly. The fact however is that practically all rivers in India have their peak flows in roughly the same period (July-October) and the 'dry' season is almost everywhere also the same. Under these circumstances, as knowledgeable observers have already opined that colossal amounts of water will need to be stored. The technical problems of storage and of transport of the water are both enormous and complex. Even if one believed that scientists and engineers can overcome them and bring the water say from the Ganges to the peninsular basins, it would be impossible, because of geographical constraints, to design a distribution network that will bring water to all segments of the recipient basins according to need. The highly uncertain benefits of this undertaking will not justify the enormous costs involved by way of investment and far more crucially the unimaginable magnitude of displacement and adverse effects on the environment.

The arguments against river linking stressed at conventions, published in newspapers and even discussed within internet groups, give an impression that the proposal for river linking is repugnant to all natural and economic logic. However, the Ministry of Water Resources, Government of India has carefully examined all the views against implementation of the project and have observed that the impediments in the implementation of the project is not insurmountable and the fear of adverse repercussion on environment is exaggerated.

There is no fallacy in the belief that with the passage of time water will be increasingly scarce globally, and at the same time the need for water, particularly for irrigation, which uses almost 90 percent of available freshwater in this country, may increase phenomenally. It is estimated that in the next fifty years, India's irrigation capacity requirement may increase by 50 per cent, and the demand to domestic and industrial water may rise at an even faster rate. A global water usage model developed by the International Water Management Institute (IWMI) shows that 50 per cent of the world's additional water needs by 2050 can be met by increasing the efficiency of water use. Most of this gain, according to IWMI, would come from the countries with large areas of irrigated rice, which singled out India and China. Together, these two countries account for half the water savings effected in their model.

Maintaining the balance between supply and demand by means of supply management does not always depend on exploiting new water resources. It can be done in many cases, as stated earlier, by improved management of the water resources made available. A rational Water Policy is therefore a crucial precondition for the framework of supply and demand management. Keeping in view the urgency of managing efficiently supply and demand of quality water for various uses and conserving this precious resource, India had formulated its water policy for the first time in 1987. However, over time, the situation further aggravated needing a fresh look to the existing policy and a new water policy was adopted in 2002. Both 1987 and 2002 Water Policies recognized water as a precious national asset. The revised policy of 2002 emphasizes integrated water resources development and management for optimal and sustainable utilization of available surface and groundwater, creation of well developed information system, use of traditional methods of water conservation, non-conventional method for water utilization and demand management. The policy also stresses involvement of people in project planning and participatory approach in water resources management. The role of voluntary organizations in taking initiatives for water development projects can hardly be over-emphasized. The need of the hour is to have a synergy between the government and the non-governmental organization in accomplishing adequate availability of quality water and its conservation. Despite many shortcomings, the National Water Policy-2002 is considered a well thought out document conveying the massage that water is an increasingly scarce and extremely valuable resource, without which sustainable development is impossible and all efforts should be made to guarantee sustainable use of freshwater.

Keeping in view the supply and demand situation of freshwater in the country, following recommendations are made for consideration and implementation.

Recommendations

1. Privatization of water supply may not be a panacea for urban water problem in every place. The solution should be found out in a community-government partnership approach. Communities can be delegated the power to manage and distribute water from the public supply and then pay the government the charges on behalf of the whole group. In turn, these charges can be collected by the community from its members according to usage.
2. Poor rural women may be empowered with adequate water rights as this would strengthen their access to water for both domestic and income-generating uses. Better access to water improves women's health and incomes and liberates them from the daily drudgery of fetching water from long distances.
3. In order to make water supply projects considerably more

effective, men's and women's complementary intra-household roles and perspectives may be incorporated in project design and implementation, upto decision making level.

4. Since a majority of the water bodies to be restored have a limited catchment areas from where rainwater will gush in to them, a body of technical and legal experts should be set up to examine how the effectiveness of these structures can be optimized.
5. The basic components of rain water harvesting in both traditional and modern systems are the roof, a storage and a gutter to canalize the flow into storage. Noting the increased pollution in the atmospheric and on the surface catchment area (especially in metropolis) and increasing in sanitary problems, special attention is needed on the quality of water from the roof.
6. To date, most increases in food crop yields took place in irrigated land. But as progress in rainfed agriculture, associated with new crops, supplemental irrigation, deficit irrigation, rainwater harvesting, and some forms of precision irrigation may also lead to fast yield increases in rainfed lands, and hence may be practiced.
7. There is a tendency to take rice, wheat and sugarcane crops as soon as water is made available. But these crops are very high water consuming, and, therefore, the productivity per unit volume of water is low as compared to the light irrigated crops, such as, mustard, groundnut, barley and pulses. Crops/cropping patterns for limited water availability area should be so selected that higher returns are obtained per unit of water applied.
8. The water-balance equation imply, in essence, that the rise or fall of the water-table depends on factors like (a) specific yield, a property of the aquifer and the textural and structural characteristics of the soil, (b) precipitation or rainfall, (c) volume of irrigation water delivered to fields, (d) volume of water discharged at the land surface, (e) evaporation from land surface and the zone of aeration respectively, (f) increments in water storage at land surface and the zone of aeration respectively. Each of these components need to be monitored in order to plan for an increase in the irrigation water to be provided in any region.

9. Mounting subsidy on irrigation influenced the cropping pattern in arid and semi-arid regions and there is a major shift in area from less water consuming crops like coarse grains, millets, oilseeds, etc. to high water consuming crops like rice, wheat, sugarcane, etc., threatening the long-term sustainability of agriculture in these regions. The subsidy policy therefore needs a re-look for rational use of irrigation water and thereby arresting the adverse impact on land and environment.
10. Rural water supply component should represent an effective marriage of initiatives from government departments and beneficiaries: locally elected governing bodies must identify groups of villagers who need improved water supplies, and these groups may take ownership and maintenance of the completed hand pumps/tubewells.
11. For sustainability of drinking water system created in the rural areas, the user community should be involved in the planning, implementation, operation and maintenance of the system. In fact, demand-based policies—*i.e.* of the type of system desired by the community or the level of service demanded, should be adopted.
12. A cost—sharing policy of drinking water system installed in rural areas should be implemented universally. Within a village, the question of how much should each family share should be settled by consensus within the community. Cross-subsidization may be necessary, since the really poor families may not be able to contribute as much as some of the others.
13. The goal of urban groundwater management should be to strike a reasonable balance between maintaining water supply availability and quality, preserving the urban infrastructure and ensuring the safe disposal of wastes. In the case of already over exploited aquifers, abstraction control may be needed. There is an urgent need to institute changes on use of groundwater and recharge groundwater via roof-top and run-off rainwater harvesting.
14. Water tariff should be set at levels that discourage excessive use. Water-efficient systems for flushing toilets should be made mandatory. Severe penalties should be imposed on those found responsible for leakage and wastage of water.

15. Use of potable water for purposes like washing of vehicles, maintenance of gardens, etc. should be prohibited. In urban concentrations, dual supply of potable and non-potable water should be undertaken. For non-potable domestic uses, tubewells should be permitted to be sunk, subject to the construction of a percolation structure in the premises. This will conserve potable water without affecting ground water availability.
16. Efforts should be made in every city and town to reuse sewage water after treatment. With tertiary treatment, water from treated sewage can be used even for air-conditioning, industrial cooling and other non-potable uses. This should be made a thrust area.
17. Groundwater should be treated as a national property and therefore there is a need for a strong institutional and legal framework for its usage, conservation and maintenance of quality. Exploitation of groundwater in an approved manner may be allowed only by granting licences.
18. Mandatory provision of roof-top rainwater harvesting structures may be extended to the old buildings, whether in residential or industrial areas. The industries should come forward and adopt the structures. Before carrying out the construction of such structures, the factors which need to be taken into account are hydrogeology, soil cover, topographical gradient, aquifer system, depth-to-water, amount and pattern of rainfall and runoff co-efficient, etc., besides the area of the roof from where the amount of rainwater is to be collected.
19. The stretches of city drains which are not polluted due to municipal waste water, sewerage run-off waste water and industrial effluents, may be taken up for harvesting of water run-off accumulating during the monsoon period. Harvested run-off water can be utilized for artificial recharge to groundwater through suitable recharge structures, such as, shafts, trenches, diffusion well, deep sink holes, etc. Abandoned dug wells and tube wells may also be utilized for the purpose.
20. Drinking water quality is evaluated on the basis of the physical, chemical, and biological parameters. However,

parameters like fluorides and nitrates that have long-term health effects need to be analyzed too.

21. For reuse of wastewater in agriculture and aquaculture, international guidelines as formulated by the World Health Organization (WHO) may be followed in order to safeguard the health of irrigators and consumers.
22. Irrigation water resource planning and management are sciences which need proper attention. The basic components of an efficient use of water should involve.
    - (i) Improving communication between suppliers and recipients of irrigation water.
    - (ii) Effective educational programmes to minimize downstream health impact;
    - (iii) Necessary protective steps to avoid problems of water logging and salination, and
    - (iv) Prevention of siltation of water reservoirs through catchment protection.
23. Integrated Water Resource Management (IWRM) system, which is a process to promote coordinated development and management of water to maximize economic and social welfare in an equitable manner, without compromising the sustainability of vital eco-system, may be implemented throughout the country.
24. The present mindset of a majority of the population is that water is government's business. This needs to be replaced by a model in which all major stakeholders participate at all levels.
25. With a view to optimally utilizing the available water resources of the country by storage and inter-basin transfer from surplus to deficit and drought prone areas, a national perspective for water resources development need to be prepared.
26. Water Users Association (WUA) should be formed in all areas where canal or tubewell irrigation systems are prevalent and water users must be involved in the governance of water resources.
27. A holistic approach to water management may be evolved. There is an urgent need for integration of surface and groundwater management, integration of upstream and downstream water related interests and integration of quantity and quality of water management.

28. In order to achieve improved water management, proper linkages have to be developed in the implementation of the programme in such a way that irrigation engineer, agricultural scientists and those engaged in managing irrigated agriculture could properly coordinate their activities in accomplishing the task.
29. Huge loss of water due to seepage from badly managed canals must be avoided.
30. Reduction of run-off and soil losses through a greater emphasis on afforestation and land management on a nation-wide scale is essential since such a strategy also constitutes our best insurance against the recurring droughts and floods.
31. Use of non-conventional sources of water, such as recycling of municipal and industrial waste water, after appropriate treatment, use of 'marginal' or saline water for irrigation combined with improved irrigation like drip and sprinkler irrigation should be practiced wherever feasible.
32. Agricultural productivity per unit of water needs to be progressively increased in order to be able to compete with higher value uses of water.
33. Water must be priced at levels that not only take care of maintenance costs and future investment, but also contribute a surplus to make the effort of managing water worthwhile. The farmers who are the actual users, must take the responsibility of maintenance, allocation and collection of water rates.
34. The government may work with the people and modify policies to facilitate community participation in developing and maintaining rainwater harvesting systems. The government should make it compulsory by framing law to prepare area-wise water harvesting structures so that not a drop of water is wasted.
35. Hard surface such as roofs and pavements that decrease groundwater percolation, constitute the catchment in the urban areas and generate enhanced runoff, has to be guided in storage reservoirs or recharged into the groundwater table through simple filtration systems for subsequent extraction.
36. As people's participation is a prerequisite for the successful

execution of watershed management programe, there is a need for critical evaluation of the successful and not-so-successful watershed development projects and identification of the underlying issues related to people's participation for sustainable development of watersheds.

37. In order that a watershed development programme gets going and achieves the envisaged results, it is imperative to enlist people's active participation in all the phases of planning and implementation of the programme. The guidelines of the programme also need to be honestly pursued ensuring people's participation.
38. As the gender issue is an important factor in people's participation, the project team should essentially include women personnel for handling gender problems and organizing them in the watershed.
39. Before finally initiating the project of inter-linking of rivers, there should be detailed agreement of its financial viability, technological capability, ecological sustainability and political feasibility.
40. There should be a national debate on the issue of such national importance like inter-linking of rivers. The government should make its detailed plans public for engineers, geographers, environmentalists, economists, agronomists, soil scientists, sociologists, social anthropologists, financial analysts and others to examine and offer their comments.
41. Efforts should be made to implement all the provisions of the New Water Policy 2002 particularly in respect of groundwater development, supply of drinking water, water rates, water quality and water sharing among the states.
42. There should be a synergy between the government and the non-governmental organizations in accomplishing adequate availability of quality water and its conservation.
43. Groundwater should be recognized as the common property resource and there is need for effective regulation of this resource. This entails mobilization of local communities to check private excesses and the specially protect the interests of the poor, dalits, adivasis and women. These efforts need legal and administrative backing from the state.

INDEX